Lecture Notes in Computer Sci

Commenced Publication in 1973
Founding and Former Series Editors:
Gerhard Goos, Juris Hartmanis, and Jan van Leeuwen

Alexey Lastovetsky Tahar Kechadi
Jack Dongarra (Eds.)

Recent Advances in Parallel Virtual Machine and Message Passing Interface

15th European PVM/MPI Users' Group Meeting
Dublin, Ireland, September 7-10, 2008
Proceedings

Volume Editors

Alexey Lastovetsky
School of Computer Science and Informatics
University College Dublin
Belfield, Dublin 4, Ireland
E-mail: alexey.lastovetsky@ucd.ie

Tahar Kechadi
School of Computer Science and Informatics
University College Dublin
Belfield, Dublin 4, Ireland
E-mail: tahar.kechadi@ucd.ie

Jack Dongarra
Computer Science Department
University of Tennessee
Knoxville, TN, USA
E-mail: dongarra@cs.utk.edu

Library of Congress Control Number: Applied for

CR Subject Classification (1998): D.1.3, D.3.2, F.1.2, G.1.0, B.2.1, C.1.2

LNCS Sublibrary: SL 2 – Programming and Software Engineering

ISSN 0302-9743
ISBN-10 3-540-87474-7 Springer Berlin Heidelberg New York
ISBN-13 978-3-540-87474-4 Springer Berlin Heidelberg New York

Springer is a part of Springer Science+Business Media

springer.com

Printed in Germany

Typesetting: Camera-ready by author, data conversion by Scientific Publishing Services, Chennai, India
Printed on acid-free paper SPIN: 12511464 06/3180 5 4 3 2 1 0

Preface

Current thinking about state-of-the-art infrastructure for computational science is dominated by two concepts: computing clusters and computational grids. Cluster architectures consistently hold the majority of slots on the list of Top 500 supercomputer sites, and computational Grids, in both experimental and production deployments, have become common in academic, government and industrial research communities around the world. The message passing is the dominant programming paradigm for high-performance scientific computing on these architectures. MPI and PVM have emerged as standard programming environments in the message-passing paradigm. The EuroPVM/MPI conference series is the premier research event for high-performance parallel programming in the message-passing paradigm. Applications using parallel message-passing programming, pioneered in this research community, are having significant impact in the areas of computational science, such as bioinformatics, atmospheric sciences, chemistry, physics, astronomy, medicine, banking and finance, energy, etc.

EuroPVM/MPI is a flagship conference for this community, established as the premier international forum for researchers, users and vendors to present their latest advances in MPI and PVM. EuroPVM/MPI is the forum where fundamental aspects of message passing, implementations, standards, benchmarking, performance and new techniques are presented and discussed by researchers, developers and users from academia and industry.

EuroPVM/MPI 2008 was organized by the UCD School of Computer Science and Informatics in Dublin, September 7–10, 2008. This was the 15th issue of the conference, which takes place each year at a different European location. Previous meetings were held in Paris (2007), Bonn (2006), Sorrento (2005), Budapest (2004), Venice (2003), Linz (2002), Santorini (2001), Balatonfured (2000), Barcelona (1999), Liverpool (1998), Krakow (1997), Munich (1996), Lyon (1995), and Rome (1994).

The main topics of the meeting were formal verification of message passing programs, collective operations, parallel applications using the message passing paradigm, one-sided and point-to-point communication, MPI standard extensions or evolution, tools for performance evaluation and optimization, MPI-I/O, mutli-core and multithreaded architectures, and heterogeneous platforms.

For this year's conference, the Program Committee Co-chairs invited seven outstanding researchers to present lectures on different aspects of the message-passing and multithreaded paradigms: George Bosilca, one of the leading members of OpenMPI, presented "The Next Frontier," Franck Cappello, one of the leading experts in the fault-tolerant message passing, presented "Fault Tolerance for PetaScale Systems: Current Knowledge, Challenges and Opportunities," Barbara Chapman, the leader of the OpenMP community, presented "Managing Multi-core with OpenMP," Al Geist, one of the authors of PVM, presented

"MPI Must Evolve or Die," William Gropp, one of the leaders of MPICH, presented "MPI and Hybrid Programming Models for Petascale Computing," Rolf Rabenseifner, one of the leading experts in optimization of collective operations, presented "Some Aspects of Message Passing on Future Hybrid Systems," and Vaidy Sunderam, one of the authors of PVM, presented "From Parallel Virtual Machine to Virtual Parallel Machine: The Unibus System."

In addition to the conference main track, the meeting featured the seventh edition of the special session "ParSim 2008 - Current Trends in Numerical Simulation for Parallel Engineering Environments." The conference also included a full-day tutorial on "Using MPI-2: A Problem-Based Approach" by Ewing Rusty Lusk and William Gropp.

The response to the call for papers was very good: we received 55 full papers submitted to EuroPVM/MPI from 22 countries including Italy, Thailand, China, Germany, India, Greece, Spain, Japan, Switzerland, Ireland, Canada, Poland, Russia, Brazil, Denmark, Belgium, Mexico, Austria, Israel, Iran, France, and USA. Out of the 55 papers, 29 were selected for presentation at the conference. Each submitted paper was assigned to four members of the Program Committee (PC) for evaluation. The PC members either reviewed the papers themselves, or solicited external reviewers. The reviewing process went quite smoothly, and almost all reviews were returned, providing a solid basis for the Program Chairs to make the final selection for the conference program. The result was a well-balanced, focused and high-quality program. Out of the accepted 29 papers, four were selected as outstanding contributions to EuroPVM/MPI 2008, and were presented at special, plenary sessions:

- "Non-Data-Communication Overheads in MPI: Analysis on Blue Gene/P" by Pavan Balaji, Anthony Chan, William Gropp, Rajeev Thakur and Ewing Lusk
- "Architecture of the Component Collective Messaging Interface" by Sameer Kumar, Gabor Dozsa, Jeremy Berg, Bob Cernohous, Douglas Miller, Joseph Ratterman, Brian Smith and Philip Heidelberger
- "X-SRQ - Improving Scalability and Performance of Multi-Core InfiniBand Clusters" by Galen Shipman, Stephen Poole, Pavel Shamis and Ishai Rabinovitz
- "A Software Tool for Accurate Estimation of Parameters of Heterogeneous Communication Models" by Alexey Lastovetsky, Maureen O'Flynn and Vladimir Rychkov

Information about the conference can be found at the conference website: `http://pvmmpi08.ucd.ie`, which will be kept available.

The EuroPVM/MPI 2008 logo was designed by Alexander Kourinniy.

The Program and General Chairs would like to thank all who contributed to making EuroPVM/MPI 2008 a fruitful and stimulating meeting, be they technical paper or poster authors, PC members, external referees, participants or sponsors. We would like to express our gratitude to all the members of the PC and the additional reviewers, who ensured the high quality of Euro PVM/MPI 2008 with their careful work.

Finally, we would like to thank the University College Dublin and the UCD School of Computer Science and Informatics for their support and efforts in organizing this event. In particular, we would like to thank Vladimir Rychkov (UCD), Alexander Ufimtsev (UCD), Angela Logue (UCD), An Nhien LeKhac (UCD) and Clare Comerford (UCD). Special thanks go to all the School of Computer Science and Informatics PhD students who helped in the logistics of the conference.

September 2008

Alexey Lastovetsky
Tahar Kechadi
Jack Dongarra

Organization

EuroPVM/MPI 2008 was organized by the School of Computer Science and Informatics, Univeristy College Dublin.

General Chair

Jack J. Dongarra	University of Tennessee, Knoxville, USA

Program Chairs

Alexey Lastovetsky	University College Dublin, Ireland
Tahar Kechadi	University College Dublin, Ireland

Program Committee

George Almasi	IBM, USA
Lamine Aouad	University College Dublin, Ireland
Ranieri Baraglia	ISTI-CNR, Italy
Richard Barrett	ORNL, USA
Gil Bloch	Mellanox, USA
George Bosilca	Univeristy of Tennesse, USA
Franck Cappello	INRIA, France
Brian Coghlan	Trinity College Dublin, Ireland
Yiannis Cotronis	University of Athens, Greece
Jean-Christophe Desplat	ICHEC, Ireland
Frederic Desprez	INRIA, France
Erik D'Hollander	University of Ghent, Belgium
Beniamino Di Martino	Second University of Naples, Italy
Jack Dongarra	University of Tennessee, USA
Edgar Gabriel	University of Houston, USA
Al Geist	OakRidge National Laboratory, USA
Patrick Geoffray	Myricom, USA
Michael Gerndt	Technische Universität München, Germany
Sergei Gorlatch	Universität Münster, Germany
Andrzej Goscinski	Deakin University, Australia
Richard L. Graham	ORNL, USA
William Gropp	Argonne National Laboratory, USA
Rolf Hempel	German Aerospace Center DLR, Germany
Thomas Herault	INRIA/LRI, France
Yutaka Ishikawa	University of Tokyo, Japan
Alexey Kalinov	Cadence, Russia

Tahar Kechadi	University College Dublin, Ireland
Rainer Keller	HLRS, Germany
Stefan Lankes	RWTH Aachen, Germany
Alexey Lastovetsky	University College Dublin, Ireland
Laurent Lefevre	INRIA, Universite de Lyon, France
Greg Lindahl	Blekko, Inc., USA
Thomas Ludwig	University of Heidelberg, Germany
Ewing Rusty Lusk	Argonne National Laboratory, USA
Tomas Margalef	Universitat Autonoma de Barcelona, Spain
Jean-François Méhaut	IMAG, France
Bernd Mohr	Forschungszentrum Jülich, Germany
John P. Morrison	University College Cork, Ireland
Matthias Müller	Dresden University of Technology, Germany
Raymond Namyst	University of Bordeaux, France
Salvatore Orlando	University of Venice, Italy
Christian Perez	IRISA, France
Neil Pundit	Sandia National Laboratories, USA
Rolf Rabenseifner	HLRS, Germany
Thomas Rauber	Universität Bayreuth, Germany
Ravi Reddy	University College Dublin, Ireland
Casiano Rodriguez-Leon	University of La Laguna, Spain
Martin Schulz	Lawrence Livermore National Laboratory, USA
Andy Shearer	NUI Galway, Ireland
Jeffrey Squyres	Cisco, Inc., USA
Jesper Larsson Träff	C&C Research Labs, NEC Europe, Germany
Carsten Trinitis	Technische Universität München, Germany
Roland Wismueller	Universitat Siegen, Germany
Felix Wolf	Forschungszentrum Jülich, Germany
Joachim Worringen	Dolphin Interconnect Solutions, Germany

Conference Organization

Alexey Lastovetsky	University College Dublin, Ireland
Tahar Kechadi	University College Dublin, Ireland
Vladimir Rychkov	University College Dublin, Ireland

External Referees

Erika Abraham
Olivier Aumage
Boris Bierbaum
Aurelien Bouteiller
Ron Brightwell
Michael Browne
Stephen Childs
Carsten Clauss
Camille Coti
Maurizio D'Arienzo
Jan Duennweber
Hubert Eichner

Markus Geimer
Ludovic Hablot
Mauro Iacono
Yvon Jégou
Julian Kunkel
Diego Latella
Pierre Lemarinier
Andreas Liehr
Stefano Marrone
Alberto F. Martín Huertas
Alastair McKinstry
Guillaume Mercier
Ruben Niederhagen
Ron Oldfield
Alexander Ploss
Marcela Printista
German Rodriguez-Herrera
John Ryan
Maraike Schellmann
Carlos Segura
Andy Shearer
Daniel Stodden
Honore Tapamo
Georg Wassen
Zhaofang Wen
Niall Wilson
Brian Wylie
Mathijs den Burger

Sponsors

The conference would have been significantly more expensive and much less pleasant to organize without the generous support of our industrial sponsors. EuroPVM/MPI 2008 gratefully acknowledges the contributions of the sponsors to a successful conference.

Platinum Level Sponsors

Gold Level Sponsors

Table of Contents

Invited Talks

Tutorial

Outstanding Papers

Applications

Collective Operations

Library Internals

Message Passing for Multi-core and Mutlithreaded Architectures

MPI Datatypes

MPI I/O

Synchronisation Issues in Point-to-Point and One-Sided Communications

Tools

Verification of Message Passing Programs

ParSim

Posters Abstracts

The Next Frontier

George Bosilca

Innovative Computing Laboratory
Electrical Engineering and Computer Science Department,
University of Tennessee, Knoxville, TN, USA
bosilca@eecs.utk.edu

Today, multi/many core systems have become prevalent, with architectures more or less exotic and heterogeneous. The overall theoretical computational power of the new generation processors has thus greatly increased, but their programmability still lacks certainty. The many changes in the newest architectures have come so rapidly that we are still deficient in taking advantage of all the new features, in terms of high performance libraries and applications. Simultaneously, application requirements grow at least at the same pace. Obviously, more computations require more data in order to feed the deepest processor pipelines. More data means either faster access to the memory or faster access to the network. But the improvement in access speed to all types of memory (network included) lags behind the increase in computational power. As a result, while extracting the right performance of the current and next generation architectures is still a challenge, it is compulsory to increase the efficiency of the current parallel programming paradigms.

Simultaneously, increasing the size of parallel machines triggers an increase in fault tolerance requirements. While the fault management and recovery topic has been thoughtfully studied over the last decade, recent changes in the number and distribution of the processor's cores have raised some interesting questions. Which fault tolerant approach fits best to the peta-scale environments is still debated, but few of these approaches show interesting performances at scale or a low degree of intrusion in the application code. Eventually, the right answer might be somewhere in between, a dynamic combination of several of these methods, strictly based on the application's properties and the hardware environment.

As expected, all these changes guarantee a highly dynamic (and exciting from a research point of view), high performance arena over the next few years. New mathematical algorithms will have to emerge in order to take advantage of these unbalanced architectures, new programming approaches will have to be established to help these algorithms, and the next generations of computer scientists will have to be fluent in understanding these architectures and competent in understanding the best programming paradigm that fits them.

How MPI will adapt to fit into this conflicting environment is still an open question. Over the last few years, MPI has been a very successful parallel programming paradigm, partially due to its apparent simplicity to express basic message exchange patterns and partially to the fact that it increases the productivity of the programmers and the parallel machines. Whatever the future of MPI will be, these two features should remain an indispensable part of its new direction of development.

A. Lastovetsky et al. (Eds.): EuroPVM/MPI 2008, LNCS 5205, p. 1, 2008.

Fault Tolerance for PetaScale Systems: Current Knowledge, Challenges and Opportunities

Franck Cappello

INRIA
fci@lri.fr

Abstract. The emergence of PetaScale systems reinvigorates the community interest about how to manage failures in such systems and ensure that large applications successfully complete. Existing results for several key mechanisms associated with fault tolerance in HPC platforms will be presented during this talk. Most of these key mechanisms come from the distributed system theory. Over the last decade, they have received a lot of attention from the community and there is probably little to gain by trying to optimize them again. We will describe some of the latest findings in this domain. Unfortunately, despite their high degree of optimization, existing approaches do not fit well with the challenging evolutions of large scale systems. There is room and even a need for new approaches. Opportunities may come from different origins like adding hardware dedicated to fault tolerance or relaxing some of the constraints inherited from the pure distributed system theory. We will sketch some of these opportunities and their associated limitations.

A. Lastovetsky et al. (Eds.): EuroPVM/MPI 2008, LNCS 5205, p. 2, 2008.

Managing Multicore with OpenMP (Extended Abstract)

Barbara Chapman

Department of Computer Science, University of Houston,
Houston, TX 77204-3010, USA
chapman@cs.uh.edu
http://www.cs.uh.edu/~chapman

High end distributed and distributed shared memory platforms with many thousands of cores will be deployed in the coming years to solve the toughest technical problems. Their individual nodes will be heterogeneous multithreading, multicore systems, capable of executing many threads of control, and with a deep memory hierarchy. For example, the petascale architecture to be put in production at the US National Center for Supercomputing Applications (NCSA) in 2011 is based on the IBM Power7 chip which uses multicore processor technology. Thousands of compute nodes with over 200,000 cores are envisioned. The Roadrunner system that will be deployed at the Los Alamos National Laboratory (LANL) is expected to have heterogneous nodes, with both AMD Opterons and IBM Cells configured, and a similar number of cores.

This brave new multicore world presents application developers with many challenges. First, the continued growth in the number of system nodes will excacerbate existing scalability problems and may introduce new ones. Second, the number of threads of control that may execute simultaneously within a node will be significantly greater than in the past. The code will need to expose a sufficient amount of parallelism. Third, the hierarchical parallelism present in the architecture will be even more pronounced than in the past, with additional resource sharing (and contention) between threads that run on the same core. There is likely to be a smaller amount of cache per thread and low bandwidth to main memory, since this is a shared resource. Last but not least, the heterogeneity within a node will need to be addressed.

Whereas MPI has proved to be an excellent means of expressing program parallelism when nodes have a small number of cores, future architectures may make this a tough proposition. In particular, MPI does not give the application developer the means to conserve memory or to directly modify the code to benefit from resource sharing and to avoid its negative implications. One possible way forward is to systematically combine MPI with OpenMP. OpenMP is a widely-supported high-level shared memory programming model that provides ease of use and maintenance simplicity. Version 3.0 of OpenMP [1] has considerably extended the scope of this API. It allows multilevel loop nest parallelism, enhances support for nested parallelism and introduces tasks, which are conceptually placed into a pool of tasks for subsequent execution by an arbitrary thread.

A. Lastovetsky et al. (Eds.): EuroPVM/MPI 2008, LNCS 5205, pp. 3–4, 2008.

In this presentation, we discuss this approach and its potential. We describe a typical implementation strategy for OpenMP (see e.g. [2]), and some immediate implications for program performance. As a pure shared memory model, OpenMP does not address the locality of data with respect to the executing threads or the system. Nor does it permit mappings of threads to the hardware resources. Yet the performance impact of these is well documented (e.g. [3]). We recently proposed a small set of extensions that enable the restriction of OpenMP's worksharing directives to a subset of the threads in a team [4]. Such a feature might make it easier to map threads to the system by making a single level of parallelism more expressive. It might also facilitate the expression of code that overlaps OpenMP computation and MPI communication. Other work has proposed enhancements to nested parallelism (e.g. [5]) and loop schedules. Industry is working to support OpenMP on heterogeneous platforms (e.g. [6,7,8,9]. We consider how these and other efforts might lead to an improved hybrid programming model.

References

1. OpenMP ARB. OpenMP application programming interface (October 2007), `http://www.openmp.org/drupal/mp-documents/spec30_draft.pdf`
2. Chapman, B., Jost, G., van der Pas, R.: Using OpenMP: Portable Shared Memory Parallel Programming. MIT Press, Cambridge (2008)
3. Antony, J., Janes, P.P., Rendell, A.P.: Exploring Thread and Memory Placement on NUMA Architectures: Solaris and Linux, UltraSPARC/FirePlane and Opteron/HyperTransport. In: Robert, Y., Parashar, M., Badrinath, R., Prasanna, V.K. (eds.) HIPC 2006. LNCS, vol. 4297, pp. 338–352. Springer, Heidelberg (2006)
4. Chapman, B.M., Huang, L., Jin, H., Jost, G., de Supinski, B.R.: Toward enhancing OpenMP's work-sharing directives. In: Europar 2006, pp. 645–654 (2006)
5. Gonzalez, M., Ayguadé, E., Martorell, X., Labarta, J., Navarro, N., Oliver, J.: NanosCompiler: supporting flexible multilevel parallelism exploitation in OpenMP. Concurrency - Practice and Experience 12(12), 1121–1218 (2000)
6. Eichenberger, A.E., O'Brien, K., Wu, P., Chen, T., Oden, P.H., Prener, D.A., Shepherd, J.C., So, B., Sura, Z., Wang, A., Zhang, T., Zhao, P., Gschwind, M.: Optimizing compiler for a cell processor. In: PACT 2005: Proceedings of the 14th International Conference on Parallel Architectures and Compilation Techniques, pp. 161–172. IEEE Computer Society, Washington (2005)
7. CAPS Enterprise. Hmpp: A hybrid multicore parallel programming platform, `http://www.caps-entreprise.com/en/documentation/caps_hmpp_product_brief.pdf`
8. Wang, P.H., Collins, J.D., Chinya, G.N., Jiang, H., Tian, X., Girkar, M., Yang, N.Y., Lueh, G.-Y., Wang, H.: Exochi: architecture and programming environment for a heterogeneous multi-core multithreaded system. In: PLDI 2007: Proceedings of the 2007 ACM SIGPLAN conference on Programming language design and implementation, pp. 156–166. ACM, New York (2007)
9. Gaster, B., Bradley, C.: Exploiting loop-level parallelism for SIMD arrays using OpenMP. In: Proceedings of IWOMP 2007 (June 2007)

MPI Must Evolve or Die

Al Geist

Oak Ridge National Laboratory,
PO Box 2008,
Oak Ridge, TN 37831-6016
gst@ornl.gov
http://www.csm.ornl.gov/~geist

Abstract. Multicore and hybrid architecture designs dominate the landscape for systems that are 1 to 20 petaflops peak performance. As such the MPI software must evolve to effectively use these types of architectures or it will die just like the vector programming models died. While applications may continue to use MPI, it is not business as usual in how communication libraries are being changed to effectively exploit the new petascale systems. This talk presents some key research in petascale communication libraries going on in the "Harness" project, which is the follow-on to the PVM research project.

The talk will cover a number of areas being explored, including hierarchical algorithm designs, hybrid algorithm designs, dynamic algorithm selection, and fault tolerance inside next generation message passing libraries. Hierarchical algorithm designs seek to consolidate information at different levels of the architecture to reduce the number of messages and contention on the interconnect. Natural places for such consolidation include the socket level, the node level, the cabinet level, and multiple-cabinet level. Hybrid algorithm designs use different algorithms at different levels of the architecture, for example, an ALL_GATHER may use a shared memory algorithm across the node and a message passing algorithm between nodes, in order to better exploit the different data movement capabilities. An adaptive communication library may dynamically select from a set of collective communication algorithms based on the number of nodes being sent to, where they are located in the system, the size of the message being sent, and the physical topology of the computer.

This talk will also describe how ORNLs Leadership computing facility (LCF) has been proactive in getting science teams to adopt the latest communication and IO techniques. This includes assigning computational science liaisons to each science team. The liaison has knowledge of both the systems and the science, providing a bridge to improved communication patterns. The LCF also has a Cray Center of Excellence and a SUN Lustre Center of Excellence on site. These centers provide Cray and SUN engineers who work directly with the science teams to improve the MPI and MPI-IO performance of their applications.

Finally this talk will take a peek at exascale architectures and the need for new approaches to software development that integrates arcitecture design and algorithm development to facilitate the synergistic evolution of both.

A. Lastovetsky et al. (Eds.): EuroPVM/MPI 2008, LNCS 5205, p. 5, 2008.

MPI and Hybrid Programming Models for Petascale Computing

William D. Gropp

Department of Computer Science
University of Illinois at Urbana-Champaign
Urbana, IL
wgropp@illinois.edu
http://www.cs.uiuc.edu/homes/wgropp

Abstract. In 2011, the National Center for Supercomputing Applications at the University of Illinois will begin operation of the Blue Waters petascale computing system. This system, funded by the National Science Foundation, will deliver a sustained performance of one to two petaflops for many applications in science and engineering.

Blue Waters will support a variety of programming models, including the "MPI everywhere" model that is the most common among today's MPI applications. In addition, it will support a variety of other programming models. The programming models may be used instead of MPI or they may be used in *combination* with MPI. Such a combined programming model is often called a *hybrid* model. The most familiar of the models used in combination with MPI is OpenMP, which is designed for shared-memory systems and is based on the use of multiple threads in each MPI process. This programming model has found mixed success to date, with many experiments showing little benefit while others show promise. The reason for this is related to the use of OpenMP within MPI programs—where OpenMP is used to complement MPI, for example, by providing better support for load-balancing adaptive computations or sharing large data tables, it can provide a significant benefit. Where it is used as an alternative to MPI, OpenMP often has difficulty achieving the performance of MPI (MPI's much-criticized requirement that the user directly manage data motion ensures that the programmer does in fact manage that memory motion, leading to improved performance). This suggests that other programming models can be productively combined with MPI as long as they complement, rather than replace, MPI.

One class of languages of great current interest are the Partitioned Global Address Space (PGAS) languages. These languages distinguish between local and remote memory (thus keeping the user aware of the performance consequences of remote memory access) but provide simpler and more efficient mechanisms for accessing or updating remote memory than are available in MPI. While some applications will undoubtedly be written entirely in these newer programming models, most applications that will run on petascale systems such as Blue Waters have already been written; at the most, some performance-critical software components may be replaced with components that are implemented using a different programming model.

A. Lastovetsky et al. (Eds.): EuroPVM/MPI 2008, LNCS 5205, pp. 6–7, 2008.

In all of these hybrid programming models, how will the different components interact? How should the different models coordinate their use of the underlying processor resources? Can these programming models share low-level infrastructure on systems such as Blue Waters? This talk will discuss some of the alternatives and suggest directions for investigation.

Future high-end systems are likely to contain hybrid computing elements. The Roadrunner system at the Los Alamos National Laboratory, which was the the first computer to exceed 1 petaflop on the HP Linpack benchmark, is an example of such a system. It combines conventional processors with a processor optimized for operations encountered in graphics processing, the Cell Broadband Engine. Like the preceding examples of programming models, such systems offer the greatest advantages when the different components complement each other. Should the programming model for such systems try to hide the differences, providing a simple, uniform view of the system, or should the programming model reflect, at an abstract level, the different strengths of the different components? This talk will look at some of the issues of using MPI in combination with other software and hardware models and discuss how MPI can remain effective at the petascale and beyond.

Some Aspects of Message-Passing on Future Hybrid Systems (Extended Abstract)

Rolf Rabenseifner

University of Stuttgart, High-Performance Computing-Center (HLRS)
70550 Stuttgart, Germany
rabenseifner@hlrs.de
www.hlrs.de

Keywords: Hybrid parallel programming, MPI, OpenMP, PGAS, Subteams, Standardization, MPI-3.

In the future, most systems in high-performance computing (HPC) will have a hierarchical hardware design, e.g., a cluster of ccNUMA or shared memory nodes with each node having several multi-core CPUs. Parallel programming must combine the distributed memory parallelization on the node inter-connect with the shared memory parallelization inside each node. There are many mismatch problems between hybrid hardware topology and the hybrid or homogeneous parallel programming models on such hardware. Hybrid programming with a combination of MPI and OpenMP is often slower than pure MPI programming. Major chances arise from the load balancing features of OpenMP and from a smaller memory footprint if the application duplicates some data on all MPI processes [1,2,3].

The master-only MPI&OpenMP style is the simplest hybrid programming paradigm style. The application only communicates outside parallel OpenMP regions. All other threads will sleep on each SMP node while the master thread is communicating. This obvious drawback can be overcome by overlapping communication and computation, e.g., the master thread communicates while all other threads are executing application code. OpenMP worksharing directives are not designed for this programming paradigm. Barbara Chapman et al. developed the idea of subteams to allow worksharing directives on a subset of threads [4].

Optimization of hybrid (and non-hybrid) parallel applications normally requires knowledge about the hardware. For example, in a torus network, the locality of MPI processes with respect to the topology of the application's domain decomposition can be an important factor in minimizing communication overhead. Knowledge about a hierarchical memory and network structure is desirable for optimizing within the application and inside the communication library. Examples are a multi-level domain decomposition inside the application to benefit from cheaper local communication, or hardware-aware optimization of collective MPI routines. Often, optimization decisions are not portable. This isn't a contradiction to MPI's goals of portability and efficiency. Already eleven years ago,

A. Lastovetsky et al. (Eds.): EuroPVM/MPI 2008, LNCS 5205, pp. 8–10, 2008.

the MPI Forum discussed *cluster attributes* to enable portable optimizations of applications on hybrid hardware [5]. Today, new challenges emerge from significant bandwidth differences between local memories (on the chip, between cores), communication through the shared memory, and between SMP nodes. It will be important not to use the memory when data is available through caches or local memories. Not only the MPI-3 standardization, but also the evolving PGAS languages (e.g., Co-Array Fortran [6,8], UPC [7,8], Chapel [9], Titanium [10,11], X10 [12]) are working to find efficient answers and programming environments for future hardware.

There is also an impact of software and benchmark standards on the future hardware development and market. In particular, micro benchmarks may exclude important aspects and focus only on special topics. For this reason, the Linpack benchmark [13] was complemented by the HPC Challenge Benchmark Suite [14,15]. With parallel I/O, benchmarks with micro-kernels tend to measure only wellformed I/O, i.e., using chunk sizes that are a multiple of some power-of-two. In contrast, real application mainly use non-wellformed I/O. For example, the b_eff_io benchmark can show significant differences between wellformed and non-wellformed junk sizes [16,17]. And as a third example, the MPI-2 Forum decided in 1997 to define a one-sided communication interface that does not really benefit from physical remote direct memory access (RDMA) although the SHMEM library [18] was already available, but only on specific hardware. Has this decision slowed down a broad development of RDMA capabilities in cluster networks? Can a new RDMA-based one-sided MPI interface help to efficiently use clusters of multi-core SMP nodes? Is it a must if mixed-model programming, like MPI & OpenMP or MPI & Co-Array Fortran, should be used? The MPI-3 Forum currently tries to address such questions.

References

1. Rabenseifner, R., Hager, G., Jost, G., Keller, R.: Hybrid MPI and OpenMP Parallel Programming. In: Half-day Tutorial No. S-10 at Super Computing 2007, SC 2007, Reno, Nevada, USA, November 10-16 (2007)
2. Rabenseifner, R.: Hybrid Parallel Programming on HPC Platforms. In: Proceedings of the Fifth European Workshop on OpenMP, EWOMP 2003, Aachen, Germany, September 22-26, pp. 185–194 (2003) , `www.compunity.org`
3. Rabenseifner, R., Wellein, G.: Communication and Optimization Aspects of Parallel Programming Models on Hybrid Architectures. International Journal of High Performance Computing Applications 17(1), 49–62 (2003)
4. Chapman, B.M., Huang, L., Jin, H., Jost, G., de Supinski, B.R.: Toward Enhancing OpenMP's Work-Sharing Directives. In: Nagel, W.E., Walter, W.V., Lehner, W. (eds.) Euro-Par 2006. LNCS, vol. 4128, pp. 645–654. Springer, Heidelberg (2006)
5. MPI-2 Journal of Development. The Message Passing Interface Forum, July 18 (1997), `http://www.mpi-forum.org`
6. Co-Array Fortran, `http://www.co-array.org/`
7. Unified Parallel C, `http://www.gwu.edu/upc/`

8. Coarfa, C., Dotsenko, Y., Mellor-Crummey, J.M., Cantonnet, F., El-Ghazawi, T.A., Mohanti, A., Yao, Y., Chavarria-Miranda, D.G.: An Evaluation of Global Address Space Languages: Co-Array Fortran and Unified Parallel C. In: Proceedings of the 10th ACM SIGPLAN Symposium on Principles and Practice of Parallel Programming (PPoPP 2005), Chicago, Illinois (June 2005)
9. Chapel: The Cascade High-Productivity Language, `http://chapel.cs.washington.edu/`
10. Titanium, `http://titanium.cs.berkeley.edu/`
11. Yelick, K.A., Semenzato, L., Pike, G., Miyamoto, C., Liblit, B., Krishnamurthy, A., Hilfinger, P.N., Graham, S.L., Gay, D., Colella, P., Aiken, A.: Titanium: A High-Performance Java Dialect. Concurrency: Practice and Experience 10(11-13) (September-November 1998)
12. The Experimental Concurrent Programming Language (X10), `http://x10.sf.net/`
13. Dongarra, J., Luszczek, P., Petitet, A.: The LINPACK Benchmark: past, present and future. Concurrency and Computation: Practice and Experience 15(9), 803–820 (2003)
14. Luszczek, P., Dongarra, J.J., Koester, D., Rabenseifner, R., Lucas, B., Kepner, J., McCalpin, J., Bailey, D., Takahashi, D.: Introduction to the HPC Challenge Benchmark Suite (March 2005) , `http://icl.cs.utk.edu/hpcc/`
15. Saini, S., Ciotti, R., Gunney, B.T.N., Spelce, T.E., Koniges, A., Dossa, D., Adamidis, P., Rabenseifner, R., Tiyyagura, S.R., Mueller, M.: Performance Evaluation of Supercomputers using HPCC and IMB Benchmarks. J. Comput. System Sci. (2007) (to be published) doi:10.1016/j.jcss.2007.07.002 (Special issue on Performance Analysis and Evaluation of Parallel, Cluster, and Grid Computing Systems)
16. Rabenseifner, R., Koniges, A.E., Prost, J.-P., Hedges, R.: The Parallel Effective I/O Bandwidth Benchmark: b_eff_io. In: Cerin, C., Jin, H. (eds.) Parallel I/O for Cluster Computing, Ch. 4., Kogan Page Ltd., Febuary 2004, pp. 107–132 (2004)
17. Saini, S., Talcott, D., Thakur, R., Adamidis, P., Rabenseifner, R., Ciotti, R.: Parallel I/O Performance Characterization of Columbia and NEC SX-8 Superclusters. In: Proceedings of the IPDPS 2007 Conference, the 21st IEEE International Parallel & Distributed Processing Symposium, Workshop on Performance Modeling, Evaluation, and Optimization of Parallel and Distributed Systems (PMEO-PDS 2007), Long Beach, California, USA, March 26-30 (2007) (to be published)
18. Message Passing Toolkit (MPT) 3.0 Man Pages, `http://docs.cray.com/`

From Parallel Virtual Machine to Virtual Parallel Machine: The Unibus System

Vaidy Sunderam

Department of Mathematics and Computer Science
Emory University
Atlanta, Georgia, USA
vss@emory.edu

Network-based concurrent computing frameworks have matured over two decades. However, in the realm of multi-domain geographically distributed computing, their true potential is yet to be realized. Popularly termed "grids", these metacomputing systems have yet to see widespread adoption and use, partly due to challenges resulting from heterogeneity, deployment issues, and dynamicity. The Unibus project proposes novel strategies for unifying and dynamically conditioning shared heterogeneous resources located within multiple administrative domains. Our approach is based on client-centric overlay software that provides unified interfaces to diverse resources, complemented by runtime systems that substantially automate setup and configuration. The overlay unifies heterogeneous resources through service drivers or mediators that operate analogously to device drivers. New resources and variations in availability are handled through self-deployment of mediators and their service daemon counterparts, enabling the overlay to adaptively present applications with coherent aggregated projections of the underlying resources. Dynamic incorporation of new resources is facilitated via automated environment conditioning that uses deployment descriptors in conjunction with site-specific data to prepare platforms for execution. We will discuss the motivations and rationale behind the Unibus approach, describe the design of the framework, and present preliminary results and experiences.

A. Lastovetsky et al. (Eds.): EuroPVM/MPI 2008, LNCS 5205, p. 11, 2008.

EuroPVM/MPI Full-Day Tutorial. Using MPI-2: A Problem-Based Approach*

William Gropp[1] and Ewing Lusk[2]

[1] University of Illinois
wgropp@illinois.edu
[2] Argonne National Laboratory
lusk@mcs.anl.gov

Abstract. MPI-2 introduced many new capabilities, including dynamic process management, one-sided communication, and parallel I/O. Implementations of these features are becoming widespread. This tutorial shows how to use these features by showing all of the steps involved in designing, coding, and tuning solutions to specific problems. The problems are chosen for their practical use in applications as well as for their ability to illustrate specific MPI-2 topics. Complete examples that illustrate the use of MPI one-sided communication, MPI parallel I/O, and hybrid programming with MPI and threads will be discussed and full source code will be made available to the attendees. Each example will include a hands-on lab session; these sessions will also introduce the use of performance and correctness debugging tools that are available for the MPI environment. Guidance on tuning MPI programs will be included, with examples and data from MPI implementations on a variety of parallel systems, including Sun, IBM, SGI, and clusters. Examples in C, Fortran, and C++ will be included. Familiarity with basic MPI usage will be assumed.

* This work was supported by the U.S. Dept. of Energy under Contract DE-AC02-06CH11357.

A. Lastovetsky et al. (Eds.): EuroPVM/MPI 2008, LNCS 5205, p. 12, 2008.

Non-data-communication Overheads in MPI: Analysis on Blue Gene/P

P. Balaji[1], A. Chan[2], W. Gropp[3], R. Thakur[1], and E. Lusk[1]

[1] Math. and Comp. Sci. Div., Argonne Nat. Lab., Argonne, IL 60439, USA
[2] Dept. of Astronomy and Astrophysics, Univ. of Chicago, Chicago, IL 60637
[3] Dept. of Computer Science, Univ. of Illinois, Urbana, IL, 61801, USA

Abstract. Modern HEC systems, such as Blue Gene/P, rely on achieving high-performance by using the parallelism of a massive number of low-frequency/low-power processing cores. This means that the local pre- and post-communication processing required by the MPI stack might not be very fast, owing to the slow processing cores. Similarly, small amounts of serialization within the MPI stack that were *acceptable* on small/medium systems can be brutal on massively parallel systems. In this paper, we study different non-data-communication overheads within the MPI implementation on the IBM Blue Gene/P system.

1 Introduction

As we move closer to the petaflop era, modern high-end computing (HEC) systems are undergoing a drastic change in their fundamental architectural model. With processor speeds no longer doubling every 18-24 months owing to the exponential increase in power consumption and heat dissipation, modern HEC systems tend to rely lesser on the performance of single processing units. Instead, they try to extract parallelism out of a massive number of low-frequency/low-power processing cores. IBM Blue Gene/L [1] was one of the early supercomputers to follow this architectural model, soon followed by other systems such as Blue Gene/P (BG/P) [5] and SiCortex [2].

While such an architecture provides the necessary ingredients for building petaflop and larger systems, the actual performance perceived by users heavily depends on the capabilities of the systems-software stack used, such as the MPI implementation. While the network itself is quite fast and scalable on these systems, the local pre- and post-data-communication processing required by the MPI stack might not be as fast, owing to the slow processing cores. For example, local processing tasks within MPI that were considered *quick* on a 3.6 GHz Intel processor, might form a significant fraction of the overall MPI processing time on the modestly fast 850 MHz cores of a BG/P. Similarly, small amounts of serialization within the MPI stack which were considered *acceptable* on a system with a few hundred processors, can be brutal when running on massively parallel systems with hundreds of thousands of cores.

In this paper, we study the non-data-communication overheads in MPI on BG/P. We identify various bottleneck possibilities within the MPI stack, with

A. Lastovetsky et al. (Eds.): EuroPVM/MPI 2008, LNCS 5205, pp. 13–22, 2008.

respect to the slow pre- and post-data-communication processing as well as serialization points, stress these overheads using different benchmarks, analyze the reasons behind such overheads and describe potential solutions for solving them.

2 BG/P Software and Hardware Stacks

BG/P has five different networks [6]. Two of them (10G and 1G Ethernet with JTAG interface) are used for file I/O and system management. The other three (3-D Torus, Global Collective and Global Interrupt) are used for MPI communication. The 3-D torus network is used for MPI point-to-point and multicast operations and connects all compute nodes to form a 3-D torus. Thus, each node has six nearest-neighbors. Each link provides a bandwidth of 425 MBps per direction (total 5.1 GBps). The global collective network is a one-to-all network for compute and I/O nodes used for MPI collective communication and I/O services. Each node has three links to this network (total 5.1 GBps bidirectional). The global interrupt network is an extremely low-latency network for global barriers and interrupts, e.g., the global barrier latency of a 72K-node partition is approximately 1.3μs. On BG/P, compute cores do not handle packets on the torus network; a DMA engine on each node offloads most of the network packet injecting and receiving work, enabling better overlap of computation and communication. However, the cores handle sending/receiving packets from the collective network.

BG/P is designed for multiple programming models. The Deep Computing Messaging Framework (DCMF) and the Component Collective Messaging Interface (CCMI) are used as general purpose libraries to support different programming models [9]. DCMF implements point-to-point and multisend protocols. The multisend protocol connects the abstract implementation of collective operations in CCMI to targeted communication networks.

IBM's MPI on BG/P is based on MPICH2 and is implemented on top of DCMF. Specifically, it borrows most of the upper-level code from MPICH2, including MPI-IO and the MPE profiler, while implementing BG/P specific details within a device implementation called `dcmfd`. The DCMF library provides low-level communication support. All advanced communication features such as allocation and handling of MPI requests, dealing with tags and unexpected messages, multi-request operations such as `MPI_Waitany` or `MPI_Waitall`, derived-datatype processing and thread synchronization are **not** handled by the DCMF library and have to be taken care of by the MPI implementation.

3 Experiments and Analysis

Here, we study the non-data-communication overheads in MPI on BG/P.

3.1 Basic MPI Stack Overhead

An MPI implementation can be no faster than the underlying communication system. On BG/P, this is DCMF. Our first measurements (Figure 1) compare

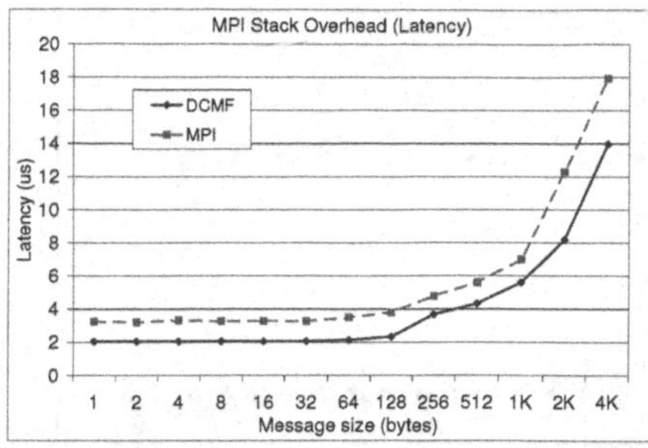

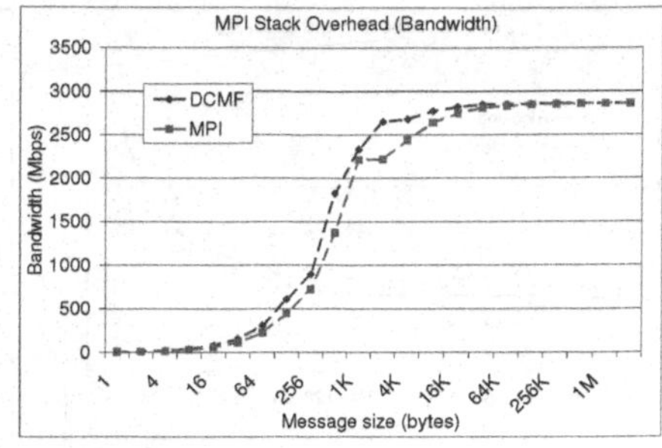

Fig. 1. MPI stack overhead

the communication performance of MPI (on DCMF) with the communication performance of DCMF. For MPI, we used the OSU MPI suite [10] to evaluate the performance. For DCMF, we used our own benchmarks on top of the DCMF API, that imitate the OSU MPI suite. The latency test uses blocking communication operations while the bandwidth test uses non-blocking communication operations for maximum performance in each case.

The difference in performance of the two stacks is the overhead introduced by the MPI implementation on BG/P. We observe that the MPI stack adds close to 1.1μs overhead for small messages; that is, close to 1000 cycles are spent for pre- and post-data-communication processing within the MPI stack. We also notice that for message sizes larger than 1KB, this overhead is much higher (closer to 4μs or 3400 cycles). This additional overhead is because the MPI stack uses a protocol switch from eager to rendezvous for message sizes larger than 1200 bytes. Though DCMF itself performs the actual rendezvous-based data communication, the MPI stack performs additional book-keeping in this mode which causes this additional overhead. In several cases, such redundant book-keeping is unnecessary and can be avoided.

3.2 Request Allocation and Queueing Overhead

MPI provides non-blocking communication routines that enable concurrent computation and communication where the hardware can support it. However, from the MPI implementation's perspective, such routines require managing `MPI_Request` handles that are needed to wait on completion for each non-blocking operation. These requests have to be allocated, initialized and queued/dequeued within the MPI implementation for each send or receive operation, thus adding overhead, especially on low-frequency cores.

In this experiment, we measure this overhead by running two versions of the typical ping-pong latency test—one using `MPI_Send` and `MPI_Recv` and the other using `MPI_Isend`, `MPI_Irecv`, and `MPI_Waitall`. The latter incurs the overhead of allocating, initializing, and queuing/dequeuing request handles. Figure 2 shows that this overhead is roughly 0.4 μs or a little more than 300 clock cycles.[1] While

[1] This overhead is more than the entire point-to-point MPI-level shared-memory communication latency on typical commodity Intel/AMD processors [7].

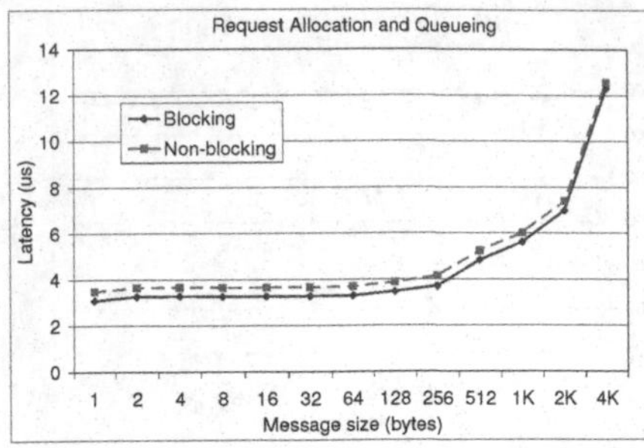

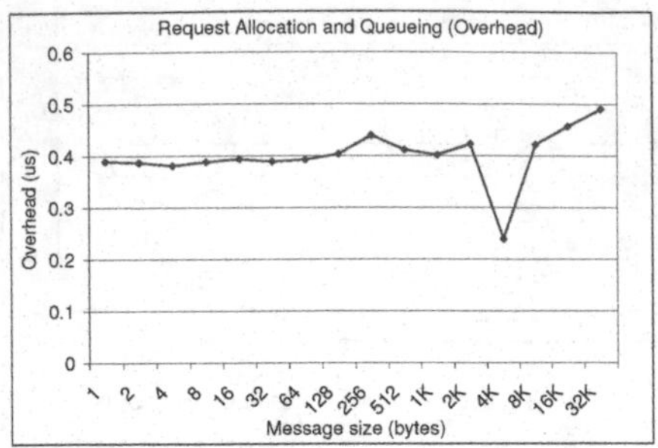

Fig. 2. Request allocation and queuing: (a) Overall performance; (b) Overhead

this overhead is expected due to the number of request management operations, carefully redesigning them can potentially bring this down significantly.

3.3 Overheads in Tag and Source Matching

MPI allows applications to classify different messages into different categories using tags. Each sent message carries a tag. Each receive request contains a tag and information about which source the message is expected from. When a message arrives, the receiver searches the queue of posted receive requests to find the one that matches the arrived message (both tag and source information) and places the incoming data in the buffer described by this request. Most current MPI implementations use a single queue for all receive requests, i.e., for all tags and all source ranks. This has a potential scalability problem when the length of this queue becomes large.

To demonstrate this problem, we designed an experiment that measures the overhead of receiving a message with increasing request-queue size. In this experiment, process P0 posts M receive requests for each of N peer processes with tag T0, and finally one request of tag T1 for P1. Once all the requests are posted (ensured through a low-level hardware barrier that does not use MPI), P1 sends a message with tag T1 to P0. P0 measures the time to receive this message not including the network communication time. That is, the time is only measured for the post-data-communication phase to receive the data after it has arrived in its local temporary buffer.

Figure 3 shows the time taken by the MPI stack to receive the data after it has arrived in the local buffer. Figures 3(a) and 3(b) show two different versions of the test—the first version keeps the number of peers to one ($N = 1$) but increases the number of requests per peer (M), while the second version keeps the number of requests per peer to one ($M = 1$) but increases the number of peers (N). For both versions, the time taken increases rapidly with increasing number of total requests ($M \times N$). In fact, for 4096 peers, which is modest considering the size BG/P can scale to, we notice that even just *one request per peer* can result in a queue parsing time of about 140000μs.

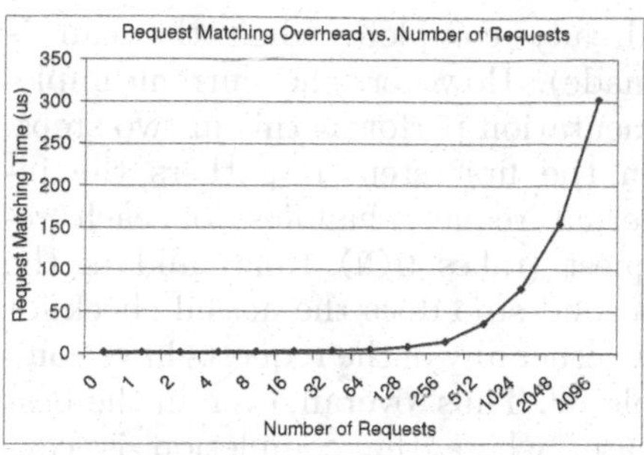

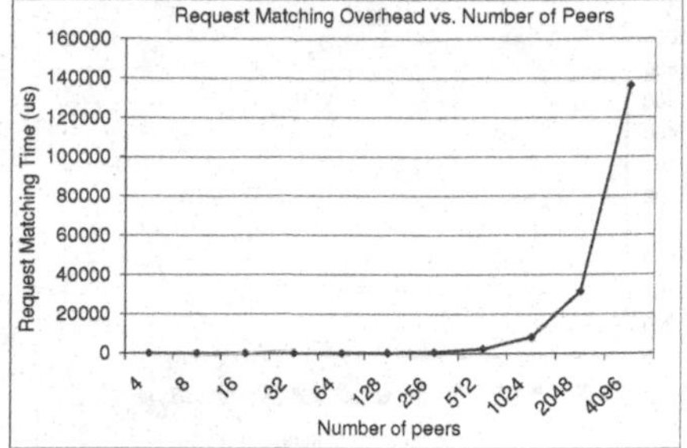

Fig. 3. Request matching overhead: (a) requests-per-peer, (b) number of peers

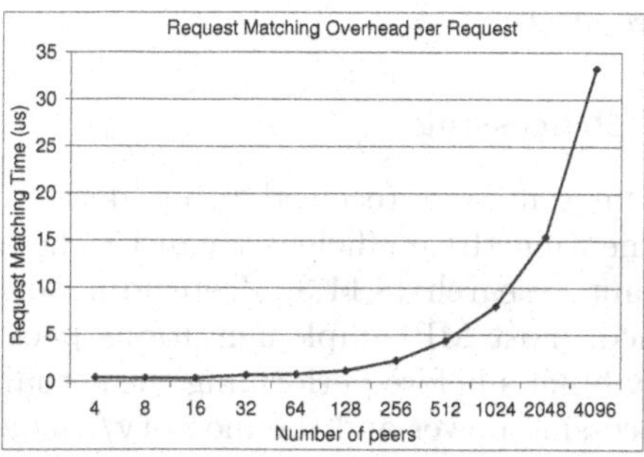

Fig. 4. Matching overhead per request

Another interesting observation in the graph is that the time increase with the number of peers is not linear. To demonstrate this, we present the average time taken per request in Figure 4—the average time per request increases as the number of requests increases! Note that parsing through the request queue should take linear time; thus the time per request should be constant, not increase. There are several reasons for such a counter-intuitive behavior; we believe the primary cause for this is the limited number of pre-allocated requests that are reused during the life-time of the application. If there are too many pending requests, the MPI implementation runs out of these pre-allocated requests and more requests are allocated dynamically.

3.4 Algorithmic Complexity of Multi-request Operations

MPI provides operations such as `MPI_Waitany`, `MPI_Waitsome` and `MPI_Waitall` that allow the user to provide multiple requests at once and wait for the completion of one or more of them. In this experiment, we measure the MPI stack's capability to efficiently handle such requests. Specifically, the receiver posts several receive requests (`MPI_Irecv`) and once all the requests are posted (ensured through a low-level hardware barrier) the sender sends just one message that matches the first receive request. We measure the time taken to receive the message, not including the network communication time, and present it in Figure 5.

We notice that the time taken by `MPI_Waitany` increases linearly with the number of requests passed to it. We expect this time to be constant since the incoming message matches the first request itself. The reason for this behavior is the algorithmic complexity of the `MPI_Waitany` implementation. While `MPI_Waitany` would have a worst-case complexity of `O(N)`, where `N` is the number of requests, its best-case complexity should be constant (when the first request is

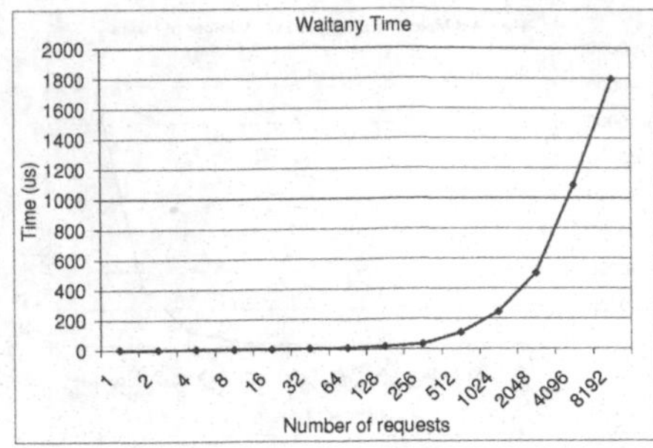

Fig. 5. `MPI_Waitany` Time

already complete when the call is made). However, the current implementation performs this in two steps. In the first step, it gathers the internal request handles for each request (takes `O(N)` time) and in the second step does the actual check for whether any of the requests have completed. Thus, overall, even in the best case, where the completion is constant time, acquiring of internal request handlers can increase the time taken linearly with the number of requests.

3.5 Overheads in Derived Datatype Processing

MPI allows non-contiguous messages to be sent and received using derived datatypes to describe the message. Implementing these efficiently can be challenging and has been a topic of significant research [8,11,3]. Depending on how densely the message buffers are aligned, most MPI implementations pack sparse datatypes into contiguous temporary buffers before performing the actual communication. This stresses both the processing power and the memory/cache bandwidth of the system. To explore the efficiency of derived datatype communication on BG/P, we looked only at the simple case of a single stride (vector) type with a stride of two. Thus, every other data item is skipped, but the total amount of data packed and communicated is kept uniform across the different datatypes (equal number of bytes). The results are shown in Figure 6.

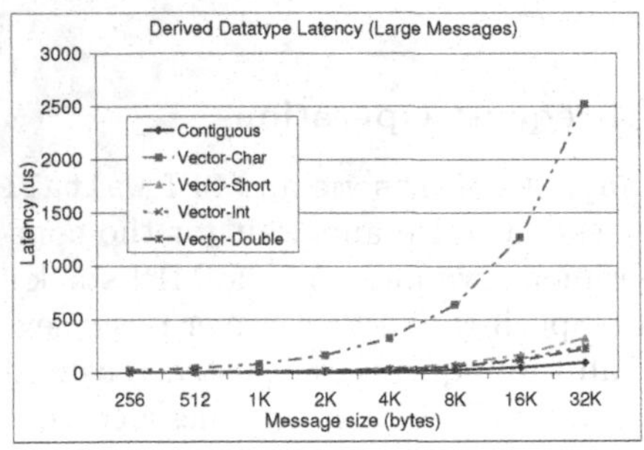

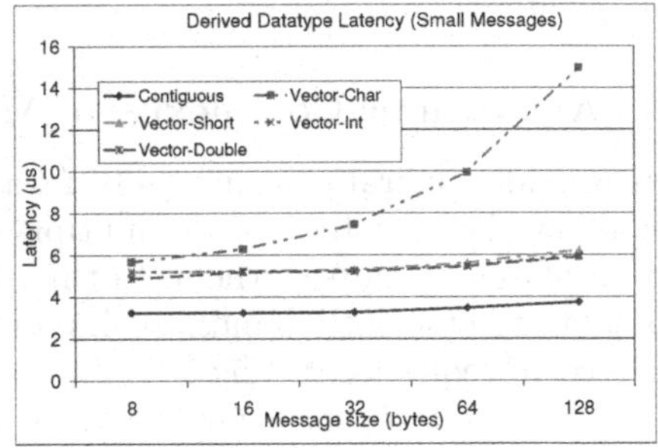

Fig. 6. Derived datatype latency: (a) long messages and (b) short messages

These results show a significant gap in performance between sending a contiguous messages and a non-contiguous message (with the same number of bytes). The situation is particularly serious for a vector of individual bytes (`MPI_CHAR`). It is also interesting to look at the behavior for shorter messages (Figure 6(b)).

This shows, roughly, a 2 μs gap in performance between contiguous send and a send of short, integer or double precision data with a stride of two.

3.6 Buffer Alignment Overhead

For operations that involve touching the data that is being communicated (such as datatype packing), the alignment of the buffers that are being processed can play a role in overall performance if the hardware is optimized for specific buffer alignments (such as word or double-word alignments), which is common in most hardware today.

In this experiment (Figure 7), we measure the communication latency of a vector of integers (4 bytes) with a stride of 2 (that is, every alternate integer is packed and communicated). We perform the test for different alignment of these integers—"0" refers to perfect alignment to a double-word boundary, "1" refers to an misalignment of 1-byte. We notice that as long as the integers are within the same double-word (0-4 byte misalignment) the performance is better as compared to when the integers span two different double-words (5-7 byte misalignment), the performance difference being about 10%. This difference is expected as integers crossing the double-word boundary require both the double-words to be fetched before any operation can be performed on them.

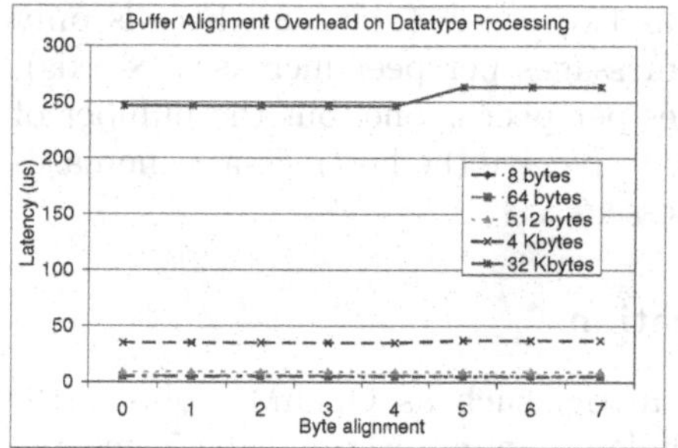

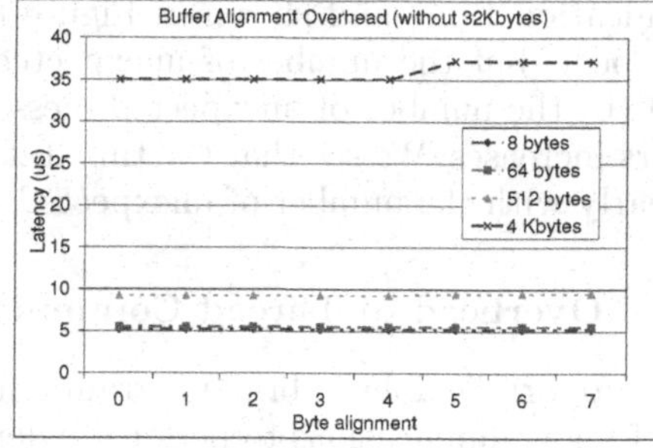

Fig. 7. Buffer alignment overhead

3.7 Unexpected Message Overhead

MPI does not require any synchronization between the sender and receiver processes before the sender can send its data out. So, a sender can send multiple messages which are not immediately requested for by the receiver. When the receiver tries to receive the message it needs, all the previously sent messages are considered *unexpected*, and are queued within the MPI stack for later requests to handle. Consider the sender first sending multiple messages of tag T0 and finally one message of tag T1. If the receiver is first looking for the message with tag T1, it considers all the previous messages of tag T0 as *unexpected* and queues them in the unexpected queue. Such queueing and dequeuing of requests (and potentially copying data corresponding to the requests) can add overhead.

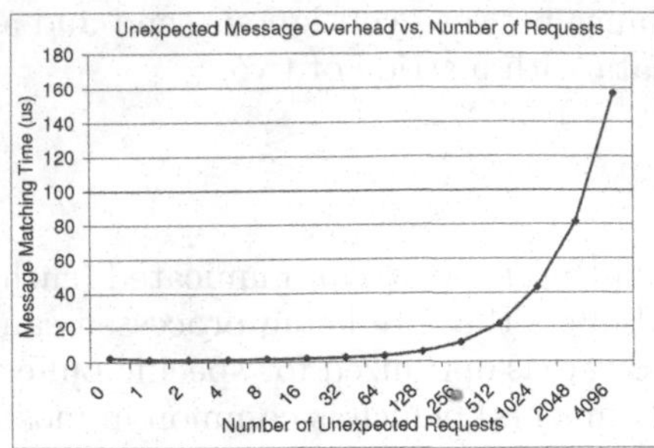

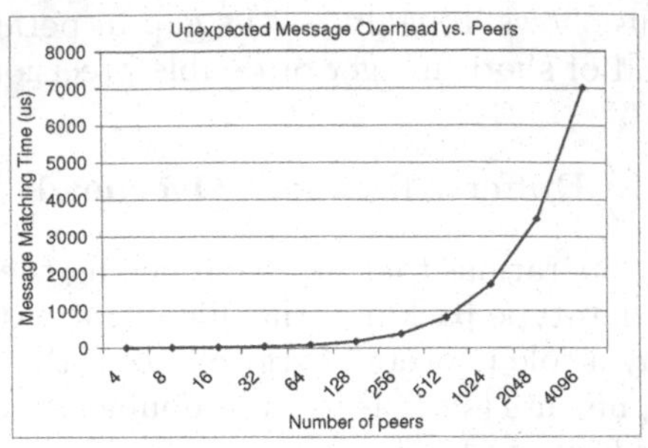

Fig. 8. Unexpected message overhead: (a) Increasing number of messages per peer, with only one peer; (b) Increasing number of peers, with only one message per peer

To illustrate this, we designed an experiment that is a symmetric-opposite of the tag-matching test described in Section 3.3. Specifically, in the tag-matching test, we queue multiple receive requests and receive one message that matches the last queued request. In the unexpected message test, we receive multiple messages, but post only one receive request for the last received message. Specifically, process P0 first receives M messages of tag T0 from each of N peer processes and finally receives one extra message of tag T1 from P1. The time taken to receive the final message (tag T1) is measured, not including the network communication time, and shown in Figure 8 as two cases: (a) when there is only one peer, but the number of unexpected messages per peer increases (x-axis), and (b) the number of unexpected messages per peer is one, but the number of peers increases. We see that the time taken to receive the last message increases linearly with the number of unexpected messages.

3.8 Overhead of Thread Communication

To support flexible hybrid programming model such as OpenMP plus MPI, MPI allows applications to perform independent communication calls from each thread by requesting for `MPI_THREAD_MULTIPLE` level of thread concurrency from the MPI implementation. In this case, the MPI implementation has to perform appropriate locks within shared regions of the stack to protect conflicts caused due to concurrent communication by all threads. Obviously, such locking has two drawbacks: (i) they add overhead and (ii) they can serialize communication.

We performed two tests to measure the overhead and serialization caused by such locking. In the first test, we use four processes on the different cores which send 0-byte messages to `MPI_PROC_NULL` (these messages incur all the overhead of the MPI stack, except that they are never sent out over the network, thus imitating an infinitely fast network). In the second test, we use four threads with `MPI_THREAD_MULTIPLE` thread concurrency to send 0-byte messages to `MPI_PROC_NULL`. In the threads case, we expect the locks to add overheads and serialization, so the performance to be lesser than in the processes case.

Figure 9 shows the performance of the two tests described above. The difference between the one-process and one-thread cases is that the one-thread case requests for the MPI_THREAD_MULTIPLE level of thread concurrency, while the one-process case requests for no concurrency, so there are no locks. As expected, in the process case, since there are no locks, we notice a linear increase in performance with increasing number of cores used. In the threads case, however, we observe two issues: (a) the performance of one thread is significantly lower than the performance of one process and (b) the performance of threads does not increase at all as we increase the number of cores used.

The first observation (difference in one-process and one-thread performance) points out the overhead in maintaining locks. Note that there is no contention on the locks in this case as there is only one thread accessing them. The second observation (constant performance with increasing cores) reflects the inefficiency in the concurrency model used by the MPI implementation. Specifically, most MPI implementations perform a global lock for each MPI operation thus allowing only one thread to perform communication at any given time. This results in virtually *zero* effective concurrency in the communication of the different threads. Addressing this issue is the subject of a separate paper [4].

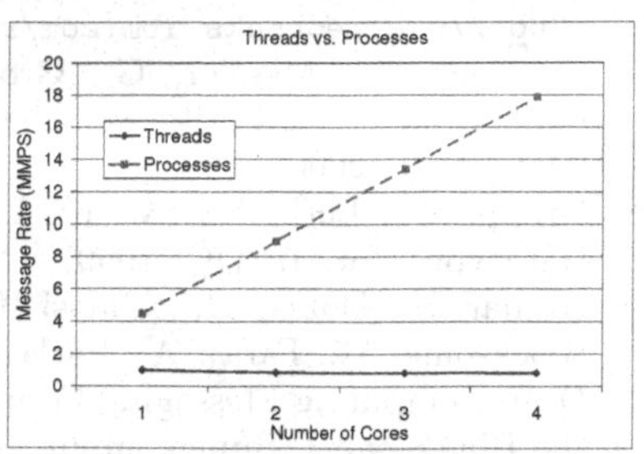

Fig. 9. Threads vs. Processes

4 Conclusions and Future Work

In this paper, we studied the non-data-communication overheads within MPI implementations and demonstrated their impact on the IBM BlueGene/P system. We identified several bottlenecks in the MPI stack including request handling, tag matching and unexpected messages, multi-request operations (such as MPI_Waitany), derived-datatype processing, buffer alignment overheads and thread synchronization, that are aggravated by the low processing capabilities of the individual processing cores on the system as well as scalability issues triggered by the massive scale of the machine. Together with demonstrating and analyzing these issues, we also described potential solutions for solving these issues in future implementations.

Acknowledgments

This work was supported by the Mathematical, Information, and Computational Sciences Division subprogram of the Office of Advanced Scientific Computing Research, Office of Science, U.S. Department of Energy, under Contract DE-AC02-06CH11357.

References

1. http://www.research.ibm.com/journal/rd/492/gara.pdf
2. http://www.sicortex.com/products/sc5832

3. Balaji, P., Buntinas, D., Balay, S., Smith, B., Thakur, R., Gropp, W.: Nonuniformly Communicating Noncontiguous Data: A Case Study with PETSc and MPI. In: IPDPS (2007)
4. Balaji, P., Buntinas, D., Goodell, D., Gropp, W., Thakur, R.: Toward Efficient Support for Multithreaded MPI Communication. Technical report, Argonne National Laboratory (2008)
5. Overview of the IBM Blue Gene/P project, `http://www.research.ibm.com/journal/rd/521/team.pdf`
6. IBM System Blue Gene Solution: Blue Gene/P Application Development, `http://www.redbooks.ibm.com/redbooks/pdfs/sg247287.pdf`
7. Buntinas, D., Mercier, G., Gropp, W.: Implementation and Shared-Memory Evaluation of MPICH2 over the Nemesis Communication Subsystem. In: Euro PVM/MPI (2006)
8. Gropp, W., Lusk, E., Swider, D.: Improving the Performance of MPI Derived Datatypes. In: MPIDC (1999)
9. Kumar, S., Dozsa, G., Almasi, G., Chen, D., Giampapa, M., Heidelberger, P., Blocksome, M., Faraj, A., Parker, J., Ratterman, J., Smith, B., Archer, C.: The Deep Computing Messaging Framework: Generalized Scalable Message Passing on the Blue Gene/P Supercomputer. In: ICS (2008)
10. D.K. Panda.: OSU Micro-benchmark Suite, `http://mvapich.cse.ohio-state.edu/benchmarks`
11. Ross, R., Miller, N., Gropp, W.: Implementing Fast and Reusable Datatype Processing. In: Euro PVM/MPI (2003)

Architecture of the Component Collective Messaging Interface

Sameer Kumar[1], Gabor Dozsa[1], Jeremy Berg[2], Bob Cernohous[2], Douglas Miller[2], Joseph Ratterman[2], Brian Smith[2], and Philip Heidelberger[1]

[1] IBM T. J. Watson Research Center,
1101 Kitchawan Rd, Yorktown Heights, NY, 10598, USA
[2] IBM Systems and Technology Group
Rochester, MN, 55901, USA

Abstract. Different programming paradigms utilize a variety of collective communication operations, often with different semantics. We present the component collective messaging interface (CCMI), that can support asynchronous non-blocking collectives and is extensible to different programming paradigms and architectures. CCMI is designed with components written in the C++ programming language, allowing it to have reuse and extendability. Collective algorithms are embodied in topological *schedules* and *executors* that execute them. Portability across architectures is enabled by the multisend data movement component. CCMI includes a programming language adaptor used to implement different APIs with different semantics for different paradigms. We study the effectiveness of CCMI on Blue Gene/P and evaluate its performance for the barrier, broadcast, and allreduce collective operations. We also present the performance of the barrier collective on the Abe Infiniband cluster.

1 Introduction

Most scientific applications use collective communication calls to optimize data exchange between processors. Performance of collective communication is critical to application scaling on large processor partitions.

Programming paradigms define different semantics for the collective operations. These collective operations can be blocking or non-blocking. In a blocking collective operation, the processor core is blocked until the collective operation is completed. Collective operations can also be synchronous or asynchronous. We define a synchronous collective as a collective communication call where progress starts only when all processors have entered the collective call. An asynchronous collective can make progress in the background as soon as one or more processors has initiated the collective call. The collective operations can be on pre-defined persistent groups called communicators or may be dynamic where they are constructed on-the-fly.

One common programming paradigm, the Message Passing Interface (MPI-2 [1]) standard defines all collective operations as blocking operations. However, non-blocking collective operations and their advantages have been explored in the LibNBC runtime [2]. MPI also defines collective operations on static pre-defined communicators. The Unified Parallel C [3] language defines non-blocking collectives on dynamic sub-groups, while

A. Lastovetsky et al. (Eds.): EuroPVM/MPI 2008, LNCS 5205, pp. 23–32, 2008.

the Charm++ programming language [4] defines non-blocking asynchronous collectives on dynamic groups called chare-array sections [5].

In this paper, we present the design of the component collective messaging interface (CCMI). We explore blocking vs non-blocking, asynchronous vs synchronous, communicator based and on-the-fly collectives. Our framework is component driven with basic building blocks written in C++. Collective operations are built using these building blocks. This approach enables most components to be reused across different architectures and parallel programming paradigms. Only the data mover component needs to be redesigned across architectures and the components specific to the semantics in a programming language for a different paradigm. We describe this approach in detail with performance results on the Blue Gene/P machine [6]. Performance is also shown for the Abe Infiniband cluster [7].

2 Architecture

The CCMI stack has four primary components that interact to define collective operations.

2.1 Multisend

Multisend is the basic data movement operation on top of which CCMI collectives are built. In CCMI, a collective is essentially a collection of multisend calls. We define three classes of data movement in our multisend call: *multicast*, *combine*, and *many-to-many*.

- In *multicast*, each processor sends the same data to several destinations. The multisend multicast also takes op-codes to specify additional hints for each destination. On architectures that provide network broadcast, the hints can provide additional information on the virtual channels and tags of the hardware multicast. On Blue Gene/P, this API allows the application to do a line broadcast in a specific direction via deposit bit packets. The deposit bit in the packet header enables the packet to be deposited on all intermediate nodes from the source to the destination.
- In a *combine* operation, processors participate in a global reduction. The combine operation can take advantage of networks that support reductions on the network. A combine operation specifies the arithmetic operation, datatype and hints to provide additional information.
- The *many-to-many* is a generalization of the all-to-all collective operation. Here each processor sends different data to many other processors.

For example, a point-to-point message is a multicast to one destination. Spanning tree broadcast can be built with a multisend multicast at each level of the spanning tree. All-to-all and scatter can be built using the many-to-many calls. The combine operation can be used to represent fast network combines supported in architectures such as Blue Gene/L [8], Blue Gene/P [6] and the Quadrics clustering interconnect [9]. Typically, each processor is involved only once in each multisend operation. So, a variety of different collective operations can be built from a collection of optimized multisends.

The multisend multicast and multisend many-to-many calls are ideally suited for low-frequency architectures such as the Blue Gene machines, as they can amortize software stack overheads across the different destinations.

As the CCMI framework supports several overlapping, non-blocking collectives, there will be several multisend messages being sent and received at the same time on each node. The different multisend messages are identified by connection ids. The sender of the multisend message chooses a connection id based on the global properties of the collective operation.

The *connection-manager* component in CCMI chooses connection identifiers for each collective operation. The number of connections may determine the maximum number of simultaneous collectives on each node. For example, a connection manger could set the connection id to be the rank of the source node. For MPI collectives, the connection manager could also use the communicator as the connection id of the collective, allowing one collective per communicator to support an optimized multi-threaded MPI environment.

2.2 Schedules

Collective operations require processors to interact in a graph topology in several phases. The *schedule* component (similar to schedules in LibNBC [2]) defines a topological graph and phases in which the processors exchange messages to accomplish a collective operation. For each phase of the collective, the schedule defines the source and destination processors and opcodes for the multisend calls to be executed.

One schedule used in CCMI is the *binomial schedule* that uses the recursive doubling algorithm [10,11] to provide tasklists for the barrier, broadcast and allreduce operations. Binomial barrier can be designed by processors exchanging messages with their neighbors in $Log_2(N)$ phases. A binomial broadcast can be designed with the root sending data to its first level neighbors in the first phase and this is propagated along the spanning tree with intermediate nodes sending data in the later phases. In the reduce operation, data moves in a direction opposite to the broadcast from the leaf nodes to the root. Allreduce can be designed either as a reduce followed by a broadcast or a direct exchange among the neighbors (we chose the latter approach). Figure 1 illustrates the phases in a binomial broadcast schedule.

As the binomial algorithm is not topology optimized, we explored *rectangle schedules* to optimize collectives on torus networks such as Blue Gene/L, Blue Gene/P and

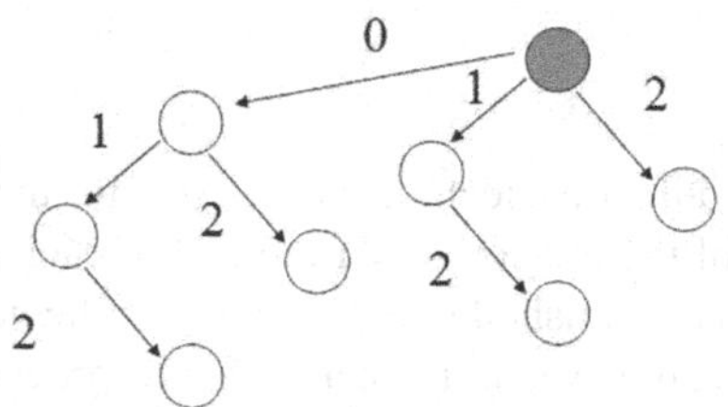

Fig. 1. Binomial schedule in CCMI

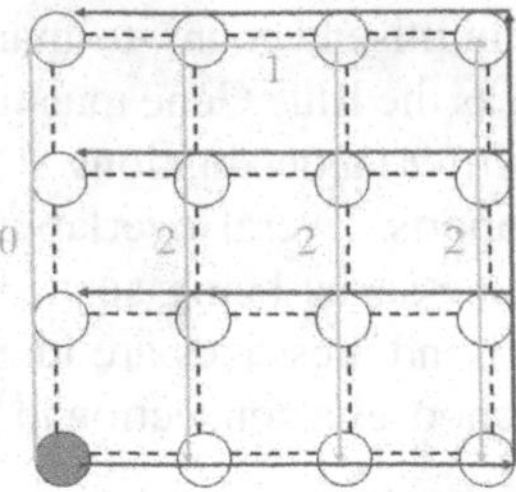

Fig. 2. Rectangle Schedules in CCMI

the Cray interconnect. On the BG/L and BG/P torus networks, each node can send packets with the deposit-bit set to deposit the packet along all the nodes in a line of the torus. The rectangle schedule takes advantage of this feature to optimize broadcast and allreduce operations.

Figure 2 shows the rectangle schedule with two independent paths on a 2-D mesh, that we have explored in CCMI. On a 2-D mesh, an X-color rectangle broadcast [12,13] (blue in figure 2) can be achieved via deposit bit broadcasts along the X+, Y+, X- dimensions in 3 phases. On a 3-D mesh, the X-color broadcast would require the broadcast messages to be sent along the X+,Y+,Z+, X- directions in four phases. When the root is in the corner on a 3-D mesh (or from any root on a 3-D torus), we can also design Y and Z color broadcasts that do not share links with each other or with the X-color broadcast.

The reduce operation traverses the schedule topology in a direction opposite to the broadcast. An allreduce operation can be supported by a reduce followed by a broadcast. For example, an X-color rectangle schedule allreduce can be designed by a reduce along the X+,Z-,Y-,X- direction followed by an X-color broadcast. Unlike broadcast, the rectangle allreduce has many more phases than the binomial scheme and is hence a throughput optimization.

We also take advantage of network accelerated global collective optimizations. For example, a fast global barrier is available on the Blue Gene/L and Blue Gene/P networks through the global interrupt network. Low-latency broadcasts and allreduce operations take advantage of the collective tree network. Each node on Blue Gene/P is a multi-core node. However, network acceleration of collectives is only available across the different nodes and local communication has to be performed in software. We have developed hybrid schedules to optimize allreduce and broadcast operations on such hybrid architectures. In this paper, we limit our exploration to network accelerated collectives.

2.3 Executors

The tasks listed in the schedule are executed by an executor that initiates the multisend operations for each phase of the collective. At the end of each phase, all the multisend messages sent and received are finished before moving to the next phase. For collective operations such as allreduce, the executor can also do pre-processing arithmetic to combine incoming messages with local state and initiate the message sends for the next phase.

There is a different executor for each collective operation to allow specific optimizations for each collective. For example, in a spanning tree broadcast there are no

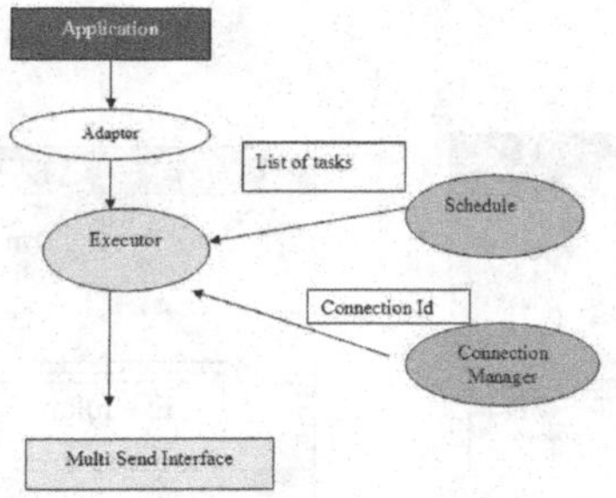

Fig. 3. CCMI Components

dependencies between phases as they all send the same data. So the broadcast executor waits for incoming data and makes a single multicast call to send the data to the next level of the spanning tree.

The spanning tree broadcast/allreduce executors can also pipeline the data to keep all the compute and network resources busy. For each chunk in the pipeline, all the phases in the schedule can be executed.

2.4 Programming Language Adaptor

The *language adaptor* component interfaces the CCMI stack to a programming language. The language adaptor can define an API suitable for that programming language. For example, the API for UPC collectives is likely to be different from MPI collectives. The implementation of the API calls will instantiate the different components in the collective operation with the specific semantics of the paradigm. A language adaptor may also define new internal components. For example, an MPI adaptor may have a communicator component which contains the pre-determined list of process ranks that participate in the collective operation. The schedules and executors can be shared across different paradigms, leaving only the language adaptor to be different for each programming paradigm. In this paper, we present performance results on two adaptors, the DCMF Adaptor and the Generic Adaptor.

The DCMF [14,15] (*Deep Computing Messaging Framework*) adaptor (Figure 4(a)) provides a non-blocking collective API in the C programming language. Optimized MPI collectives on Blue Gene/P are implemented on top of this API. It uses a multisend built on top of the DCMF messaging API. The *generic adaptor* (Figure 4(b)) enables C++ applications to directly construct and access C++ components of CCMI such as schedules and executors. A smart compiler can inline the C++ interface functions of schedules and executors resulting in good performance. The generic adaptor provides an implementation of multisend that uses MPI operations, allowing CCMI to be portable across different architectures.

3 Collective Operations and Algorithms

We have explored several collective operations in the CCMI framework, such as barrier, broadcast, allreduce and all-to-all. For each collective operation, we have designed one

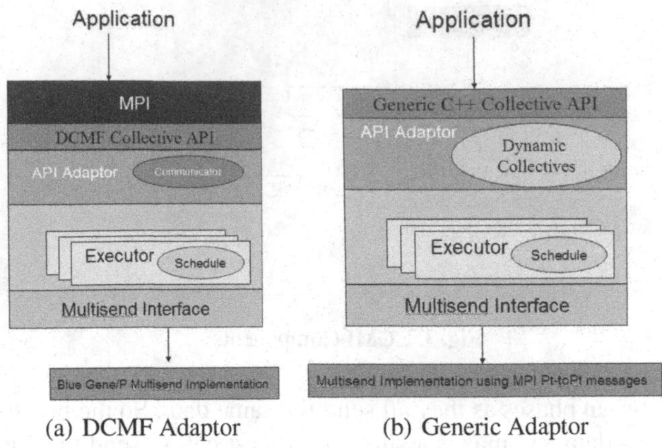

(a) DCMF Adaptor (b) Generic Adaptor

Fig. 4. Stack views of CCMI adaptors

or more schedules typically reusing the same executor. For example, the barrier executor with the binomial schedule can provide an implementation of the binomial barrier algorithm.

Barrier: The barrier executor keeps a vector of counters with one slot for each of the phases returned from the schedule. For each message received the phase-counter is incremented. When messages have been sent and received in all the phases the barrier is complete and the counters are reset.

The barrier executor can read the schedule during construction and cache the list of sources and destinations to communicate with to optimize performance. In programming paradigms where the communicator is static and does not change often, subsequent barrier calls will only look at the schedule cache to do the message passing. During each message passing operation, the multisend interface is invoked to send messages to one or more destinations.

Broadcast: The broadcast executor is asynchronous, non-blocking, and supports on-the-fly group creation. This means the broadcast payload data can move on the network in the background and context information is saved to ensure all nodes eventually make progress. Groups can be created as the multisend propagates down the spanning tree. The broadcast executor on the root immediately sends data to the intermediate nodes on the spanning tree using the multisend interface.

The DCMF adaptor has two interfaces for asynchronous broadcast, (i) an active message based callback interface with callbacks to allocate space for the broadcast message and the schedule-executor and (ii) a non-blocking broadcast interface where each participating process posts a broadcast with the communicator object. The DCMF adaptor registers a callback to process the first packet from the multisend interface. In case (i), the adaptor delegates the buffer allocation to an application callback. In case (ii), the DCMF adaptor keeps a queue of posted broadcasts and unexpected broadcasts. The first incoming packet is matched with the local posted broadcast and remaining packets are copied into the posted buffer. If the broadcast has not been posted, when the first packet arrives, the adaptor will allocate a buffer for the broadcast data and construct a schedule/executor pair for that collective operation. In both cases the executor will read the schedule and perform the remaining phases of the broadcast.

Allreduce: The allreduce executor can execute schedules that define an allreduce operation based on the gossiping algorithm or the reduce followed by broadcast algorithm. In the gossiping scheme, all participating processors communicate with atleast $log_2(P)$ processors, after which every processor has the final result. In the reduce followed by broadcast scheme, the data is reduced over a spanning tree to a root processor where it is then broadcast via the spanning tree. This means that the allreduce executor can even execute a broadcast schedule.

4 Performance Results

We measured the the latency of a 1 byte multicast with the DCMF runtime (on the BG/P torus via the DMA) to be $2\mu s$ for a single destination, but only $0.34\mu s$ for each additional destination, showing the effectiveness of the multisend interface. The performance of MPI Barrier with the binomial and global-interrupt schedules in SMP mode is presented in Figure 5(a). The figure also shows the portability of the CCMI stack with results on the Abe Infiniband cluster [7], with the generic adaptor and a multisend implementation developed on top of MPI. From the results, CCMI has a similar barrier latency as the MVAPICH library [16].

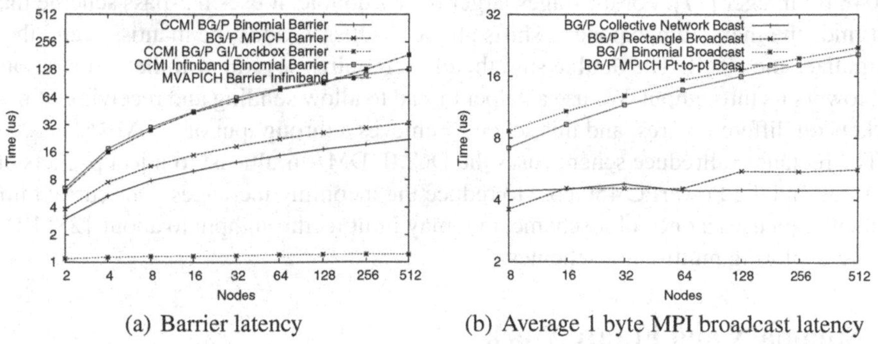

(a) Barrier latency

(b) Average 1 byte MPI broadcast latency

Fig. 5. Performance of collective operations on BG/P and Infiniband

Figure 5(b) shows the average latency of a 1 byte broadcast operation using different optimizations on Blue Gene/P. This benchmark measures the latency of several broadcast operations from the same root, allowing successive broadcast calls to pipeline each other. These runs were performed on a booted partition with 512 nodes, and hence the collective network latency is unchanged for smaller configurations. The CCMI rectangle optimization, (a software optimization), has comparable performance to the collective network. This is mainly due to the multicast operation that supports deposit bit broadcasts and allows the payload to be multicast to many destinations with low overheads. The rectangle broadcast latency also does not increase much with partition size.

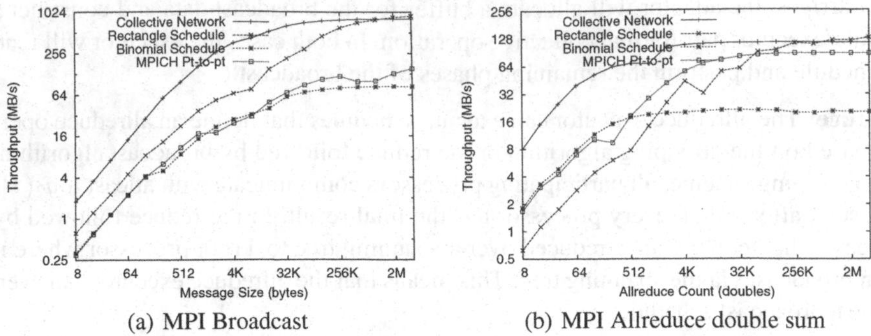

(a) MPI Broadcast (b) MPI Allreduce double sum

Fig. 6. BG/P collectives throughput on 512 nodes in SMP mode

Figure 6(a) shows the throughput of the broadcast operation on 512 nodes of BG/P in SMP mode. The collective network achieves 96% of peak throughput. Rectangle torus broadcast uses three colors allowing the broadcast message to be sent simultaneously on three different routes. For messages larger than 2MB its performance is better than the collective network.

On Blue Gene/P, the collective network can only perform integer and fixed point allreduce operations. The throughput of allreduce floating point double sum on 512 nodes of BG/P in SMP mode is shown by Figure 6(b). For short messages, the collective network double sum operation uses a 1-pass scheme that expands a 64 bit floating number into a 2048 bit integer [17]. For messages larger than 1 double, it uses a 2-pass scheme that first finds the maximum exponent, shifts all the mantissas, adds the mantissas and then normalizes the result. The double sum therefore requires several arithmetic operations that lowers its throughput. We use a helper thread to allow sending and receiving of tree packets on different cores, and this scheme achieves a throughput of 120MB/s.

The rectangle allreduce scheme uses the DCMF DMA multicast to inject packets on the torus and the PowerPC 450 core to reduce the incoming messages. Our current implementation uses a one color scheme, that may limit its throughput to about 128MB/s. We are exploring multi-color schemes.

5 Summary and Future Work

We presented the architecture of the CCMI stack that defines reusable components to support asynchronous non-blocking collectives. This software stack is the open-source production runtime on the Blue Gene/P machine. We showed the effectiveness of the multisend component that allows software overheads of message passing to be amortized across all the processors the data is sent to. The performance gain is demonstrated by the rectangle broadcast optimization on BG/P. To extend CCMI to a new architecture only the multisend component needs to be designed and implemented, and the remaining components can be reused. We showed the performance of CCMI with the generic adaptor on an Infiniband cluster.

To extend CCMI to a new programming language, only a new language adaptor may need to be designed. This adaptor will provide an API with semantics suitable for that programming language. In this paper, we only presented performance results for MPI collectives, but we plan to explore other adaptors for other programming paradigms such as Charm++ and UPC. We also plan to explore higher dimensional multinomial schedules that can take advantage of relatively low overhead for each additional destination in a multisend multicast operation on BG/P.

Acknowledgments

We would like to thank Gheorghe Almasi, Charles Archer, Michael Blocksome, Dong Chen, Jose G. Castanos, Ahmad Faraj, Thomas Gooding, Mark E. Giampapa, John Gunnels, Carl Obert, Jeff Parker, Craig Stunkel and Robert Wisniewski for their support in the design, development and optimization of collectives and CCMI on the Blue Gene/P machine.

The work presented in this paper was funded in part by the US Government contract No. B554331.

References

1. Forum, M.P.I.: MPI-2: Extensions to the message-passing interface (1997), http://www.mpi-forum.org/docs/mpi-20-html/mpi2-report.html
2. Hoefler, T., Lumsdaine, A., Rehm, W.: Implementation and Performance Analysis of Non-Blocking Collective Operations for MPI. In: Proceedings of SC 2007. IEEE Computer Society/ACM (2007)
3. Nishtala, R., Almasi, G., Cascaval, C.: Performance without Pain = Productivity. In: Proceedings of PPOPP 2008 (2008)
4. Kale, L.V., Krishnan, S.: Charm++: Parallel Programming with Message-Driven Objects. In: Wilson, G.V., Lu, P. (eds.) Parallel Programming using C++, pp. 175–213. MIT Press, Cambridge (1996)
5. Department of Computer Science,University of Illinois at Urbana-Champaign, Urbana, IL, The CHARM (5.9) Programming language manual (2007)
6. IBM Blue Gene Team, Overview of the Blue Gene/P project, IBM J. Res. Dev., vol. 52 (January 2008), http://www.research.ibm.com/journal/rd/521/team.html
7. NCSA Abe Cluster, http://www.ncsa.uiuc.edu/UserInfo/Resources/Hardware/Intel64Cluster
8. Gara, A., Blumrich, M.A., Chen, D., Chiu, G.L.-T., Coteus, P., Giampapa, M.E., Haring, R.A., Heidelberger, P., Hoenicke, D., Kopcsay, G.V., Liebsch, T.A., Ohmacht, M., Steinmacher-Burow, B.D., Takken, T., Vranas, P.: Overview of the Blue Gene/L System Architecture. IBM Journal of Research and Development 49(2/3), 195–212 (2005)
9. Petrini, F., chun Feng, W., Hoisie, A., Coll, S., Frachtenberg, E.: The quadrics network: high-performance clustering technology. IEEE Micro. 22(1), 46–57 (2002)
10. Hensgen, D., Finkel, R., Manber, U.: Two algorithms for barrier synchrofization. International Journal of Parallel Programming 17(1) (1988)
11. Thakur, R., Rabenseifner, R., Gropp, W.: Optimization of Collective Communication Operations in MPICH. International Journal of High Performance Computing Applications 19, 49–66 (2005)
12. Almasi, G., et al.: Design and implementation of message-passing services for the Blue Gene/L supercomputer. IBM J. Res. Dev. 49, 393–406 (2005)

13. Almasi, G., Archer, C., Erway, C., Heidelberger, P., Martorell, X., Moreira, J., Steinmacher-Burow, B., Zheng, Y.: Optimization of MPI Collective Communication on BlueGene/L Systems. In: Proceedings of the 19th annual international conference on Supercomputing ICS 2005, pp. 253–262. ACM Press, New York (2005)
14. Kumar, S., Dozsa, G., Almasi, G., Chen, D., Heidelberger, P., Giampapa, M.E., Blocksome, M., Faraj, A., Parker, J., Ratterman, J., Smith, B., Archer, C.: The Deep Computing Messaging Framework: Generalized Scalable Message passing on the BlueGene/P Supercomputer. In: International Conference on Supercomputing ICS 2008 (to appear, 2008)
15. DCMF (2008), `http://dcmf.anl-external.org/wiki`.
16. Huang, W., Santhanaraman, G., Jin, H.-W., Gao, Q., Panda, D.K.x.D.K.: Design of High Performance MVAPICH2: MPI2 over InfiniBand. In: CCGRID 2006: Proceedings of the Sixth IEEE International Symposium on Cluster Computing and the Grid, pp. 43–48. IEEE Computer Society, Washington (2006)
17. Almasi, G., Dozsa, G., Erway, C.C., Steinmacher-Burow, B.D.: Efficient Implementation of Allreduce on BlueGene/L Collective Network. In: Proceedings of Euro PVM/MPI, pp. 57–66 (2005)

X-SRQ - Improving Scalability and Performance of Multi-core InfiniBand Clusters

Galen M. Shipman[1], Stephen Poole[1], Pavel Shamis[2], and Ishai Rabinovitz[2]

[1] Oak Ridge National Laboratory*, Oak Ridge, TN, USA
{gshipman,spoole}@ornl.gov
[2] Mellanox Technologies, Yokneam, Israel
{pasha,ishai}@mellanox.co.il

Abstract. To improve the scalability of InfiniBand on large scale clusters Open MPI introduced a protocol known as B-SRQ [2]. This protocol was shown to provide much better memory utilization of send and receive buffers for a wide variety of benchmarks and real-world applications.

Unfortunately B-SRQ increases the number of connections between communicating peers. While addressing one scalability problem of InfiniBand the protocol introduced another. To alleviate the connection scalability problem of the B-SRQ protocol a small enhancement to the reliable connection transport was requested which would allow multiple shared receive queues to be attached to a single reliable connection. This modified reliable connection transport is now known as the extended reliable connection transport.

X-SRQ is a new transport protocol in Open MPI based on B-SRQ which takes advantage of this improvement in connection scalability. This paper introduces the X-SRQ protocol and details the significantly improved scalability of the protocol over B-SRQ and its reduction of the memory footprint of connection state by as much as 2 orders of magnitude on large scale multi-core systems. In addition to improving scalability, performance of latency-sensitive collective operations are improved by up to 38% while significantly decreasing the variability of results. A detailed analysis of the improved memory scalability as well as the improved performance are discussed.

1 Introduction

The widespread availability of commodity multi-core CPUs from both Intel and AMD is changing the landscape of near-commodity clusters. Compute nodes with 8 cores (2 quad core CPUs) and even 16 cores (4 quad core CPUs) are becoming more common and 8 or more cores in a single socket are expected in the next 12-18 months. A number of these multi-core clusters are connected with InfiniBand (IB), thereby increasing the need to examine the scalability of MPI in such environments.

* Research sponsored by the Mathematical, Information, and Computational Sciences Division, Office of Advanced Scientific Computing Research, U.S. Department of Energy, under Contract No. DE-AC05-00OR22725 with UT-Battelle, LLC.

A. Lastovetsky et al. (Eds.): EuroPVM/MPI 2008, LNCS 5205, pp. 33–42, 2008.

Open MPI [1] supports the IB interconnect using the reliable connected (RC) transport layer. RC in IB currently requires a connection to be established between each communicating pair of processes and consumes one page (commonly $4KB$) of system memory for each connection. Multi-core systems increase the number of dedicated processes per node and therefore increase the number of connections per node. This additional memory consumed on the node may be substantial in a large scale multi-core system. Furthermore, maintaining a fixed amount of memory per core is becoming increasingly difficult as memory prices remain high relative to the falling price of a CPU core. Pressure on memory will increase as applications are migrated to multi-core machines.

This paper describes Open MPI's use of the extended reliable connection (XRC) which alleviates some of the memory pressure in multi-core environments. In addition to reducing overall memory consumption in Open MPI, the use of XRC in conjunction with B-SRQ improves performance. This conjunction will be referred to as X-SRQ throughout this paper.

The rest of this paper is organized as follows. Section 2 provides a brief discussion of previous work in this area as well as an overview of the XRC architecture. Section 3 describes the new protocol, including necessary modifications to our on-demand connection wire-up scheme. Section 4 describes the test platform followed by performance analysis of the results. Section 5 summarizes relevant results and concludes with a discussion of areas of possible future work.

2 Background

The InfiniBand specification details 5 transport layers:

1) **Reliable Connection** (RC)**:** connection-oriented and acknowledged
2) **Reliable Datagram** (RD)**:** multiplexed and acknowledged
3) **Unreliable Connection** (UC)**:** connection-oriented and unacknowledged
4) **Unreliable Datagram** (UD)**:** connectionless and unacknowledged
5) **Raw Datagram:** connectionless and unacknowledged

RC provides a connection-oriented transport between two queue pairs (QPs). Work requests posted to a QP are implicitly addressed to the remote peer. The scalability limitations of connection-oriented transports are well known [2], [3] requiring $\binom{N}{2}$ connections for N peers.

Fig. 1 illustrates two nodes, each with two cores connected via RC. In this example each core is running a single process and is connected to each of the remote processes. If we assume that shared memory is used for intra-node MPI communication then the total number of QPs is 4 per node.

RD allows using a single QP to send and receive from any other addressable RD QP. RD was designed to provide a number of desirable scalability features but in practice RD has proven difficult to implement with no manufacturer currently supporting this transport layer.

While some are examining the use of UD to enhance scalability [4], the additional costs of user-level reliability and implementation complexity are still being examined.

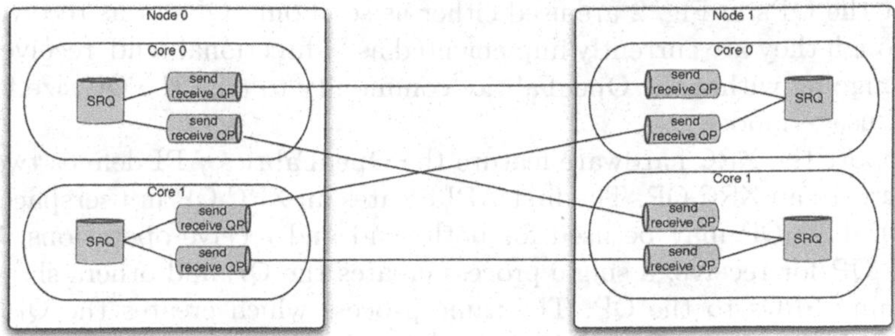

Fig. 1. RC - 2 Nodes

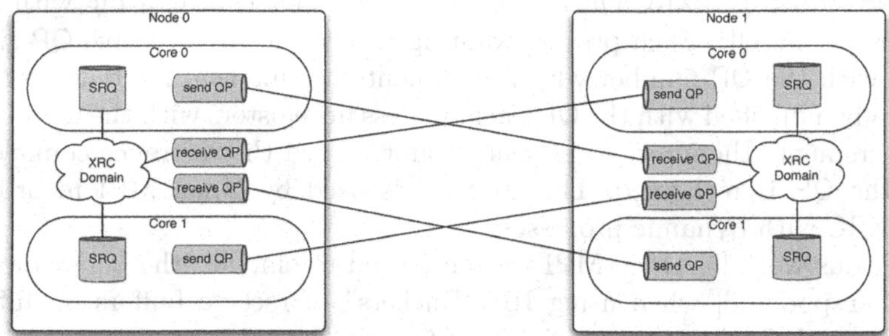

Fig. 2. XRC - 2 Nodes

To improve the scalability of InfiniBand in multi-core environments Mellanox has introduced XRC - the new transport layer. Requiring changes both in the HCA and in the software stack; XRC allows a single receive QP to be shared by multiple shared receive queues (SRQs) across one or more processes. Note that these SRQs can exist in any process which shares the same XRC domain as the receive QP. The SRQ "destination" is specified in the work queue entry (WQE) at the send QP. The receive QP consumes a buffer from the specified SRQ and enques a WQE to the completion queue (CQ) connected to this SRQ. This mechanism allows each process to maintain a single send QP to each host rather than to each remote process. A receive QP is established per remote send QP. These receive QPs can be shared among all the processes on the host.

Fig. 2 illustrates two nodes with two processes per node (PPN) using XRC for communication. Note that each receive QP is a machine resource and exists within an XRC domain. SRQs are setup in each process and are connected to one or more receive QPs. Send QPs are a per process resource (as in RC), however each process can use a single send QP to communicate with any process on the remote machine by specifying the SRQ of the target process. If we assume that shared memory is used for intra-node MPI communication, then each node uses 4 QPs. In general, XRC can be used to reduce the number of QPs by a factor of $PPN-1$. Thus for applications running P processes on N nodes, XRC decreases the number of required transport connections from $P^2*(N-1)$ to $P*2*(N-1)$.

Note that the QPs in Fig. 2 are used either as send only QPs or as receive only QPs although they are currently implemented as bidirectional send/receive QPs. Work is ongoing within the OpenFabrics community to trim the QP size for the new XRC usage model.

To support the XRC hardware feature the OpenFabrics API defines two new APIs to create an XRC QP. The first API creates an XRC QP in userspace (just as in RC). This QP may be used for both send and receive operations. When using the QP for receive, a single process creates the QP and others share this by attaching SRQs to the QP. The same process which creates the QP must eventually destroy the QP, but it must defer the destruction of the QP until after all other processes have finished using the QP.

The second API creates a receive XRC QP at the kernel level. This allows a process to open the XRC QP and later exit without coordinating with other consumers of the QP. Each process wanting to use the XRC receive QP simply registers with the QP number which increments the internal reference count of the QP. When finished with the QP each process unregisters with the QP number which decrements the internal reference count. When the reference count drops to zero the QP is reclaimed. This method is used by Open MPI in order to support XRC with dynamic processes.

In previous work [5] Open MPI was enhanced to emulate the behavior of receive buffer pools [6] when using IB. "Buckets" of receive buffers of different sizes, with each bucket using a single SRQ, are created in each process. Each SRQ is then associated with a single QP. The sender can achieve better memory utilization on both sender and receiver sides by simply sending data on the QP with the minimum buffer of adequate size. This protocol was shown to significantly enhance memory utilization across a wide variety of applications and in some cases it enhance performance.

As part of this work a generic mechanism was created to allow the user or system administrator to specify any number of QPs (up to system limits) with various size buffers associated with them. Each QP can be specified to use either an SRQ or per-peer receive buffers posted to the QP directly. This mechanism allows for greater flexibility in tuning and experimentation.

Our current work involves modifying this generic mechanism to allow XRC QPs to be used in a similar fashion. Multiple SRQs can then be used to increase memory utilization without the need for creating a separate QP for each SRQ.

3 Protocol Description

One substantial drawback to the B-SRQ protocol is that each SRQ requires a separate QP. This limits the overall scalability of the protocol as node and core counts continue to increase. The X-SRQ protocol removes this limitation. Fig. 3 illustrates two nodes with two PPN using X-SRQ for communication. Note that each process maintains a number of SRQs each with a different buffer size. These SRQs are then attached to a single node level receive QP per remote process.

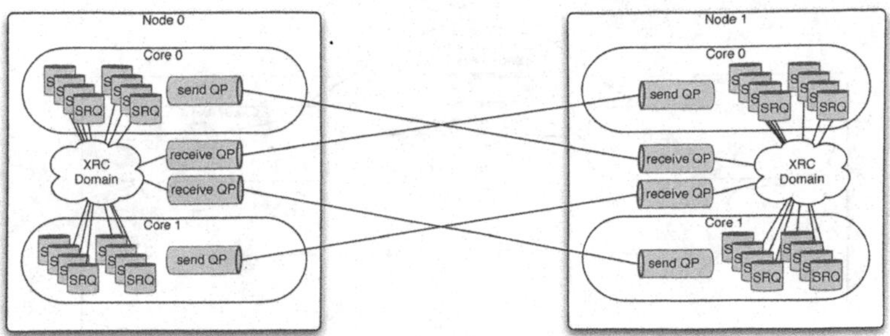

Fig. 3. X-SRQ - 2 Nodes

A number of changes in Open MPI were required in order to make use of XRC. Instead of addressing processes by simply using the implicit addressing of an RC QP, XRC requires specifying the remote SRQ number in the work request prior to enqueuing a send on an XRC QP. A few other minor changes were required such as extending our QP specification syntax to allow specifying XRC QPs via Open MPI's Modular Component Architecture parameter system [7].

Of special interest is the connection establishment mechanism within Open MPI which supports XRC. The XRC QP establishment is considerably different from the usual RC QP wireup [3]. During process initialization, all processes perform the following:

1. exchange Local Identifiers (LIDs)
2. create an XRC domain
3. create SRQ entries

After the initialization phase each process keeps a table of LIDs for all remote processes and the connection status to the remote process. Fig. 4(a) illustrates a unidirectional connection establishment from process 0 to process 2. When process 0 initiates a send to process 2, process 0 checks the connection status to process 2. If the connection is closed then process 0 will create an XRC send QP (user level), and will send a connection request to process 0 (step 1 in Fig. 4(a)). Process 2 will create all the SRQ receive buffers as specified by the configuration (step 2 in Fig. 4(a)). Process 2 will then open an XRC receive QP (kernel level) and respond with an SRQ number for each SRQ as well as the XRC receive QP number (step 3 in Fig. 4(a)). Process 0 will receive the remote QP number and SRQ numbers and will connect the send QP to process 2 (step 4 in Fig. 4(a)). Process 0 will then update the connection status (both IB and MPI) of process 2 to "connected". Process 3's table entry is updated as well, but the MPI connection status remains "disconnected". At the end of the handshake there will be a unidirectional QP from process 0 on node 0 to process 2 on node 1.

Fig. 4(b) illustrates process 0 initiating a unidirectional connection to process 3 on node 1. Process 0 first checks the IB status and sees that process 0 already has an XRC send QP to node 1 (process 3). Process 0 then sends a connection request with an XRC receive QP number on node 1 to process 3 (step 1 in

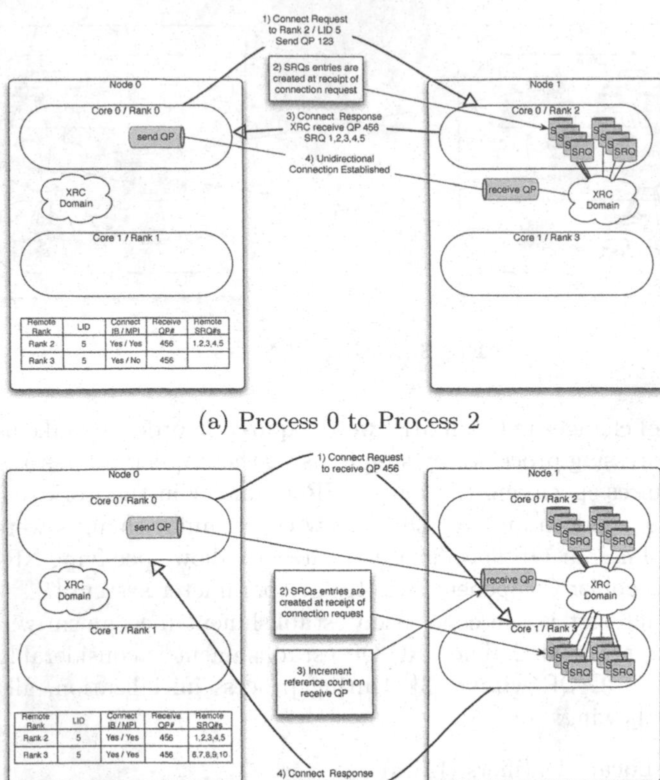

(a) Process 0 to Process 2

(b) Process 0 to Process 3

Fig. 4. XSRQ Connection Establishment

Fig. 4(b)). Process 3 creates all the SRQ receive buffers as specified by the configuration (step 2 in Fig. 4(b)). Process 3 will increase a reference counter on the XRC receive QP with the requested QP number (step 3 in Fig. 4(b)) and respond to process 0 with the SRQ numbers (step 4 in Fig. 4(b)). Process 0 receives the SRQ numbers and changes the connection status of process 3 to MPI "connected".

Other aspects of the X-SRQ protocol remain similar to that of the B-SRQ protocol. As previously discussed, Open MPI supports mixing QPs with receive buffers posted directly (per-peer) on the QP with QPs using SRQs . This flexibility allows using flow-control mechanisms over the per-peer QP while using SRQs for scalability. Currently Open MPI does not support mixing XRC QPs with non-XRC QPs; this is left for future work.

4 Results

All experiments were conducted on a 32 node cluster located at Mellanox Technologies, USA. Each node consisted of dual quad-core Intel Xeon X5355 CPUs running at 2.66GHz with 8GB memory and a Mellanox ConnectX HCA running firmware 2.3.0. Each node ran Redhat Enterprise Linux with the 2.6.9-42 SMP kernel, OFED version 1.3 (release candidate 5) and Open MPI trunk subversion r17144. All nodes were connected via a DDR switch.

The most notable result was the significant reduction in the memory footprint of the Open MPI library when using multiple SRQs per process. X-SRQ has much better memory scaling characteristics than B-SRQ as the number of QPs is significantly smaller. The number of QPs for B-SRQ is governed by the following: PPN - number of processes per node; N - number of nodes; $NSRQ$ - number of SRQs; and NQP - number of QPs.

For B-SRQ, the number of QPs created is

$$NQP = PPN^2 * NSRQ * (N - 1)$$

For X-SRQ, the number of QPs created is

$$NQP = PPN * 2 * (N - 1)$$

Note that for X-SRQ the $NSRQ$ parameter is dropped. Instead of squaring PPN, we multiplied it by 2 to account for the separate send QP and receive QP. Fig. 5(a),5(b) illustrate the impact of increasing the number of processes per node - as is often the case for multi-core clusters. At 1024 nodes and 8 PPN, QP memory resources peak at $256MB$ when using RC as opposed to only $64MB$ for XRC .

Fig. 6(a),6(b) illustrate the impact of increasing the number of SRQs per process at 8 PPN. At 1024 nodes and 8 SRQs per process, QP memory resources peak at $2GB$ as opposed to only $64MB$ for XRC

In addition to significant memory savings, XRC improves performance. As the number of QPs is decreased, the HCA needs to manage fewer interconnect context resources. Consequently, the HCA is better able to cache context resource data and thereby to avoid a lookup on host memory. To evaluate the performance improvements of XRC, the SkaMPI collectives benchmarks [8] were used. MPI_BARRIER, MPI_REDUCE (8 bytes), MPI_ALLREDUCE (8 bytes) and MPI_ALLGATHER (8 bytes) collectives where chosen in order to evaluate the performance of X-SRQ. Overall performance of X-SRQ was better than that of B-SRQ . Fig. 7(a),7(b) illustrate that X-SRQ performance improvements reach up

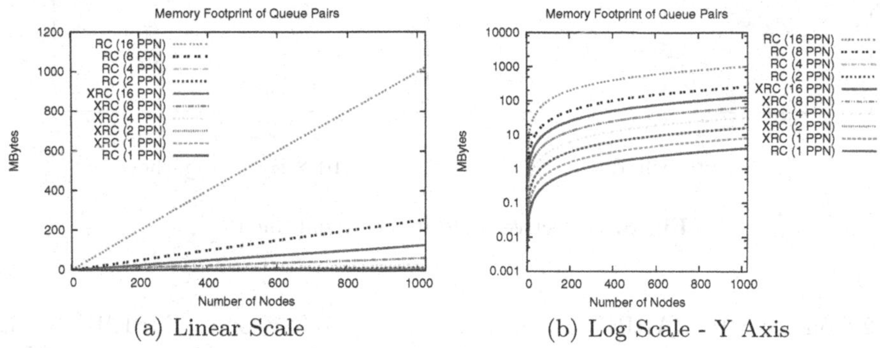

(a) Linear Scale (b) Log Scale - Y Axis

Fig. 5. Memory footprint of QPs (Varying PPN)]

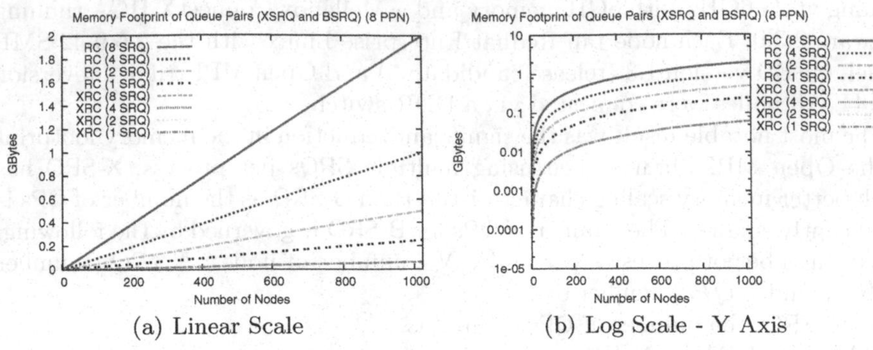

(a) Linear Scale (b) Log Scale - Y Axis

Fig. 6. Memory footprint of QPs (Varying SRQ Count)]

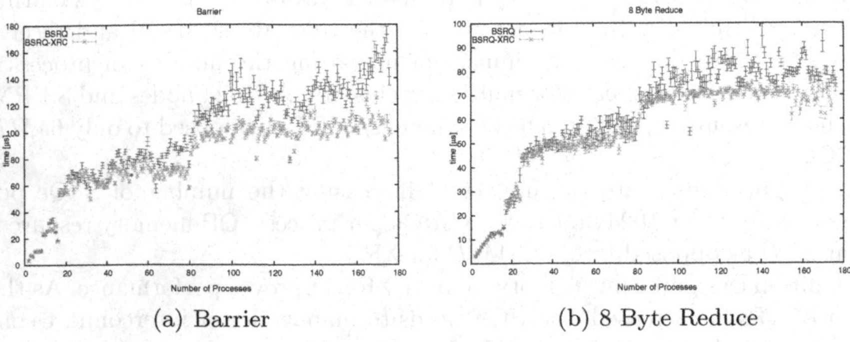

(a) Barrier (b) 8 Byte Reduce

Fig. 7. Collective performance

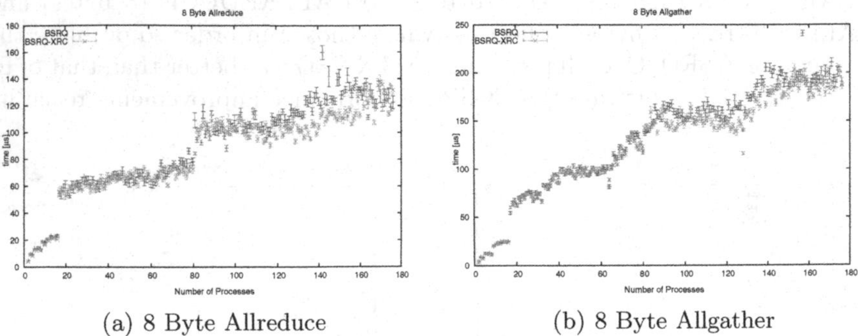

(a) 8 Byte Allreduce (b) 8 Byte Allgather

Fig. 8. Collective performance (continued)

to 42% on the MPI_BARRIER benchmark and up to 38% on the MPI_REDUCE benchmark. The standard deviation of the results was also much lower for X-SRQ when compared to B-SRQ.

Fig. 8(a), 8(b) also show some improvement of MPI_ALLREDUCE and MPI_ALLGATHER benchmarks at larger process counts, although not as significant as that of MPI_BARRIER and MPI_REDUCE benchmarks. Larger overheads for these collectives may minimize the overall performance improvement. The standard deviation of the results was again lower for X-SRQ compared to B-SRQ.

5 Conclusions

Through a novel use of the XRC transport layer, both the scalability and performance of Open MPI have improved. The X-SRQ protocol improves both send and receive buffer utilization while significantly improving the scalability of QP connections. While not limited to multi-core systems, these scalability improvements are significant in larger multi-core environments. Current trends point towards increased core counts for the foreseeable future thereby making these scalability enhancements a necessity for clusters using IB.

In addition to improved scalability, X-SRQ improves performance on latency-sensitive operations due to more efficient use of network resources. These performance improvements are consistent with the HCA architecture and are relevant not only for larger clusters, but for any multi-core cluster (as small as 32 nodes) using InfiniBand.

Open MPI does not currently support the use of XRC QPs and RC QPs concurrently. Future work will involve allowing these different QP "types" to be used concurrently within a single Open MPI job.

References

[1] Garbriel, E., Fagg, G., Bosilica, G., Angskun, T., Squyres, J.J.D.J., Sahay, V., Kambadur, P., Barrett, B., Lumsdaine, A., Castain, R., Daniel, D., Graham, R., Woodall, T.: Open MPI: goals, concept, and design of a next generation MPI implementation. In: Proceedings, 11th European PVM/MPI Users' Group Meeting (2004)

[2] Brightwell, R., Maccabe, A.B.: Scalability limitations of VIA-based technologies in supporting MPI. In: Proceedings of the Fourth MPI Devlopers' and Users' Conference (2000)

[3] Shipman, G.M., Woodall, T.S., Graham, R.L., Maccabe, A.B., Bridges, P.G.: Infiniband scalability in Open MPI. In: International Parallel and Distributed Processing Symposium (IPDPS 2006) (2006)

[4] Koop, M.J., Sur, S., Gao, Q., Panda, D.K.: High performance mpi design using unreliable datagram for ultra-scale infiniband clusters. In: ICS 2007: Proceedings of the 21st annual international conference on Supercomputing, pp. 180–189. ACM, New York (2007)

[5] Shipman, G.M., Brightwell, R., Barrett, B., Squyres, J.M., Bloch, G.: Investigations on infiniband: Efficient network buffer utilization at scale. In: Proceedings, Euro PVM/MPI, Paris, France (2007)

[6] Brightwell, R., Maccabe, A.B., Riesen, R.: Design, implementation, and performance of MPI on Portals 3.0. International Journal of High Performance Computing Applications 17(1) (2003)
[7] Squyres, J.M., Lumsdaine, A.: The component architecture of open MPI: Enabling third-party collective algorithms. In: Getov, V., Kielmann, T. (eds.) Proceedings, 18th ACM International Conference on Supercomputing, Workshop on Component Models and Systems for Grid Applications, St.Malo, France, pp. 167–185. Springer, Heidelberg (2004)
[8] Reussner, R., Sanders, P., Prechelt, L., Müller, M.: Skampi: A detailed, accurate mpi benchmark. In: Alexandrov, V.N., Dongarra, J. (eds.) PVM/MPI 1998. LNCS, vol. 1497, pp. 52–59. Springer, Heidelberg (1998)

A Software Tool for Accurate Estimation of Parameters of Heterogeneous Communication Models

Alexey Lastovetsky, Vladimir Rychkov, and Maureen O'Flynn

School of Computer Science and Informatics, University College Dublin, Belfield, Dublin 4, Ireland
{alexey.lastovetsky,vladimir.rychkov,maureen.oflynn}@ucd.ie
http://hcl.ucd.ie

Abstract. Analytical communication performance models play an important role in prediction of the execution time of parallel applications on computational clusters, especially on heterogeneous ones. Accurate estimation of the parameters of the models designed for heterogeneous clusters is a particularly challenging task due to the large number of parameters. In this paper, we present a set of communication experiments that allows us to get the accurate estimation of the parameters with minimal total execution time, and software that implements this solution. The experiments on heterogeneous cluster demonstrate the accuracy and efficiency of the proposed solution.

Keywords: Heterogeneous cluster, heterogeneous communication performance model, MPI, communication model estimation.

1 Introduction

Heterogeneous computational clusters have become a popular platform for parallel computing with MPI as their principle programming system. Unfortunately, many MPI-based applications that were originally designed for homogeneous platforms do not have the same performance on heterogeneous platforms and require optimization. The optimization process is typically based on the performance models of heterogeneous clusters, which are used for prediction of the execution time of different configurations of the application, including its computation and communication costs. The accuracy of the performance models is very influential in determining the efficiency of parallel applications. The optimization of communications is an important aspect of the optimization of parallel applications. The performance of MPI collective operations, the main constituent of MPI, may degrade on heterogeneous clusters. The implementation of MPI collective operations can be significantly improved, by taking the communication performance model of the executing platform into account.

Traditionally, communication performance models for high performance computing are analytical and built for homogeneous clusters. The basis of these

A. Lastovetsky et al. (Eds.): EuroPVM/MPI 2008, LNCS 5205, pp. 43–54, 2008.

models is a point-to-point communication model characterized by a set of integral parameters, having the same value for each pair of processors. Collective operations are expressed as a combination of the point-to-point parameters, and the collective communication execution time is analytically predicted for different message sizes and numbers of processors. The core of this approach is the choice of such a point-to-point model that is the most appropriate to the targeted platform, allowing for easy and natural expression of different algorithms of collective operations. For homogeneous clusters, the point-to-point parameters are found statistically from the measurements of the execution time of communications between any two processors. When such a homogeneous communication model is applied to a cluster of heterogeneous processors, its point-to-point parameters are found by averaging values obtained for every pair of processors. Thus, in this case, the heterogeneous cluster will be treated as homogeneous in terms of the performance of communication operations.

When some processors or links in the heterogeneous cluster significantly differ in performance, predictions based on the homogeneous communication model may become inaccurate. More accurate performance models would not average the point-to-point communication parameters. On the other hand, the taking into account the parameters for each pair of processors will make the total number of point-to-point parameters and the amount of time required to estimate them significantly larger. In [1], [2], we proposed an analytical heterogeneous communication model designed for prediction of the execution time of MPI communications on heterogeneous clusters based on a switched network. The model includes the parameters that reflect the contributions of both links and processors to the communication execution time, and allows us to represent the aspects of heterogeneity for both links and processors. At the same time, the design of communication experiments for accurate and efficient estimation of the parameters of this model is not a trivial task.

Usually, to estimate the point-to-point parameters, different variations of sending/receiving messages between two processors are used. As regards the heterogeneous model proposed in [1], [2], with point-to-point communications only, we cannot collect enough data to estimate the parameters, and therefore must conduct some additional independent experiments. We design these additional communication experiments as a combination of scatter and gather. The observation of scatter and gather on the clusters based on a switched network show that the execution time may be non-linear and non-deterministic, especially if the MPI software stack includes the TCP/IP layer. Therefore, in our design we take into account all the irregularities, which might make the estimation inaccurate, and carefully select the message size.

The statistical methods of finding the point-to-point parameters, normally used in the case of homogeneous communication models, will result in unacceptably large number of measurements if applied as they are to the heterogeneous communication model. Therefore, another issue that has to be addressed is the minimization of the number of measurements necessary to accurately find the point-to-point parameters. We managed to reduce the number of measurements with the same accuracy as the exhaustive statistical analysis.

To the best of the authors' knowledge, there are no other publications describing heterogeneous communication performance models of computational clusters and the accurate estimation of the parameters of such models. In this paper, we present the software tool that automates the estimation of the heterogeneous communication performance model of clusters based on a switched network. The software tool can also be used in the high-level model-based optimization of MPI collective operations. This is particularly important for heterogeneous platforms where the users typically have neither authority nor knowledge for making changes in hardware or basic software settings.

This paper is organized as follows. In Section 2, related work on estimation of the parameters of communication performance models is discussed. In Section 3, we describe the point-to-point model of heterogeneous clusters based on a switched network and the design of communication experiments required to estimate its parameters. Section 4 presents the software tool for the estimation of the parameters of the heterogeneous communication performance model and the experimental results that demonstrate the accuracy and efficiency of the proposed solution.

2 Related Work

In this section, we discuss how the parameters of existing communication performance models are estimated. As all these models are built for homogeneous platforms, their parameters are the same for all processors and links. Therefore, to estimate them, it is sufficient to perform a set of communication experiments between any two processors.

The Hockney model [3] of the execution time of point-to-point communication is $\alpha + \beta m$, where α is the latency, β is the bandwidth and m is the message size. There are two ways to obtain a statistically reliable estimation of the Hockney parameters:

- To perform two series of roundtrips with empty messages (to get the latency parameter from the average execution time), and with non-empty ones (to get the bandwidth), or
- To perform a series of roundtrips with messages of different sizes and use results in a linear regression which fits the execution time into a linear combination of the Hockney parameters and a message size.

The LogP model [4] predicts the time of network communication for small fixed-sized messages in terms of the latency, L, the overhead, o, the gap per message, g, and the number of processors, P. The gap, g, is the minimum time between consecutive transmissions or receptions; it is the reciprocal value of the end-to-end bandwidth between two processors, so that the network bandwidth can be expressed as L/g. According to LogP, the time of point-to-point communication can be estimated by $L + 2o$. In [5], the estimation of the LogP parameters is presented, with the sending, o_s, and receiving, o_r, overheads being distinguished. The set of communication experiments used for estimation of the LogP parameters is as follows:

- To estimate the sending overhead parameter, o_s, a small number of messages are sent consecutively in one direction. The averaged sending time measured on the sender side will approximate o_s.
- The receiving overhead, o_r, is found directly from the time of receiving a message in the roundtrip. In this experiment, after completion of the send operation, the sending processor waits for some time, sufficient for the reply to reach the receiving processor, and only then posts a receive operation. The execution time of the receive operation is assumed to approximate o_r.
- The latency is found from the execution time of the roundtrip with small message $L = RTT/2 - o_s - o_r$.
- To estimate the gap parameter, g, a large number of messages are sent consecutively in one direction. The gap is estimated as $g = T_n/n$, where n is a number of messages and T_n is the total execution time of this communication experiment measured on the sender processor. The number of messages is chosen to be large to ensure that the point-to-point communication time is dominated by the factor of bandwidth rather than latency. This experiment, also known as a saturation, reflects the nature of the gap parameter but takes a long time.

In contrast to the Hockney model, LogP is not designed for the communications with arbitrary messages, but there are some derivatives, such as the LogGP model [6], which takes into account the message size by introducing the gap per byte parameter, G. The point-to-point communication time is estimated by $L + 2o + (m - 1)G$. The gap per byte, G, can be assessed in the same way as the gap parameter of the LogP model, saturating the link with large messages M, $G = g/M$.

In the PLogP (parameterized LogP) model [10], all parameters except for latency are piecewise linear functions of the message size, and the meaning of parameters slightly differs from LogP. The meaning of latency, L, is not intuitive; rather it is a constant that combines all fixed contribution factors such as copying to/from the network interfaces and the transfer over the network. The send, $o_s(m)$, and receive, $o_r(m)$, overheads are the times that the source and destination processors are busy for the duration of communication. They can be overlapped for sufficiently large messages. The gap, $g(m)$, is the minimum time between consecutive transmissions or receptions; it is the reciprocal value of the end-to-end bandwidth between two processors for messages of a given size m. The gap is assumed to cover the overheads: $g(m) \geq o_s(m)$ and $g(m) \geq o_r(m)$. According to the PLogP model, the point-to-point execution time is equal to $L + g(m)$ for the message of m bytes. The estimation of the PLogP parameters includes the experiments which are similar to the LogP ones but performed for different message sizes. Although this model is adaptive in nature, because of the number and location of breaks of piecewise linear functions are determined while the model is being built, the total number of parameters may become too large.

There are two main approaches to modeling the performance of communication operations for heterogeneous clusters. The first one is to apply traditional homogeneous communication performance models to heterogeneous clusters. In this case, the parameters of the models are estimated for each pair of processors and the average values for all pairs are then used in modelling. The second approach is to use dedicated heterogeneous models, where different pairs of heterogeneous processors are characterized by different parameters. While simpler in use, the homogeneous models are less accurate. When some processors or links in the heterogeneous cluster significantly differ in performance, predictions based on the homogeneous models may become quite inaccurate. The number of communication experiments required for the accurate estimation of both homogeneous and heterogeneous models will be of the same order, $O(n^2)$.

The traditional models use a small number of parameters to describe communication between any two processors. The price to pay is that such a traditional point-to-point communication model is not intuitive. The meaning of its parameters is not clear. Different sources of the contribution into the execution time are artificially and non-intuitively mixed and spread over a smaller number of parameters. This makes the models difficult to use for accurate modelling of collective communications. For example, the Hockney model uses only two parameters to describe communication between two processors. The parameters accumulate contributions of the participating processors and the communication layer into the constant and variable delays respectively. In order to model, say, the scatter operation on a switched cluster in an intuitive way, we need separate expressions for the contribution of the root processor, the communication layer and each of the receiving processors. Otherwise, we cannot express the serialization of outgoing messages on the root processor followed by their parallel transmission over the communication layer and parallel processing on the receiving processors. The use of the Hockney model as it is results in either ignoring the serialization or ignoring the parallelization. In the former case, the predictions will be too optimistic. In the latter case, the predictions will be too pessimistic. In both cases, they are not accurate. While using more parameters, the LogGP model faces the same problem because it does not separate the contribution of the processors and the communication layer into the variable delay. The traditional way to cope with this problem is to use an additional (and non-intuitive) fitting parameter, which will make the overall model even less clear. While this approach can somehow work for homogeneous models, it becomes hardly applicable to heterogeneous models. The point is that a heterogeneous model would need multiple fitting parameters making it fully impractical.

The alternative approach is to use original point-to-point heterogeneous models that allow for easy and intuitive expression of the execution time of collective communication operations such as the LOM model [1], [2] designed for switched heterogeneous clusters. While easy and intuitive in use, these models encounter a new challenging problem. The problem is that the number of point-to-point parameters describing communication between a pair of processors becomes larger than the number of independent point-to-point communication experiments traditionally used for estimation of the parameters. In this paper, we describe the set of communication experiments sufficient for the accurate and efficient estimation of the parameters and present the software tool that implements this approach.

3 Heterogeneous Communication Performance Model and Its Estimation

The LOM model [1] includes both link-specific and processor-specific parameters. Like most of point-to-point communication models, its point-to-point parameters represent the communication time by a linear function of the message size. The execution time of sending a message of M bytes from processor i to processor j in a heterogeneous cluster $i \xrightarrow{M} j$ is estimated by $C_i + t_i M + C_j + t_j M + \frac{M}{\beta_{ij}}$, where C_i, C_j are the fixed processing delays; t_i, t_j are the delays of processing of a byte; β_{ij} is the transmission rate. The delay parameters, which are attributed to each processor, reflect the heterogeneity of the processors. The transmission rates correspond to each link and reflect the heterogeneity of communications; for networks with a single switch, it is realistic to assume $\beta_{ij} = \beta_{ji}$.

To estimate the parameters of such a model, an approach with roundtrip point-to-point experiments is not enough. For a network consisting of n processors, there will be $2n + C_n^2$ unknowns: n fixed processing delays, n variable processing delays, and C_n^2 transmission rates. The execution time of the roundtrip, namely sending M_1 bytes and receiving M_2 bytes between nodes $i \overset{M_2}{\underset{M_1}{\longleftrightarrow}} j$, is equal to $T_{ij}(M_1, M_2) = (C_i + t_i M_1 + C_j + t_j M_1 + \frac{M_1}{\beta_{ij}}) + (C_i + t_i M_2 + C_j + t_j M_2 + \frac{M_2}{\beta_{ij}})$. The roundtrip experiments will give us only C_n^2 equations. Therefore, the first challenge we face is to find a set of experiments that gives a sufficient number of linearly independent linear equations, whose variables represent the unknown point-to-point parameters.

First, we measure the execution time of the roundtrips with empty messages between each pair of processors $i < j$ (C_n^2 experiments). The fixed processing delays can be found from $T_{ij}(0) = 2C_i + 2C_j$ solved for every three roundtrips $i \overset{0}{\underset{0}{\longleftrightarrow}} j$, $j \overset{0}{\underset{0}{\longleftrightarrow}} k$, $k \overset{0}{\underset{0}{\longleftrightarrow}} i$ ($i < j < k$): $\{T_{ij}(0) = 2C_i + 2C_j, T_{jk}(0) = 2C_j + 2C_k, T_{ki}(0) = 2C_k + 2C_i\}$.

In order to find the rest $n + C_n^2$ parameters, we might use the roundtrips with non-empty message, but it would give us only C_n^2 linearly independent equations. Instead, we use the additional experiments, which include communications from one processor to two others and backward, and express the execution time of the communication experiments in terms of the heterogeneous point-to-point communication performance model. As will be shown below, the set of point-to-point and point-to-two communication experiments is enough to find the fixed processing delay and transmission rates, but there is one more important issue to be addressed. The point-to-two experiments are actually a particular combination of scatter and gather. The scatter and gather operations may have some irregular behaviour on the clusters based on a switched network, especially if the MPI software stack includes the TCP/IP layer. Therefore, the message sizes for the additional experiments have to be carefully selected to avoid these irregularities.

We observed the leap in the execution time of scatter for large messages and the non-deterministic escalations of the execution time of gather for medium-sized messages (see Fig. 1). It prompted us introduce the particular threshold parameters to categorize the message size ranges where distinctly different behaviour of the collective MPI operations is observed, and to apply different formula for these regions to express the execution time with the heterogeneous point-to-point parameters.

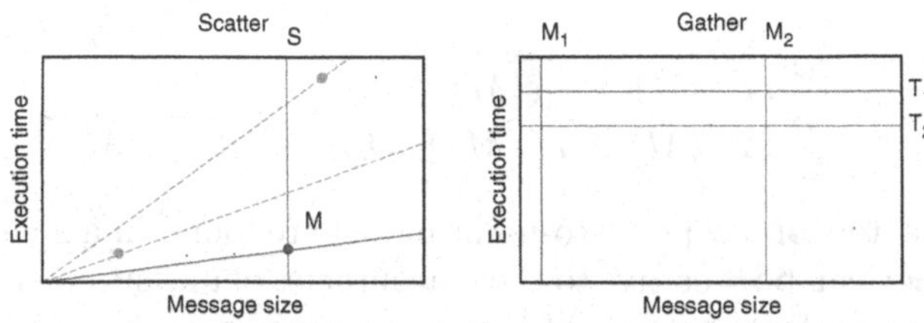

Fig. 1. The execution time of collective communications against the message size

The estimated time of scattering messages of size M from node 0 to nodes $1, ..., n$ is given by $n(C_0 + t_0 M) + \max_{1 \le i \le n}\{C_i + t_i M + \frac{M}{\beta_{0i}}\}$, if $M \le S$, and $n(C_0 + t_0 M) + \sum_{1 \le i \le n}\{C_i + t_i M + \frac{M}{\beta_{0i}}\}$, if $M > S$, where C_0, t_0, C_i, t_i are the fixed and variable processing delays on the source node and destinations. This reflects the parallel communication for small messages and the serialized communication for large messages. The threshold parameter S corresponds to the leap in the execution time, separating small and large messages. It may vary for different combinations of clusters and MPI implementations.

For the gather operation, we separate small, medium and large messages by introducing parameters M_1 and M_2. For small messages, $M < M_1$, the execution time has a linear response to the increase of message size. Thus, the execution time for the many-to-one communication involving n processors ($n \le N$, where N is the cluster size) is estimated by $n(C_0 + t_0 M) + \max_{1 \le i \le n}\{C_i + t_i M + \frac{M}{\beta_{0i}}\} + \kappa_1 M$, where $\kappa_1 = const$ is a fitting parameter for correction of the slope. For large messages, $M > M_2$, the execution time resumes a linear predictability with increasing message size. Hence, this part is similar in design but has a different slope of linearity that indicates greater values due to overheads: $n(C_0 + t_0 M) + \sum_{1 \le i \le n}\{C_i + t_i M + \frac{M}{\beta_{0i}}\} + \kappa_2 M$. The additional parameter $\kappa_2 = const$ is a fitting constant for correction of the slope. For medium messages, $M_1 \le M \le M_2$, we observed a small number of discrete levels of escalation, that remain constant as the message size increases.

Thus, following the model of scatter and gather, in our experiments we gather zero-sized messages in order to avoid the non-deterministic escalations. For scatter, the message size M is taken less than the value of the threshold parameter S. The wrong selection of the message size can make the estimation of the point-to-point parameters inaccurate, which is shown in Fig. 1. c In order to find variable processing delays t_i and transmission rates β_{ij}, we measure the execution time of the C_n^2 experiments $i \xleftrightarrow[0]{M} j$ $(i < j)$, the roundtrips with empty replies, and the C_n^3 experiments $i \xleftrightarrow[0]{M} j, k$ $(i < j < k)$, where the source processor sends the messages of the same size to two processors and receives zero-sized messages from them. The execution time $T_i(M)$ of one-to-two communications with root i can be expressed by $T_i(M) = 4C_i + 2t_iM + \max(2C_j + t_jM + \frac{M}{\beta_{ij}}, 2C_k + t_kM + \frac{M}{\beta_{ik}})$. The execution times of these experiments are used in the following formula to get the values of the variable processing delays and then the values of transmission rates:

$$t_i = \begin{cases} \frac{T_i(M) - T_{ij}(M) - 2C_i}{M}, T_{ij}(M) > T_{ik}(M) \\ \frac{T_i(M) - T_{ik}(M) - 2C_i}{M}, T_{ik}(M) > T_{ij}(M) \end{cases} \quad \frac{1}{\beta_{ij}} = \frac{T_{ij}(M) - 2C_i - 2C_j}{M} - t_i - t_j$$

As the parameters of our point-to-point model are found in a small number of experiments, they can be sensitive to the inaccuracies of measurement. Therefore, it makes sense to perform a series of the measurements for one-to-one and one-to-two experiments and to use the averaged execution times in the corresponding linear equations. Minimization of the total execution time of the experiments is another issue that we address. The advantage of the proposed design is that these series do not have to be lengthy (typically, up to ten in a series) because all the parameters have been already averaged with the process of their finding.

The procedure of the estimation of the point-to-point parameters is preceded by the estimation of the threshold parameters. To estimate the threshold parameters, we use the scatter and gather benchmarks for different message sizes. The data rows for scatter and gather consist of the message sizes taken with some stride and the measured execution time $\{M^i, T^i\}$, $M^{i+1} = M^i + stride$. Typical data rows for heterogeneous clusters based on a switched network are shown in Fig. 1. One can see that:

- the execution time of scatter can be approximated by the piecewise linear function with one break that correspond to the threshold parameter S to be found;
- the execution time of gather has the regions of linearity for small, $M < M_1$, and large, $M > M_2$, messages and can also be approximated by the two linear functions.

To find the threshold parameters, we use the algorithm proposed in [8]. It considers the statistical linear models with multiple structural changes and uses dynamic programming to identify optimal partitions with different numbers of segments. The algorithm allows us to locate the break in the execution time of scatter, S, and the range of large messages for gather, M_2.

Then we perform the linear regression of the execution time of gather on this range to estimate the slope correction parameter κ_2, that is used to adjust the prediction of many-to-one execution time for large messages. The linear regression gives us two values c_0 and c_1: $T \approx c_0 + c_1 M$, $M > M_2$. The slope correction parameter κ_2 is found as follows: $\kappa_2 = c_1 - \sum_{i=1}^{n}(t_i + \frac{1}{\beta_{0i}})$. We find M_1 as $M_1 \approx M_k$, $k = min\{i : T^{i+1}/T^1 > 10\}$. The linear regression on the data row $\{M^i, T^i\}$, $i = 1, ..., k$ is performed to obtain the linear parameters for the small messages, $T \approx c_0 + c_1 M$, $M > M_1$, and to calculate the slope correction parameter $\kappa_1 = c_1 - max_{1 \leq i \leq n}\{t_i + \frac{1}{\beta_{0i}}\}$.

4 The Software Design and Experimental Results

To the best of the authors' knowledge, there are no available software tools for the estimation of heterogeneous communication models of computational clusters. In this section, we present such a software tool, and describe its features and design.

We design the software tool in the form of a library implemented on top of MPI. In addition to the library, the software tool provides a command line utility that can be used for one-time estimations. The utility uses the library to estimate the parameters of the heterogeneous communication performance model with the given accuracy and saves the data in a file that can be used later. One-estimation may be done during the installation of the software tool, or each time the parallel platform or MPI implementation has been changed. The estimation can also be performed in the user application at runtime, with the invocation of the library functions. The library consists of three modules:

1. The Measurement module is responsible for the measurement of the execution time of the communication experiments required to estimate the parameters of the heterogeneous model. It uses the MPIBlib benchmarking library [7], namely, the point-to-point, scatter and gather benchmarks. In addition, the Measurement module includes the function for measuring the execution time of the point-to-two communication experiments, $i \underset{0}{\overset{M}{\longleftrightarrow}} j, k$, required to find the variable processing delays and transmission rates. The point-to-point and point-to-two experiments are optimized for clusters with a single switch. As network switches are capable of forwarding packets between sources and destinations appropriately, several point-to-point or point-to-two communications can be run in parallel, with each process being involved in no more than one communication. This decreases the execution time the benchmark takes, giving quite accurate results.
2. The Model module provides the API, which allows the user to estimate the parameters of the heterogeneous communication performance model inside their application. This module uses the results of benchmarks provided by the Measurement module and the MPIBlib library, builds and solves the systems of equations described in the previous section. For estimation of the threshold parameters required to select the message size for point-to-two experiments, the **strucchange** library of the R statistical package is used [9]. It automates the detection of the structural changes in the linear regression models. The statistical analysis is performed with help of GSL

(GNU scientific library). More specifically, the parameters of the heterogeneous communication performance model are estimated within a confidence interval that indicates the reliability of estimation, which is implemented with help of GSL. For linear regression, the software tool uses GSL routines for performing least squares fits to experimental data.

3. The Optimization module provides a set of the optimized implementations of collective operations, such as scatter and gather, which use the parameters of the heterogeneous model [10].

To demonstrate the accuracy provided by the software tool, we compare the execution time of a single point-to-point communication observed for different message sizes with the predictions provided by the logp_mpi package [6] and by our software tool (Fig. 2). The logp_mpi package was used for the predictions of the PLogP and LogGP models. The experiments were carried out between two processors of the 16-node heterogeneous cluster, which has the following characteristics: 11 x Intel Xeon 2.8/3.4/3.6, 2 x P4 3.2/3.4, 1 x Celeron 2.9, 2 x AMD Opteron 1.8, Gigabit Ethernet, LAM 7.1.3. The PLogP model is more accurate but much more costly. The accuracy is due to the use of the functional parameters, each of which is approximated by a large number of scalar parameters. The linear predictions of LogGP and our point-to-point models are practically the same.

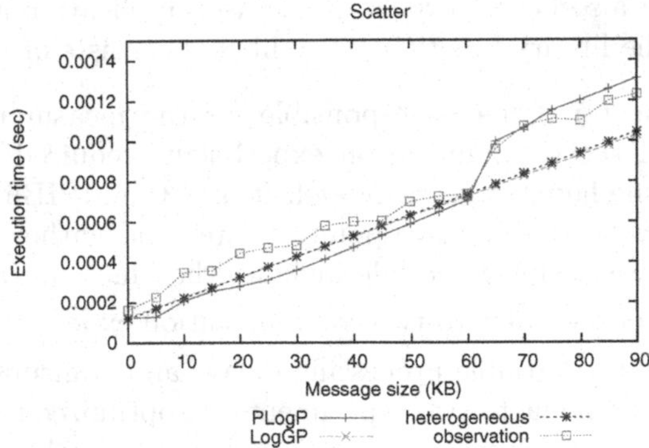

Fig. 2. The observed and predicted execution times of the point-to-point communication on the 16-node heterogeneous cluster

The point-to-point parameters estimated by the software tool are used in the analytical models of collective communication operations for prediction of their execution time. Therefore, the accuracy of estimation of these parameters can be validated by the comparison of the observed execution time of the collectives and the one predicted by the analytical models using the values of the point-to-point parameters estimated by the software tool. For the experiment, we use the linear scatter and gather, the analytical models of which are presented in Section 3. Fig. 3 shows the results of this experiment. One can see that the execution time of scatter is predicted with high accuracy. The same is true for gather, given

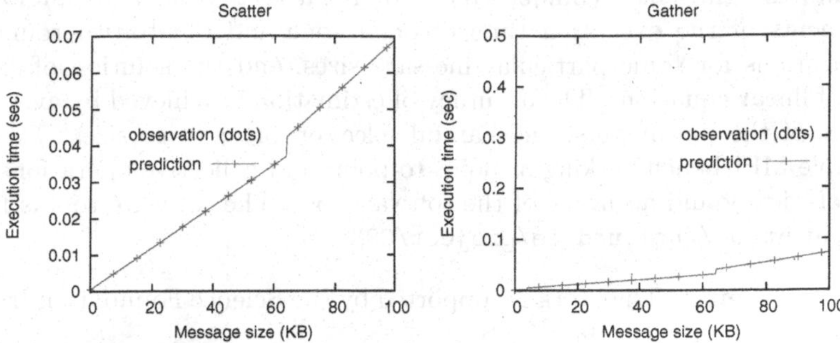

Fig. 3. The observed and predicted execution time of scatter and gather on the 16-node heterogeneous cluster

that the analytical model is not supposed to predict irregular escalations of the execution time for medium-sized messages.

Usually, the statistically reliable estimation is achieved by averaging the results of numerous repetitions of the same experiment. The software tool has an addition level of the averaging of the experimental results. Namely, each individual experiment produces multiple estimates of the same parameter that are also averaged. Consider, for example, the experiment estimating the fixed processing delays. When the execution time of the empty roundtrips between all pairs of processors has been measured, the fixed processing delay of a processor can be found in an identical manner from C_{n-1}^2 systems of equations, one for each of the C_{n-1}^2 triplets of the processors that include this processor. Therefore, the first approximation of the fixed processing delay will be calculated by averaging these C_{n-1}^2 values. For more accurate estimation, this communication experiment can be then repeated several times, giving several estimates of the fixed processing delay which can be further averaged. As a result, the number of the repetitions will be much smaller.

In total, the following series of repetitions are performed:

- a series of the k_0 repetitions for the experiment including C_n^2 empty roundtrips,
- a series of the k_1 repetitions for the experiment including C_n^2 one-to-one communications, and
- a series of the k_2 repetitions for the experiment including $3C_n^3$ one-to-two communications.

In our experiments on the 16-node heterogeneous cluster, no more than ten repetitions in a series were needed to achieve the acceptable accuracy. The estimation of the parameters took just fractions of a second, which allows us to use the library for the runtime estimation in user applications.

5 Conclusion

This paper has described the software tool for accurate estimation of parameters of the heterogeneous communication performance model. The software tool implements the efficient technique that requires a relatively small number of measurements of the execution time of one-to-one and one-to-two roundtrip communications for some particular message sizes, and the solution of simple systems of linear equations. The accuracy of estimation is achieved by averaging the values of the parameters, and careful selection of message sizes. The fast and reliable MPI benchmarking of point-to-point and collective operations also support efficiency and accuracy of the software tool. The software tool is freely available at `http://hcl.ucd.ie/project/CPM`

Acknowledgments. This work is supported by the Science Foundation Ireland and in part by the IBM Dublin CAS.

References

1. Lastovetsky, A., Mkwawa, I., O'Flynn, M.: An Accurate Communication Model of a Heterogeneous Cluster Based on a Switch-Enabled Ethernet Network. In: Proc. of ICPADS 2006, vol. 2, pp. 15–20. IEEE Computer Society Press, Los Alamitos (2006)
2. Lastovetsky, A., O'Flynn, M.: A Performance Model of Many-to-One Collective Communications for Parallel Computing. In: Proceedings of IPDPS 2007 (2007)
3. Hockney, R.: The communication challenge for MPP: Intel Paragon and Meiko CS-2. Parallel Computing 20, 389–398 (1994)
4. Culler, D., Karp, R., Patterson, D., Sahay, A., Schauser, K., Santos, E., Subramonian, R., von Eicken, T.: LogP: Towards a realistic model of parallel computation. In: Proceedings of PPoPP 1993, pp. 1–12. ACM, New York (1993)
5. Culler, D., Liu, L., Martin, R., Yoshikawa, C.: LogP Performance Assessment of Fast Network Interfaces. IEEE Micro. 16(1), 35–47 (1996)
6. Alexandrov, A., Ionescu, M., Schauser, K., Scheiman, C.: LogGP: Incorporating long messages into the LogP model. In: Proc. of SPAA 1995, pp. 95–105. ACM, New York (1995)
7. Kielmann, T., Bal, H., Verstoep, K.: Fast measurement of LogP parameters for message passing platforms. In: Rolim, J. (ed.) IPDPS-WS 2000. LNCS, vol. 1800, pp. 1176–1183. Springer, Heidelberg (2000)
8. Lastovetsky, A., Rychkov, V., O'Flynn, M.: MPIBlib: Benchmarking MPI Communications for Parallel Computing on Homogeneous and Heterogeneous Clusters. In: Lastovetsky, A., Kechadi, T., Dongarra, J. (eds.) EuroPVM/MPI 2008. LNCS, vol. 5205. Springer, Heidelberg (2008)
9. Bai, J., Perron, P.: Computation and Analysis of Multiple Structural Change Models. J. of Applied Econometrics 18, 1–22 (2003)
10. Zeileis, A., Leisch, F., Hornik, K., Kleiber, C.: Strucchange: An R package for testing for structural change in linear regression models. J. of Statistical Software 7(2), 1–38 (2002)
11. Lastovetsky, A., O'Flynn, M., Rychkov, V.: Optimization of Collective Communications in HeteroMPI. In: Cappello, F., Herault, T., Dongarra, J. (eds.) PVM/MPI 2007. LNCS, vol. 4757, pp. 135–143. Springer, Heidelberg (2007)

Sparse Non-blocking Collectives in Quantum Mechanical Calculations

Torsten Hoefler[1], Florian Lorenzen[2], and Andrew Lumsdaine[1]

[1] Open Systems Lab, Indiana University, Bloomington IN 47405, USA
{htor,lums}@cs.indiana.edu
[2] Institut für Theoretische Physik, FU Berlin, Arnimalle 14, 14195 Berlin, Germany
lorenzen@physik.fu-berlin.de

Abstract. For generality, MPI collective operations support arbitrary dense communication patterns. However, in many applications where collective operations would be beneficial, only sparse communication patterns are required. This paper presents one such application: Octopus, a production-quality quantum mechanical simulation. We introduce new sparse collective operations defined on graph communicators and compare their performance to MPI_Alltoallv. Besides the scalability improvements to the collective operations due to sparsity, communication overhead in the application was reduced by overlapping communication and computation. We also discuss the significant improvement to programmability offered by sparse collectives.

1 Introduction

Ab-initio quantum mechanical simulations play an important role in nano and material sciences as well as many other scientific areas, e. g., the understanding of biological or chemical processes. Solving the underlying Schrödinger equation for systems of hundreds or thousands of atoms requires a tremendous computational effort that can only be mastered by highly parallel systems and algorithms.

Density functional theory (DFT) [10,11] is a computationally feasible method to calculate properties of quantum mechanical systems like molecules, clusters, or solids. The basic equations of DFT are the static and time-dependent Kohn-Sham equations:[1]

$$\boldsymbol{H}\boldsymbol{\varphi}_j = \varepsilon_j \boldsymbol{\varphi}_j \qquad\qquad i\frac{\partial}{\partial t}\boldsymbol{\varphi}_j(t) = \boldsymbol{H}(t)\boldsymbol{\varphi}_j(t) \tag{1}$$

The electronic system is described by the Hamiltonian operator

$$\boldsymbol{H} = -\frac{1}{2}\nabla^2 + \boldsymbol{V}, \tag{2}$$

where the derivative accounts for kinetic energy and $\boldsymbol{V}$ for the atomic potentials and electron-electron interaction. The vectors $\boldsymbol{\varphi}_j$, $j = 1, \ldots, N$, are the Kohn-Sham orbitals each describing one of N electrons.

[1] i denotes the imaginary unit $i = \sqrt{-1}$ and t is the time parameter.

A. Lastovetsky et al. (Eds.): EuroPVM/MPI 2008, LNCS 5205, pp. 55–63, 2008.

The scientific application `octopus` [3] solves the eigenvalue problem of Eq. (1, left) by iterative diagonalization for the lowest N eigenpairs $(\varepsilon_j, \boldsymbol{\varphi}_j)$ and Eq. (1, right) by explicitly evolving the Kohn-Sham orbitals $\boldsymbol{\varphi}_j(t)$ in time. The essential ingredient of iterative eigensolvers as well as of most real-time propagators [2] is the multiplication of the Hamiltonian with an orbital $\boldsymbol{H}\boldsymbol{\varphi}_j$. Since `octopus` relies on finite-difference grids to represent the orbitals, this operation can be parallelized by dividing the real-space mesh and assigning a certain partition (domain) to each node as shown in Fig. 1(a).

The potential $\boldsymbol{V}$ is a diagonal matrix, so the product $\boldsymbol{V}\boldsymbol{\varphi}_j$ can be calculated locally on each node. The Laplacian operator of (2) is implemented by a finite-difference stencil as shown in Fig. 1(b). This technique requires to send values close to the boundary (gray shading in Fig. 1(b)) from one partition (orange) to a neighboring one (green).

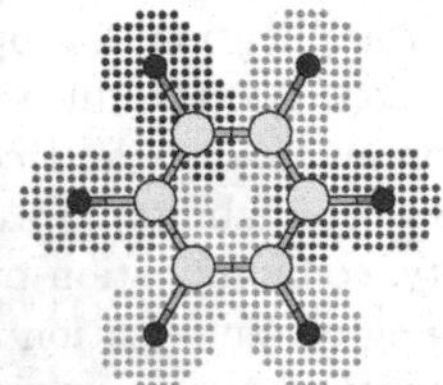

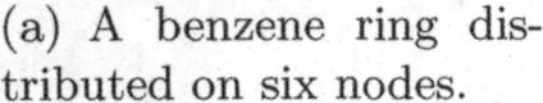

(a) A benzene ring distributed on six nodes.

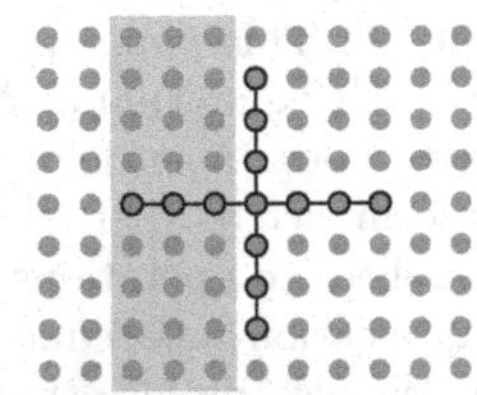

(b) Boundary values for a third order stencil.

Fig. 1. Partitions of `octopus`' real-space finite-difference mesh

The original implementation of $\boldsymbol{H}\boldsymbol{\varphi}_j$ is:

1. Exchange boundary values between partitions
2. $\boldsymbol{\varphi}_j \leftarrow -\frac{1}{2}\nabla^2\boldsymbol{\varphi}_j$ (apply kinetic energy operator)
3. $\boldsymbol{\varphi}_j \leftarrow \boldsymbol{\varphi}_j + \boldsymbol{V}\boldsymbol{\varphi}_j$ (apply potential)

In this article, we describe a simplified and efficient way to implement and optimize the neighbor exchange with non-blocking collective operations that are defined on topology communicators.

2 Parallel Implementation

This section gives a detailed analysis of the communication and computation behavior of the domain parallelization and presents alternative implementations using non-blocking and topology-aware colletives that provide higher performance and better programmability.

2.1 Domain Parallelization

The application of the Hamiltonian to an orbital $\boldsymbol{H}\boldsymbol{\varphi}_j$ can be parallelized by a partitioning of the real-space grid. The best decomposition depends on the distribution of grid-points in real-space which depends on the atomic geometry of the system under study. We use the library METIS [9] to obtain partitions that are well balanced in the number of points per node.

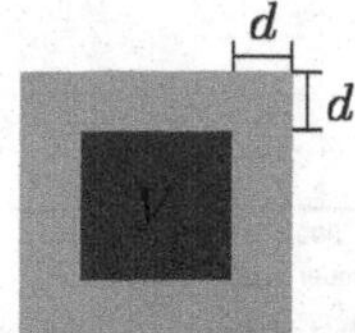

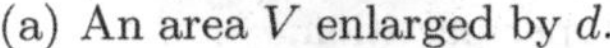

(a) An area V enlarged by d.

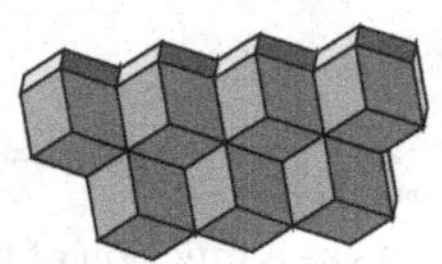

(b) Rhombic dodecahedrons, consisting of twelve rhombic faces, stacked to fill up space.

Fig. 2. Message sizes and number of neighbors

The communication overhead of calculating $\boldsymbol{H}\varphi_j$ is dominated by the neighbor exchange operation on the grid. To determine a model to assess the scaling of the communication time which can be used to predict the application's running time and scalability, we need to assess the message-sizes, and the average number of neighbors of every processor. Both parameters are influenced by the discretization order d that affects how far the stencil leaks into neighbouring domains, and by the number of points in each partition. Assuming a nearly optimal domain decomposition, NP points in total, and P processors we can consider the ratio $V = NP/P$ as "volume" per node. The number of communicated points is $p(P) = V_d - V$ with V_d being the volume V increased by the discretization order d and reads

$$p(P) = \alpha d^3 + \beta d^2 \sqrt{V(P)} + \gamma d \sqrt[3]{V(P)^2} \tag{3}$$

with coefficients α, β, γ depending on the actual shape of the partitions. The derivation of (3) is sketched schematically in Fig. 2(a) for a 2D situation: an area V is increased by d in each direction. The enlargement is proportional to d^2 (green) and $d\sqrt{V}$ (red). In 3D, the additional dimension causes these terms to be multiplied by d and leads to one more term proportional to $d\sqrt[3]{V^2}$. Fig. 3(a) shows the number of exchanged points measured for a cylindrical grid of 1.2 million points and the analytical expression (3) fitted to the data-points.

Since the average number of neighbors (ν) depends on the structure of the input system, we cannot derive a generic formula for this quantity but instead give the following estimate: METIS minimizes edge-cut which is equivalent to minimization of surfaces. This can be seen in Fig. 1(a) where the partition borders are almost between the gray Carbon atoms, the optimum in this case. In general, the minimal surface a volume can take on is spherical. Assuming the partitions to be stacked rhombic dodecahedrons as approximation to spheres, shown in Fig. 2(b), we conclude that, for larger P, ν is clearly below P because each dodecahedron has at maximum twelve neighbors. This consideration, of course, assumes truly minimum surfaces that METIS can only approximate. In

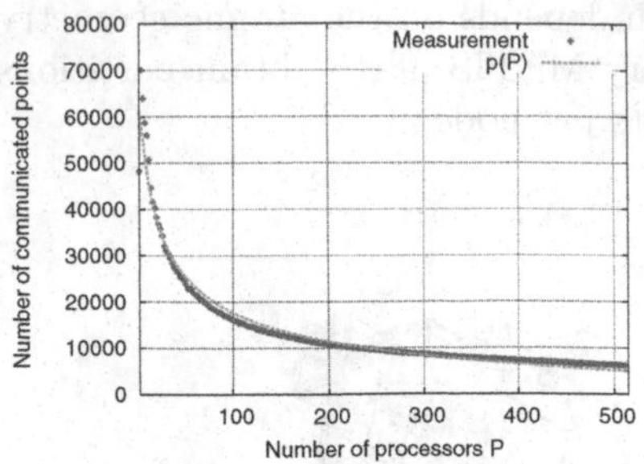

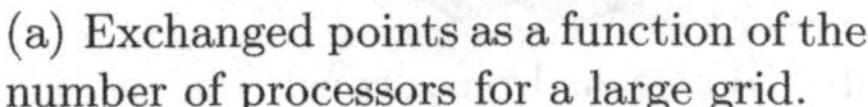
(a) Exchanged points as a function of the number of processors for a large grid.

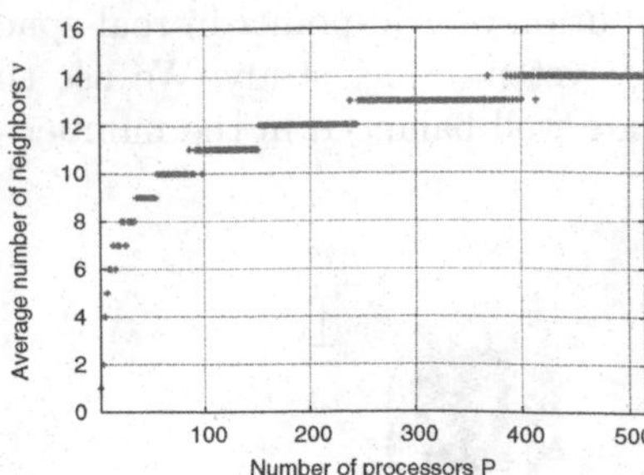

(b) Average and maximum number of neighbors ν per partition.

Fig. 3. Communicated points and neighbor-count for different numbers of processors

practice, we observe an increasing number of neighbors for larger P, see Fig. 3(b). Nevertheless, the number of neighbors is an order of magnitude lower than the number of processors.

Applying the well-know LogGP model [4] to our estimations of the scaling of the message sizes and the number of neighbors ν, we can derive the following model of the communication overhead (each point is represented by an 8 byte double value):

$$t_{comm} = L + o\nu + g(\nu - 1) + G(\nu \cdot 8\, p(P)) \tag{4}$$

We assume a constant number of neighbors ν at large scale. Thus, the communication overhead scales with $O\left(\sqrt{NP/P}\right)$ in P. The computational cost of steps 2 and 3 that determines the potential to overlap computation and communication scales with NP/P for the potential term and $\alpha d^3 + \beta d^2 \sqrt{NP/P} + \gamma d \sqrt[3]{(NP/P)^2} + \delta NP/P$ for the kinetic term.[2] We observe that our computation has a similar scaling behaviour as the communication overhead, cf. Eq. (4). We therefore conclude that overlapping the neighbor exchange communication with steps 2 and 3 should show a reasonable performance benefit at any scale.

Overlapping this kind of communication has been successfully demonstrated on a regular grid in [1]. We expect the irregular grid to achieve similar performance improvements which could result in a reduction of the communication overhead.

Practical benchmarks show that there are two calls that dominate the communication overhead of `octopus`. On 16 processors, about 13% of the application time is spent in many 1 real or complex value MPI_Allreduce calls caused by dot-products and the calculation of the atomic potentials. This communication can not be optimized or overlapped easily and is thus out of the scope of this article. The second biggest source of communication overhead is the neighbor

[2] The derivation of this expression is similar to (3) except that we shrink the volume by the discretzation order d.

communication which causes about 8.2% of the communication overhead. Our work aims at efficiently implementing the neighbor exchange and reducing its communication overhead with new non-blocking collective operations that act on a process topology.

2.2 Optimization with Non-blocking Collective Operations

Non-blocking collective operations that would support efficient overlap for this kind of communication are not available in the current MPI standard. We used the open-source implementation LibNBC [6] that offers a non-blocking interface for all MPI-defined collective operations.

Implementation with NBC_Ialltoallv. The original implementation used MPI_Alltoallv for the neighbor exchange. The transition to the use of non-blocking collective operations is a simple replacing of MPI_Alltoall with NBC_Ialltoallv and the addition of a handle. Furthermore, the operation has to be finished with a call to NBC_Wait before the communicated data is accessed.

However, to achieve the best performance improvement, several additional steps have to be performed. The first step is to maximize the time to overlap, i. e., to move the NBC_Wait as far behind the respective NBC_Ialltoallv as possible in order to give the communication more time to proceed in the background. Thus, to overlap communication and computation we change the original algorithm to:

1. Initiate neighbor exchange (NBC_Ialltoallv)
2. $\varphi_j \leftarrow \boldsymbol{v}\varphi_j$ (apply potential)
3. $\varphi_j \leftarrow \varphi_j - \frac{1}{2}\nabla^2\varphi_j^{inner}$ (apply kinetic energy operator to inner points)
4. Wait for the neighbor exchange to finish (NBC_Wait)
5. $\varphi_j \leftarrow \varphi_j - \frac{1}{2}\nabla^2\varphi_j^{edge}$ (apply kinetic energy operator to edge points)

We initiate the exchange of neighboring points (step 1) and overlap it with the calculation of the potential term (step 2) and the inner part of the kinetic energy, which is the derivative of all points that can be calculated solely by local points (step 3). The last step is waiting for the neighbor-exchange to finish (step 4) and calculation of the derivatives for the edge points (step 5).

A usual second step to optimize for overlap is to introduce NBC_Test() calls that give LibNBC the chance to progress outstanding requests. This is not necessary if the threaded version of LibNBC is running on the system. We have shown in [5] that the a naively threaded version performs worse, due to the loss of a computational core. However, for this work, we use the InfiniBand optimized version of LibNBC [7] which does not need explicit progression with NBC_Test() if there is only a single communication round (which is true for all non-blocking operations used in `octopus`).

As shown in Sec. 2.1, the maximum number of neighbors is limited. Thus, the resulting communication pattern for large-scale runs is sparse. The MPI_Alltoallv function, however, is not suitable for large-scale sparse communication patterns because it is not scalable due to the four index arrays which have to be filled for every process in the communicator regardless of the communication pattern. This results in arrays mostly filled with zeros that still have to be generated, stored and processed in the MPI call and is thus a performance bottleneck at large-scale. Filling those arrays correctly is also complicated for the programmer and a source of common programming errors. To tackle the scalability and implementation problems, we propose new collective operations [8] that are defined on the well known MPI process topologies. The following section describes the application of one of the proposed collective operations to the problem described above.

Topological Collective Operations. We define a new class of collective operations defined on topology communicators. The new collective operation defines a neighbor exchange where the neighbors are defined by the topology. MPI offers a regular (cartesian) topology as well as a graph topology that can be used to reflect arbitrary neighbor relations. We use the graph communicator to represent the neighborship of partitions generated by METIS for the particular input system. MPI_Graph_create is used to create the graph communicator. We implemented our proposal in LibNBC, the functions NBC_Get_neighbors_count and NBC_Comm_neighbors return the neighbor count and the order of ranks for the send/receive buffers respectively. The operation NBC_Ineighbor_xchg performs a non-blocking neighbor exchange in a single step.

Programmability. It seems more natural to the programmer to map the output of a graph partitioner (e. g., an adjacency list that represents topological neighbors) to the creation of a graph communicator and simply perform collective communication on this communicator rather than performing the Alltoallv communication. To emphasize this, we demonstrate pseudocodes that perform a similar communication operation to all graph neighbors indicated in an undirected graph (`list[i][0]` represents the source and `list[i][1]` the destination vertex of edge `i` and is sorted by source node).

Listing 1. NBC_Ialltoall Implementation

```
rdpls = malloc(p*sizeof(int)); sdpls = malloc(p*sizeof(int));
rcnts = malloc(p*sizeof(int)); scnts = malloc(p*sizeof(int));
for(i=0; i<p; i++) { scnts[i] = rcnts[i] = 0; }
for(i=0; i<len(list); i++) if(list[i][0] == rank)
    scnts[list[i][1]] = count; rcnts[list[i][1]] = count;
sdispls[0] = rdispls[0] = 0;
for(i=1; i<p; i++) {
   sdpls[i] = sdpls[i-1] + scnts[i];
   rdpls[i] = rdpls[i-1] + rcnts[i]; }
NBC_Ialltoallv(sbuf, scnts, sdpls, dt, rcnts, rdpls, dt, comm, req);
/* computation goes here */
NBC_Wait(req, stat);
```

Listing 1 shows the NBC_Ialltoall implementation which uses four different arrays to store the adjacency information. The programmer is fully responsible for administering those arrays. Listing 2 shows the implementation with our newly proposed operations that acquire the same information from the MPI library (topology communicator layout). The processes mapping in the created graph communicator might be rearranged by the MPI library to place tightly coupled processes on close processors (e. g. on the same SMP system). The collective neighbor exchange operation allows other optimizations (e. g. starting off-node communication first to overlap local memory copies of on-node communication). Due to the potentially irregular grid (depending on the input system), the number of points communicated with each neighbor might vary. Thus, we used the vector variant NBC_Ineighbor_xchgv to implement the neighbor exchange for `octopus`.

Listing 2. NBC_Ineighbor_xchg Implementation

```
last = list[0][0]; counter = 0; // list is sorted by source
for(i=0; i<len(list); i++) {
 if(list[i][0] != last) index[list[i][0]] = counter;
 edges[counter++] = list[i][1];
}
MPI_Graph_create(comm, nnodes, index, edges, 1, topocomm);
NBC_Ineighbor_xchg(sbuf, count, dt, rbuf, count, dt, topocomm, req);
/* computation goes here */
NBC_Wait(req, stat);
```

3 Performance Analysis

We benchmarked our implementation on the CHiC supercomputer system, a cluster computer consisting of nodes equipped with dual socket dual-core AMD 2218 2.6 GHz CPUs, connected with SDR InfiniBand and 4 GB memory per node. We use the InfiniBand-optimized version of LibNBC [7] to achieve highest performance and overlap. Each configuration was ran three times on all four cores per node (4-16 nodes were used) and the average values are reported.

Fig. 4(a) shows the microbenchmark results for the overhead of NBC_Ialltoallv and NBC_Ineigbor_xchgv of *NBCBench* [6] with 10 neighbors under th assumption that the whole communication time can be overlapped. The overhead of the new neighbor exchange operation is slightly lower than the NBC_Ialltoallv overhead because the implementation does not evaluate arrays of size P. Fig. 4(b) shows the communication overhead of a fixed-size ground state calculation of a chain of Lithium and Hydrogene atoms. The overhead varies (depending on the technique used) between 22% and 25% on 16 processes. The bars in Fig. 4(b) show the total communication overhead and the tackled neighbor exchange overhead (lower part). We analyze only the overhead-reduction and easier implementation of the neighbor exchange in this work. The application of non-blocking neighbor collective operations efficiently halves the neighbor exchange overhead and thus improves the performance of `octopus` by about 2%. The improvement is smaller

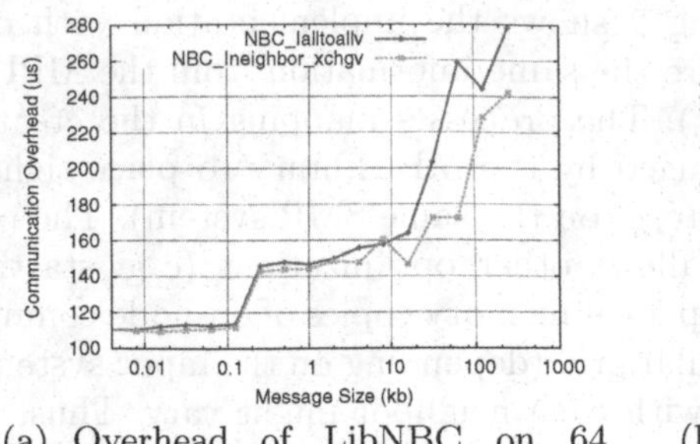

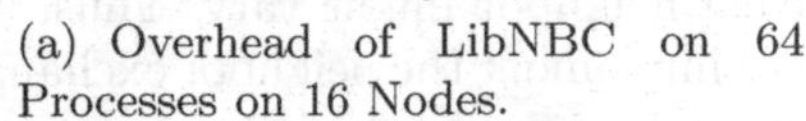
(a) Overhead of LibNBC on 64 Processes on 16 Nodes.

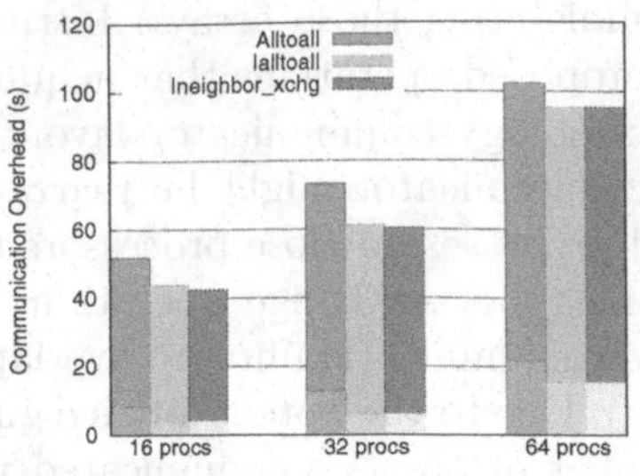

(b) Octopus Communication Overhead on 4-16 Nodes.

Fig. 4. LibNBC and octopus communication overhead

on 64 processes because the time to overlap is due to the strong scaling problem much smaller than in the 32 or 16 process case. The gain of using the nearest neighbor exchange collective is marginal at this small scale. Memory restrictions prevented bigger strong-scaling runs.

4 Conclusions and Future Work

We proposed a new class of collective operations that enable collective communication on a processor topology defined by an MPI graph communicator and thus simplify the implementation significantly. We showed the application of the new operations to the quantum mechanical simulation program `octopus`. The communication overhead of the neighbor exchange operation was efficiently halved by overlapping of communication and computation improved the application performance.

Acknowledgements. The authors want to thank Frank Mietke for support with the CHiC cluster system and Xavier Andrade for support regarding the implementation in `octopus`. This work was partially supported by a grant from the Lilly Endowment, National Science Foundation grant EIA-0202048 and a gift the Silicon Valley Community Foundation on behalf of the Cisco Collaborative Research Initiative.

References

1. Hoefler, T., et al.: Optimizing a Conjugate Gradient Solver with Non-Blocking Collective Operations. Journal of Parallel Computing 33(9), 624–633 (2007)
2. Castro, A., Marques, M.A.L., Rubio, A.: Propagators for the time-dependent kohn-sham equations. The Journal of Chemical Physics 121(8), 3425–3433 (2004)
3. Castro, A.: Octopus: a tool for the application of time-dependent density functional theory. Phys. stat. sol (b) 243(11), 2465–2488 (2006)
4. Alexandrov, A., et al.: LogGP: Incorporating Long Messages into the LogP Model. Journal of Parallel and Distributed Computing 44(1), 71–79 (1995)

5. Hoefler, T., et al.: A Case for Standard Non-Blocking Collective Operations. In: Recent Advances in Parallel Virtual Machine and Message Passing Interface 2007, vol. 4757, pp. 125–134. Springer, Heidelberg (2007)
6. Hoefler, T., et al.: Implementation and Performance Analysis of Non-Blocking Collective Operations for MPI. In: 2007 International Conference on High Performance Computing, Networking, Storage and Analysis, SC 2007, IEEE Computer Society/ACM (November 2007)
7. Hoefler, T., et al.: Optimizing non-blocking Collective Operations for InfiniBand. In: 22nd International Parallel & Distributed Processing Symposium (April 2008)
8. Hoefler, T., Lorenzen, F., Gregor, D., Lumsdaine, A.: Topological Collectives for MPI-2. Technical report, Open Systems Lab, Indiana University (February 2008)
9. Karypis, G., Kumar, V.: A fast and high quality multilevel scheme for partitioning irregular graphs. SIAM J. Sci. Comput. 20(1), 359–392 (1998)
10. Kohn, W., Sham, L.J.: Self-consistent equations including exchange and correlation effects. Phys. Rev. 140, A1133 (1965)
11. Runge, E., Gross, E.K.U.: Density-functional theory for time-dependent systems. Phys. Rev. Lett. 52(12), 997 (1984)

Dynamic Load Balancing on Dedicated Heterogeneous Systems*

Ismael Galindo[1], Francisco Almeida[1], and José Manuel Badía-Contelles[2]

[1] Department of Statistics and Computer Science
La Laguna University Spain
[2] Department of Computer Science and Engineering
Jaume I University Spain

Abstract. Parallel computing in heterogeneous environments is drawing considerable attention due to the growing number of these kind of systems. Adapting existing code and libraries to such systems is a fundamental problem. The performance of this code is affected by the large interdependence between the code and these parallel architectures. We have developed a dynamic load balancing library that allows parallel code to be adapted to heterogeneous systems for a wide variety of problems. The overhead introduced by our system is minimal and the cost to the programmer negligible. The strategy was validated on several problems to confirm the soundness of our proposal.

1 Introduction

The spread of heterogeneous architectures is likely to increase in the coming years due to the growing trend toward the institutional use of multiple computing resources (usually heterogeneous) as the sole computing resource [1]. The performance of this kind of system is very conditioned by the strong dependence that exists between parallel code and architecture [2]. Specifically, the process of allocating tasks to processors often becomes a problem requiring considerable programmer effort [3].

We have devised a library that allows dynamic task balancing within a parallel program running on a dedicated heterogeneous system, while adapting to system conditions during execution. This library facilitates the programmer the task of tailoring parallel code developed for homogeneous systems to heterogeneous ones [4]. The library has been implemented in a way that does not require changing any line of code in existing programs, thus minimizing code intrusion. All that is required is to use three new functions:

- Library start: `ULL_MPI_init_calibratelib()`
- Library end: `ULL_MPI_shutdown_calibratelib()`
- Balancing function: `ULL_MPI_calibrate(...)`

* This work has been supported by the EC (FEDER) and the Spanish MEC with the I+D+I contract number: TIN2005-09037-C02-01.

A. Lastovetsky et al. (Eds.): EuroPVM/MPI 2008, LNCS 5205, pp. 64–74, 2008.

We validated our proposal on three test problems: matrix product [5], the Jacobi method for solving linear systems [5] and resource allocation optimization via dynamic programming algorithms [6]. The computational results show that the benefits yielded by using our balancing library offer substantial time reductions in every case. The efficiency level obtained, considering the minimum code intrusion, makes this library a useful tool in the context of heterogeneous platforms.

This paper is structured as follows: in Section 2 we introduce some of the issues that motivated this research and the main goals to achieve. Section 3 shows how to use our library and the advantages our approach yields. In Section 4 we describe the balancing algorithm used by the library and Section 5 shows the validation performed on the selected problems. We close with some conclusions and future research directions.

2 Background and Objectives

Programming on heterogeneous parallel systems is obviously architecture dependent and the performance obtained is strongly conditioned by the set of machines performing the computation. This means that, in most cases, the techniques used on homogeneous parallel systems must be reworked to be applied to systems which are not necessarily homogeneous [3,7].

Specifically, we set out to solve the problem of synchronizing parallel programs in heterogeneous architectures. Given a program developed for a homogeneous system, we hope to obtain a version that makes use of the system's heterogeneous abilities by allocating tasks according to the computational ability of each processing element. The simplest way to approach the problem consists on manually adapting the code as required by the architectural characteristics[8]. This approach usually implies at least a knowledge of said characteristics, such that the parallel program's tasks can be allocated according to the computational capacity of each processor. A more general approach can be obtained in the context of self-optimization strategies based on a run time model [4,9]. In this approach, an analytical model that parametrizes the architecture and the algorithm is instantiated for each specific case so as to optimize program execution. This strategy is considerably more general than the previous one, though more difficult to apply since the modeling process is not trivial [10,11], nor is its subsequent instantiation and minimization for each case. A search of the literature yields some generic tools such as mpC [12,13] and HeteroMPI [14,15] which provide the mechanisms that allow algorithms to be adapted to heterogeneous architectures, but which also require more input from the user and are more code intrusive. Adaptive strategies have been also proposed in AMPI [16] and Dyn-MPI [17]. AMPI is built on Charm++ [18] and allows automatic load balancing based on process virtualization. Although it is an interesting generic tool, it involves a complex runtime environment.

Our objective is to develop a simple and efficient dynamic adaptation strategy of the code for heterogeneous systems that minimizes code intrusion, so that the program can be adapted without any prior knowledge of the architecture and

```
// procs = Number of processors; miid = Process ID; n = Problem size
...
despl = (int *) malloc( nprocs * sizeof(int));
count = (int *) malloc( nprocs * sizeof(int));
nrows = n/nprocs; despl[0] = 0;
for (i = 0; i < nprocs; i++) {
   count[i] = nrows;
   if (i) despl[i] = despl[i-1] + count[i-1];
}
while (it < maxit) {
  fin = despl[miid] + count[miid];
  resi_local = 0.0;
  for (i = despl[miid]; i < fin; i++) {
    sum = 0.0;
    for (j = 0; j < n; j++)
      sum += a[i][j] * x[j];
    resi_local += fabs(sum - b[i]);
    sum += -a[i][i] * x[i];
    new_x[i] = (b[i] - sum) / a[i][i];
  }
  MPI_Allgatherv (&new_x[despl[miid]], count[miid], MPI_DOUBLE,
                  x, count, despl, MPI_DOUBLE, new_com);
  it++;
}
```

Listing 1.1. Basic algorithm of an iterative scheme

without the need to develop analytical models. We intend to apply the technique to a wide variety of problems, specifically to parallel programs which can be expressed as a series of synchronous iterations. To accomplish this, we have developed a library with which to instrument specific sections in the code. The instrumentation required is minimal, as it is the resulting overhead. Using this instrumentation, the program will dynamically adapt itself to the destination architecture. This approach is particularly effective in SPMD applications with replicated data. DynMPI is perhaps a tool closer to our library in terms of the objectives but it is focussed on non dedicated clusters. DynMPI has been implemented as a MPI extension and has a wider range of applicability. However, is more code intrusive since data structures, code sections and communication calls must be instrumented. It uses daemons to monitor the system, what means extra overhead, and the standard MPI execution script must be replaced by the extended version.

Our library's design is directed at solving the time differences obtained when executing the parallel code without the necessity of extra monitoring daemons. It is based on an iterative scheme, such as that appearing in Listing 1.1, which shows a parallel version of the iterative Jacobi algorithm to solve linear systems. The code involves a main loop that executes $maxit$ iterations where a calculation operation is performed for each iteration. Each processor performs calculations in accordance with the size of the task allocated, $n/nprocs$. Following this calculation, a collective communication operation is carried out during which all the processors synchronize by gathering collecting data before proceeding to the next iteration.

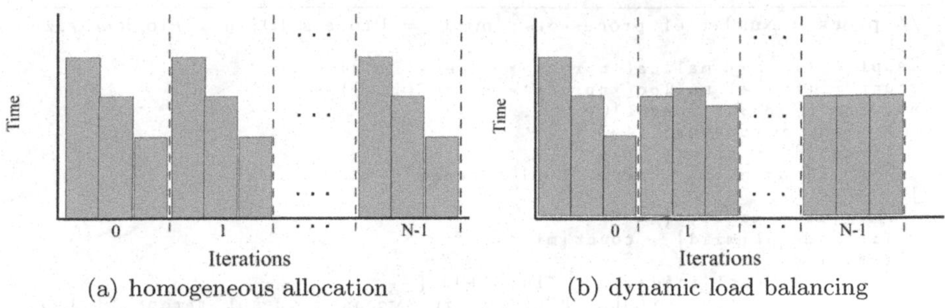

(a) homogeneous allocation (b) dynamic load balancing

Fig. 1. Time diagrams on heterogeneous systems. Each bar corresponds to the execution time of a processor on an iteration.

```
for (i = 0; i <= N; i++) {
    for (j = 0; j <= M; j++) { // irregular loop. Iteration j is O(j)
        G[i][j] = (*f)(i, 0);
        for (x = 0; x <= j; x++) {
            fij = G[i - 1][j - x] + (*f)(i, x);
            if (G[i][j] < fij)
               G[i][j] = fij;
        }
    }
}
```

Listing 1.2. Sequential algorithm for the resource allocation problem

Let's suppose that a code like that showed in Listing 1.1 is executed on a heterogeneous cluster made up of, for example, 3 processors, such that processor 2 is twice as fast as processor 1, and processor 2 is four times as fast as processor 0. Then, implementing a homogeneous task allocation, where the same problem size is assigned to each node results in an execution time which is directly dependent on that of the slowest processor. Figure 1(a) shows the results with a problem size 1500 and with subproblems of size 500. In this case the slowest processor that determines the execution time is processor 0.

A typical solution to this problem consists of allocating on each processor a static load proportional to the its computational capacity. However, several reasons brought us to consider the dynamic strategy. The allocation of tasks according to the computational power of the processors depends on the processors and also on the application. This fact involves some benchmarking to determine the computational power of the processors and usually it is highly code intrusive. On the other hand, when facing the parallelization of codes with non-regular loops (see code of the resource allocation in Listing 1.2), the static proportional allocation is not a trivial task and if performed at runtime, the overhead introduced may not be negligible.

The next section details the strategy used for balancing task allocation with a low overhead for the execution time of each processor.

```
// procs = Number of processors; miid  = Process ID; n = Problem size
...
despl = (int *) malloc( nprocs * sizeof(int));
count = (int *) malloc( nprocs * sizeof(int));
nrows = n/nprocs; despl[0] = 0;
for (i=0; i< nprocs; i++) {
   count[i] = nrows;
   if (i) despl[i] = despl[i-1] + count[i-1];
}
while (it < maxit) {
  fin = despl[miid] + count[miid];
  resi_local = 0.0;
  for (i = despl[miid]; i < fin; i++) {
    ULL_MPI_calibrate(ULL_MPI_INIT, it, &count, &despl,threshold,1,n)
    sum = 0.0;
    for (j = 0; j < n; j++)
      sum += a[i][j] * x[j];
    resi_local += fabs(sum - b[i]);
    sum += -a[i][i] * x[i];
    new_x[i] = (b[i] - sum) / a[i][i];
  }
  ULL_MPI_calibrate(ULL_MPI_END, it, &count, &despl, threshold, 1, n)
  MPI_Allgatherv (&new_x[despl[miid]], count[miid], MPI_DOUBLE,
                         x, count, despl, MPI_DOUBLE, new_com);
  it++;
}
```

Listing 1.3. Calibrated version of the basic algorithm of an iterative scheme

3 Dynamic Task Allocation

The library we developed allows for dynamic balancing with the introduction of just two calls to the `ULL_MPI_calibrate`() function in the section of code that is to be balanced, as shown by the code in Listing 1.3. A call is introduced at the beginning and end of the section to be balanced, so that each processor can know on runtime how long it will take to execute the assigned task. The balanced load results from a comparison of this execution time for each processor and the subsequent task redistribution.

Listing 1.4 shows the interface of the calibrating function. The following arguments are input to the balancing function:

- **section:** The section is used to determine the entry point where the routine is used. It can take the following two values:
  - **ULL_MPI_INIT:** Beginning of section to balance.
  - **ULL_MPI_END:** End of section to balance.

```
int ULL_MPI_calibrate (ULL_MPI_Section section, int iteration,
                       int **counts, int **displs,
                       int threshold,
                       int size_object, int size_problem);
```

Listing 1.4. Prototype of the ULL calibrating function

- **iteration:** Indicates the iteration to be balanced. A 0 value indicates whether the program is on its first or subsequent iterations. The first iteration has a particular treatment.
- **counts[], displs[]:** Indicates the task size to be computed by each processor. `counts[]` is an integer array containing the amount of work that is processed by each processor. `displs[]` specifies the distance (relative to the work data vector) at which to place the data processed by each processor.
- **threshold:** Corresponds to a number of microseconds that indicate wheather to balance or not. The behaviour per iteration is as follows:
  - Let T_i be the time processor i takes to execute the task assigned.
  - $T_{max} = Maximum(T_i)$
  - $T_{min} = Minimum(T_i)$
  - If $(T_{max} - T_{min}) > threshold$ then balance. If not, the system has already balanced the workload.
- **size_objects:** The size of the data type manipulated during computation expressed as the number of elements to be communicated in the communication routine, i.e, in the example of Listing 1.3, size objects is 1, since the elements of the matrix are double and in the communication routine they are communicated as `MPI_DOUBLE` data types.
- **size_problem:** Corresponds to the total problem size to be computed in parallel, so the calculations of the new task sizes are consistent with the tasks allocated to each processor `counts[]`, `displs[]`.

Running the synthetic code again on the previous three-processor cluster with a problem size equal to 1500 and a 100-microsecond threshold yields the following values for problem size (counts[]) and execution times (T_i):

- Iteration i = 0. The algorithm begins with a homogeneous task allocation:
  - counts[proc0] = 500, $T_0 = 400\ us$
  - counts[proc1] = 500, $T_1 = 200\ us$
  - counts[proc2] = 500, $T_2 = 100\ us$
  - if $((T_{max} = 400) - (T_{min} = 100)) > (threshold = 100)$ then balance(counts[])
- Iteration i = 1. A balancing operation is performed automatically:
  - counts[proc0] = 214, $T_0 = 171\ us$
  - counts[proc1] = 428, $T_1 = 171\ us$
  - counts[proc2] = 858, $T_2 = 171\ us$

Figure 1(b) shows a diagram of the iterations required to correct the load imbalance. For this synthetic code the load distribution is exactly proportional to the hypothetical loads, but this is not necesarily true in practice.

Note the library's ease of use and the minimum code intrusion. The only change necessary is to add calls to the functions at the beginning and end of the code to initialize and clear the memory (`ULL_MPI_init_calibratelib()`, `ULL_MPI_shutdown_calibratelib()`).

Table 1. Heterogenous platform used for the tests. All the processors are Intel (R) Xeon (TM).

Cluster 1		Cluster 2		Cluster 3	
Processor	Frequency	Processor	Frecuency	Processor	Frecuency
0	3.20 GHz	0, 1	3.20 GHz	0, 1	3.20 GHz
1	2.66 GHz	2, 3	2.66 GHz	2, 3, 4, 5	2.66 GHz
2	1.40 GHz	4, 5	1.40 GHz	6, 7	1.40 GHz
3	3.00 GHz	6, 7	3.00 GHz	8, 9, 10, 11	3.00 GHz

4 The Balancing Algorithm

The call to the `ULL_MPI_calibrate(...)` function must be made by all the processors and implements the balancing algorithm. Although a large number of balancing algorithms can be found in the literature [19], we opted for a simple and efficient strategy that yielded satisfactory results. The methodology chosen, however, allows for the implementation of balancing algorithms which may be more efficient. All processors perform the same balancing operations as follows:

- The time required by each processor to carry out the computation in each iteration has to be given to the algorithm. A collective operation is performed to share these times among all processors.
  - $T[]$ = vector where each processor gathers all the times (T_i).
  - $size\_problem$ = the size of the problem to be computed in parallel.
  - $counts[]$ = holds the sizes of the tasks to be computed on each processor.
- The first step is to verify that the threshold is not being exceeded
  if (MAX($T[]$) - MIN($T[]$)) $> THRESHOLD$ then, BALANCE
- The relative power $RP[]$ is calculated for each processor and corresponds to the relationship between the time $T[i]$ invested in performing the computation for a size $counts[i]$ versus the time taken for a computational unit as a function of the problem size, $size\_problem$:
  $RP[i] = \frac{counts[i]}{T[i]}, 0 \leq i \leq Num\_procs - 1; SRP = \sum_{i=0}^{Num\_procs-1} RP[i]$
- Finally, the sizes of the new $counts$ are calculated for each processor:

$$counts[i] = size\_problem * \frac{RP[i]}{SRT}$$

Once the $counts$ vector is computed, the $displs$ vector is also updated. Using this method, each processor fits the size of the task allocated according to its own

Table 2. Overhead of the calibration running on an homogeneous system with 8 processors Intel (R) Xeon (TM) 3.2 GHz

	Matrix Product		Jacobi		Resource Allocation Problem	
Size Problem	Parallel	Calibrated	Parallel	Calibrated	Parallel	Calibrated
1152	4.15	4.16	5.68	6.18	2	1.77
2304	96.11	93.44	19.23	19.76	14.23	9.19
4608	782.56	757.56	72.09	72.75	112.72	65.17

computational capacity. The system could be extended to run on heterogeneous non dedicated systems and on systems with dynamic load. For that purpose, the array $T[]$ must be fed not only with the execution times but with the loading factor on each processor.

To test the overhead introduced by our tool, we have executed classical parallel codes and the calibrated instrumented versions on an homogeneous system with 8 Intel 3.20 GHz processors. The parallel codes perform block assignments of the tasks with blocks of the same size for each processor. Since we are dealing with an homogeneous system, no performance improvement should be achieved and the differences in running times represent the overhead introduced. Table 2 shows the running times in each case. The overhead introduced by the tool is negligible. Note that in the resource allocation problem, the performance is improved by the code using the calibration tool. This is due to the fact that an homogeneous block data distribution is not the best choice in this case.

5 Computational Results

To check the advantages of the proposed method, we carried out a comprehensive computational experiment where the three aforementioned applications were balanced on different heterogeneous systems. The tests were run on three clusters (Table 1) to check the library's response to an increase in the number of processors with varying computational capacities. For the sake of the simplicity, the clock frecuency is the indicator of the level of heterogeneity, however it is a well known fact that better adjustments can be done by executing representative samples of the applications to determine the speeds of the processors. We will first analyze the performance of the resource allocation problem. Four algorithms were implemented: sequential and parallel homogeneous, heterogenous and calibrated. All parallel versions of the algorithm were run on the three clusters, giving the following results:

- T_{seq}: Time in seconds of the sequential version.
- T_{par}: Time in seconds of the parallel version, homogeneous data distribution.
- T_{het}: Time in seconds of the parallel version, static heterogeneous data distribution proportional to the computational load.

Table 3. Results for the Resource Allocation Problem

Cluster	$Size$	T_{sec}	T_{par}	T_{het}	T_{cal}	G_{Rpar}	G_{Rhet}
1	1152	7.59	7.04	4.41	3.38	51.98	23.31
	2304	60.44	55.36	33.35	22.62	59.14	32.18
	4608	483.72	430.37	257.98	168.22	60.91	34.79
2	1152	7.59	4.35	6.1	3.92	9.68	35.7
	2304	60.54	30.70	20.64	23.94	22.01	-15,97
	4608	483.72	237.75	139.06	121.77	48.19	12,43
3	1152	7.59	5.32	4,68	4.30	19.17	8,18
	2304	60.54	24.81	20.41	18.90	23.82	7.38
	4608	483.72	167.04	113.54	95.89	42.59	15.55

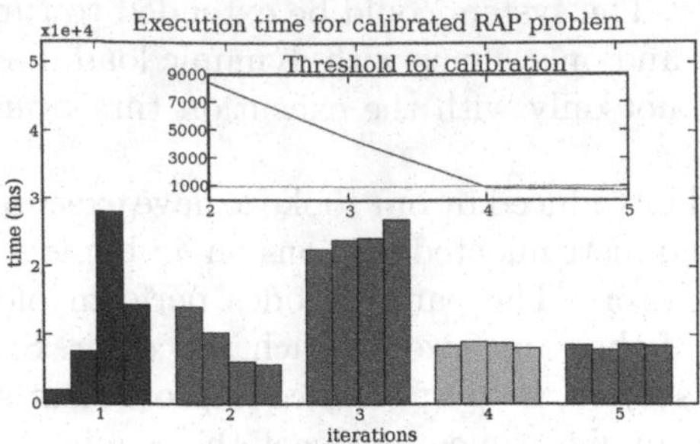

Fig. 2. Execution time per iteration for the Resource Allocation Problem on cluster 1 (four processors) with a size of 2304

- T_{cal}: Time in seconds of the balanced parallel version.
- $G_{Rpar} = \frac{T_{par}-T_{cal}}{T_{par}} * 100$: Gain relative to the homogeneous version.
- $G_{Rhet} = \frac{T_{het}-T_{cal}}{T_{het}} * 100$: Gain relative to the heterogeneous version.

In the calibrated version, the calibrating function was only added where appropriate, without altering the code from the parallel version. The sequential version was executed on the fastest processor within each cluster. The threshold is problem dependent and for testing purposes has been stated experimentally. The results are shown in Table 3 and are expressed in seconds. A 100-microsecond threshold was used for the calibrating algorithm. We observe important performance gains when using our tool. Only in one case our tool worsened the performance, and that is likely due to the threshold used in this case.

Figure 2 shows the results obtained after each iteration on cluster 1 (four processors) with a problem size of 2304. Each bar represents the execution time for each processor. Note that the times in iteration 0 of processors 2 and 3 are much higher than the rest due to the unbalanced execution. The calibration redistributes the workload, placing a higher load on processors 0 and 1 and decreasing the load on processors 2 and 3. The problem gets calibrated at iteration 4 when using a 1000–milliseconds threshold.

Table 4. Results for the matrix product and for the Jacobi method

	Matrix Multiplication					Jacobi Method				
Cluster	*Size*	T_{sec}	T_{par}	T_{cal}	G_R	*Size*	T_{sec}	T_{par}	T_{cal}	G_R
1	1152	30.98	47.21	18.94	59.8	1152	34.76	30.26	14.77	51.18
	2304	720.81	400.49	248.46	37.9	2304	138.74	116.44	50.71	56.44
	4608	5840.44	3344.19	2035.84	39.1	4608	553.46	463.89	190.74	58.88
2	1152	30.98	29.92	15.36	48.6	1152	34.76	17.54	17.05	2.79
	2304	720.81	247.98	184.62	25.5	2304	138.74	63.43	53.44	15.74
	4608	5840.44	2239.31	1639.96	26.7	4608	553.46	256.80	220.49	14.13
3	1152	30.98	20.009	15.42	22.9	1152	34.76	19.79	14.37	27.38
	2304	720.81	165.40	134.15	18.8	2304	138.74	54.02	39.38	27.10
	4608	5840.44	1487.51	1093.37	26.5	4608	553.46	178.81	162.53	9.10

For the matrix product and Jacobi cases the tests used square matrices of size $Size * Size$. A threshold of 2000 microseconds was chosen for the balancing algorithm. Problem size used was a multiple of the number of processors selected. The results are shown in Table 4. Note the significant gain resulting from the dynamic balancing, which in some cases exceeds 50%. For the Jacobi method a 100-microsecond threshold was chosen for the calibrating algorithm.

6 Conclusions and Future Research

We have developed a library to perform dynamic load balancing in heterogeneous systems. The library can be applied to a wide range of problems and the effort required by the programmer is minimal, since the approach taken involves minimum intrusion in the user's code. In future work we plan to widen the library's applicability to other types of programs and systems.

References

1. Top500 Org: Systems under development (2006), `http://www.top500.org/orsc/2006/comes.html`
2. Dongarra, J., Bosilca, G., Chen, Z., Eijkhout, V., Fagg, G.E., Fuentes, E., Langou, J., Luszczek, P., Pjesivac-Grbovic, J., Seymour, K., You, H., Vadhiyar, S.S.: Self-adapting numerical software (sans) effort. IBM Journal of Research and Development 50(2-3), 223–238 (2006)
3. Kalinov, A., Lastovetsky, A.L., Robert, Y.: Heterogeneous computing. Parallel Computing 31(7), 649–652 (2005)
4. Cuenca, J., Giménez, D., Martinez, J.P.: Heuristics for work distribution of a homogeneous parallel dynamic programming scheme on heterogeneous systems. Parallel Comput. 31(7), 711–735 (2005)
5. Wilkinson, B., Allen, M.: Parallel Programming: Techniques and Applications Using Networked Workstations and Parallel Computers. Prentice Hall, Englewood Cliffs (2004)
6. Alba, E., Almeida, F., Blesa, M.J., Cotta, C., Díaz, M., Dorta, I., Gabarró, J., León, C., Luque, G., Petit, J.: Efficient parallel lan/wan algorithms for optimization. The mallba project. Parallel Computing 32(5-6), 415–440 (2006)
7. Kalinov, A.: Scalability of heterogeneous parallel systems. Programming and Computer Software 32(1), 1–7 (2006)
8. Aliaga, J.I., Almeida, F., Badía-Contelles, J.M., Barrachina-Mir, S., Blanco, V., Castillo, M.I., Dorta, U., Mayo, R., Quintana-Ortí, E.S., Quintana-Ortí, G., Rodríguez, C., de Sande, F.: Parallelization of the gnu scientific library on heterogeneous systems. In: ISPDC/HeteroPar, pp. 338–345. IEEE Computer Society, Los Alamitos (2004)
9. Almeida, F., González, D., Moreno, L.M.: The master-slave paradigm on heterogeneous systems: A dynamic programming approach for the optimal mapping. Journal of Systems Architecture 52(2), 105–116 (2006)
10. Wu, X.: Performance Evaluation, Prediction and Visualization of Parallel Systems. Kluwer Academic Publishers, Dordrecht (1999)

11. Al-Jaroodi, J., Mohamed, N., Jiang, H., Swanson, D.R.: Modeling parallel applications performance on heterogeneous systems. In: IPDPS, p. 160. IEEE Computer Society, Los Alamitos (2003)
12. Lastovetsky, A.: Adaptive parallel computing on heterogeneous networks with mpc. Parallel computing 28, 1369–1407 (2002)
13. mpC: parallel programming language for heterogeneous networks of computers, `http://hcl.ucd.ie/Projects/mpC`
14. Lastovetsky, A., Reddy, R.: Heterompi: Towards a message-passing library for heterogeneous networks of computers. Journal of Parallel and Distributed Computing 66, 197–220 (2006)
15. HeteroMPI: Mpi extension for heterogeneous networks of computers, `http://hcl.ucd.ie/Projects/HeteroMPI`
16. Huang, C., Lawlor, O., Kale, L.: Adaptive mpi (2003)
17. Weatherly, D., Lowenthal, D., Lowenthal, F.: Dyn-mpi: Supporting mpi on non dedicated clusters (2003)
18. charm++ System, `http://charm.cs.uiuc.edu/research/charm/index.shtml#Papers`
19. Bosque, J.L., Marcos, D.G., Pastor, L.: Dynamic load balancing in heterogeneous clusters. In: Hamza, M.H. (ed.) Parallel and Distributed Computing and Networks, pp. 37–42. IASTED/ACTA Press (2004)

Communication Optimization for Medical Image Reconstruction Algorithms

Torsten Hoefler[1], Maraike Schellmann[2], Sergei Gorlatch[2], and Andrew Lumsdaine[1]

[1] Open Systems Lab, Indiana University, Bloomington IN 47405, USA
{htor,lums}@cs.indiana.edu
[2] Institute of Computer Science, University of Münster, Germany
{schellmann,gorlatch}@uni-muenster.de

Abstract. This paper presents experiences and results obtained in optimizing the parallel communication performance of a production-quality medical image reconstruction application. The fundamental communication operations in the application's principal algorithm are collective reductions. The overhead of these operations was reduced by transforming the algorithm to overlap its computation and communication. Several different approaches to communication progress were studied, both user-directed and asynchronous. Experimental results comparing the new approach to the previous implementation show overall application performance improvements of up to 8%, when run on 32 nodes.

1 Introduction

Modern medical methods for diagnosis and treatment require very accurate, high-resolution 3D images of the inside of a human body. In order to provide the required accuracy and resolution, reconstruction algorithms in medical imaging are becoming more complex and time-consuming. In this paper, we study Positron Emission Tomography (*PET*) reconstruction, where one of the most popular, but also most time-consuming algorithms—the list-mode OSEM algorithm—requires several hours on a common PC in order to compute a 3D reconstruction. With advanced algorithms that incorporate more physical aspects of the PET process, computation times are rising even further [1]. This motivates the parallelization of the algorithm on multiprocessor machines [2].

Our current parallel implementation uses Message Passing Interface (MPI) [3] collective operations and OpenMP. Collective operations allow the programmer to express high-level communication patterns in a portable way, such that implementers of communication libraries provide machine-optimized algorithms for those complex communications. Our earlier work showed that many parallel algorithms can be implemented with exclusive use of collective communications and that portability, readability, programmability, code maintenance and performance are often improved in this case [4].

In this paper, we use non-blocking collective operations to reduce the communication overhead of the parallel list-mode OSEM algorithm. Non-blocking

A. Lastovetsky et al. (Eds.): EuroPVM/MPI 2008, LNCS 5205, pp. 75–83, 2008.

collective operations are a new class of collective operations that combines all benefits of collective operations with the ability to overlap communication and computation [5]. They also relax the tight bond between computation and communication by performing communication tasks in the background. In our case study, the scalability for the fixed problem size (strong scaling) is limited by a collective data reduction operation in which the message size is independent of the number of MPI processes (in our example 48 MiB). To reduce the communication overhead, we transform our code to leverage non-blocking collective operations offered by LibNBC [6], which provide—additionally to the overlapping of communication with computation— high-level communication offload using the InfiniBand network. We analyze the code transformations and provide an analytical runtime model that identifies the overlap potential of our approach.

The rest of the paper is organized as follows: we start with an introduction to the list-mode OSEM algorithm in Section 1.1 and describe its current parallelization in Section 1.2. In Section 2, we show the necessary changes to our implementation that allow overlapping of communication and computation, and in Section 3, we discuss the optimization of LibNBC to maximize overlap. Conclusions are presented in Section 4.

1.1 List-Mode OSEM Algorithm

PET is a medical imaging technique that displays metabolic processes in a human or animal body. PET acquisition proceeds as follows: A slightly radioactive substance which emits positrons when decaying is applied to the patient who is then placed inside a *scanner*. The detectors of the scanner measure so-called events: When the emitted positrons of the radioactive substance collide with an electron residing in the surrounding tissue near the decaying spot (up to 3 mm from the emission point), they are annihilated. During annihilation two gamma rays emit from the annihilation spot in opposite directions and form a line, see Fig. 1. These gamma rays are registered by the involved detectors; one such registration is called *event*.

During one investigation, typically 10^7 to $5 \cdot 10^8$ events are registered, from which a reconstruction algorithm computes a 3D image of the substance's distribution in the body.

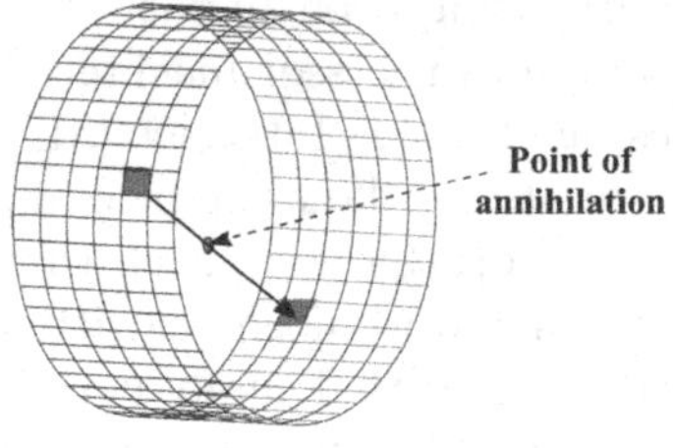

Fig. 1. Detectors register an event in a PET-scanner with 6 detector rings

```
for each(iteration k){
 for each(subiteration l){
  for (event i ∈ S_l){
    compute A_i
    compute c_l+ = (A_i)^t 1/(A_i f_l^k) }
  f_{l+1}^k = f_l^k c_l}
 f_0^{k+1} = f_{l+1}^k}
```

Listing 1. Sequential list-mode OSEM algorithm

In this work, we focus on the very accurate, but also quite time-consuming list-mode OSEM (Ordered Subset Expectation Maximization) reconstruction algorithm [7] which computes the image f from the m events saved in a list.

The algorithm works block-iteratively: in order to speed up convergence, a complete iteration over all events is divided into s subiterations (see Listing 1). Each subiteration processes one block of events, the so-called subset. The starting image vector is $f_0 = (1, ..., 1) \in \mathbb{R}^N$, where N is the number of voxels in the image being reconstructed. For each subiteration $l \in 0, ..., s-1$, the events in subset l are processed in order to compute a new, more precise reconstruction image f_{l+1}, which is used again for the next subiteration as follows:

$$f_{l+1} = \underbrace{\frac{1}{A^t_{norm}\mathbf{1}}}_{:=a} f_l c_l; \quad c_l = \sum_{i \in S_l} (A_i)^t \frac{1}{A_i f_l}, \tag{1}$$

where S_l are the indices of events in subset l, $\mathbf{1} = (1, ..., 1)$. For the i-th row A_i of the so-called system-matrix $A \in \mathbb{R}^{m \times N}$, element a_{ik} denotes the length of intersection of the line between the two detectors of event i with voxel k. The so-called normalization vector $a = \frac{1}{A^t_{norm}\mathbf{1}}$ is independent of the current subiteration and can thus be precalculated. In the computation of f_{l+1} the multiplication of $a f_l c_l$ is performed element by element.

After one iteration over all subsets, the reconstruction process can either be stopped, or the result can be improved with further iterations over all subsets (see pseudocode in Listing 1). Note that the optimal number of events per subset $m_s = m/s$ only depends on the scanner geometry and is thus fixed (for our scanner [8], it is $m_s = 10^6$).

1.2 Algorithm Parallelization Concept

Two strategies to parallelize the list-mode OSEM algorithm exist: PSD (Projection Space Decomposition) and ISD (Image Space Decomposition). In [2] we showed that PSD outperforms ISD in almost all cases and we therefore chose the PSD strategy that distributes the events among the processes for our parallelization: Since f_{l+1} depends on f_l we parallelize the computations within one subset. We decompose the input data, i.e., the events of one subset into P (=number of nodes) blocks and process each block simultaneously. The calculations for one subset includes four steps on every node j ($\forall j = 1, \ldots, P$) (cf. Fig. 2):

1. read m_s/P events
2. compute $c_{l,j} = \sum_{i \in S_{l,j}} (A_i)^t \frac{1}{A_i f_l}$. This includes the on-the-fly computation of A_i for each event in $S_{l,j}$.
3. sum up $c_{l,j} \in \mathbb{R}^N$ ($\sum_j c_{l,j} = c_l$) with MPI_Allreduce
4. compute $f_{l+1} = f_l c_l$

We implemented steps 1 and 3 (i.e., the reading of data and the actual communication of the parallel algorithm) using MPI_File_Read and blocking MPI_Allreduce. We start one process per node and use the SMP node in a cluster by additionally parallelizing steps 2 and 4 using OpenMP.

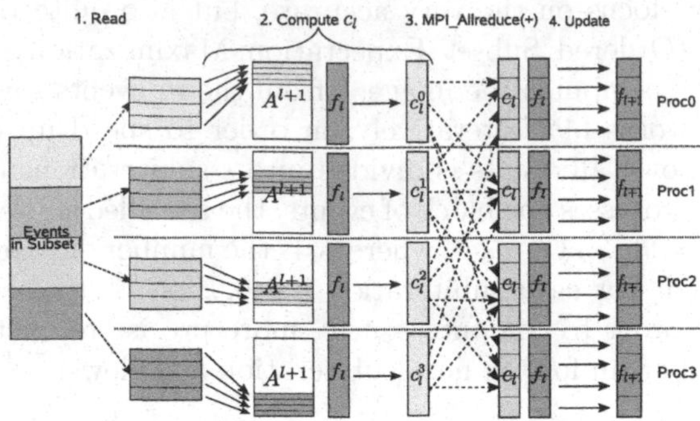

Fig. 2. Parallel list-mode OSEM algorithm on four nodes with the blocking MPI_Allreduce using four OpenMP threads per node

2 Parallel Algorithm with Non-blocking Collectives

In order to optimize the parallel algorithm, we reduce the overhead arising from the allreduce step by overlapping its communication with computations that are independent of the communicated data. We use LibNBC's [6] non-blocking version of MPI_Allreduce called NBC_Iallreduce, and the MPI_Wait counterpart NBC_Wait.

We overlap the reading of events for subset l and the computation of the corresponding sub-matrix A^l (which is composed of rows $i \in S_l$) with the communication of c_{l-1} of the preceding subset (see Fig. 3). Hence, the non-blocking parallel algorithm on nodes j ($\forall\, j = 1, \ldots, P$) reads as follows:

1. read m_s/P events in the first subset
2. compute $c_{l,j} = \sum_{i \in S_{l,j}} (A_i)^t \dfrac{1}{A_i f_l}$. This includes the on-the-fly computation of A_i for each event in $S_{l,j}$ in the first subset. Beginning from the second subset, rows A_i have already been computed in parallel with NBC_Iallreduce
3. start NBC_Iallreduce for $c_{l,j}$ ($\sum_j c_{l,j} = c_l$)
4. in every but the last subset, each node reads the m_s/P events for subset $l+1$ and computes A_i for subset $l+1$
5. perform NBC_Wait to finish NBC_Iallreduce
6. compute $f_{l+1} = f_l c_l$

Note that in this approach, A^l has to be kept in memory. If not enough memory is available, one part A^l can be computed as in the original version in step 2 and the other part in step 4. Also, since A_i is precomputed, the computation of $c_l += (A_i)^t \dfrac{1}{A_i f_l}$ could cause CPU cache misses that influence the performance.

2.1 Analyzing the Overlap Potential

In order to identify the overlap potential of our approach, we develop an analytical runtime model for the overlappable computations. We denote the sequential

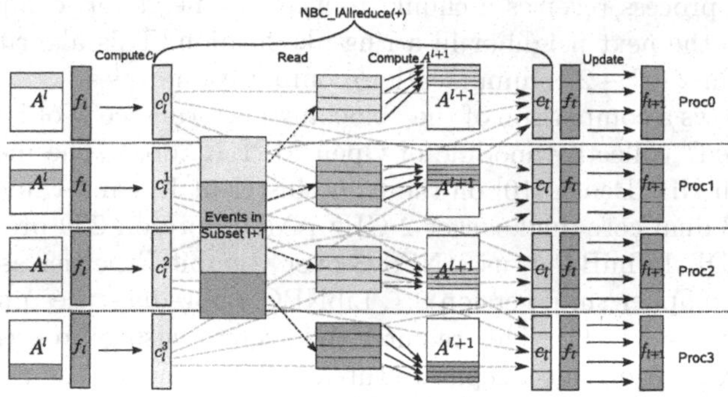

Fig. 3. Parallel list-mode OSEM algorithm on four nodes with the non-blocking NBC_Iallreduce using four OpenMP threads per node

time to compute the m_s rows of A^l by $t^1_{A_l}(m_s)$ and the time to read each node's m_s/P events by $t^P_{read}(m_s/P)$. If we assume that $t^P_{read}(m_s/P) \approx t^P_{read}(m_s)/P$, we obtain a computational overlap time per subset with one thread on each of the P nodes of

$$t^P_{CompOver} = t^P_{read}(m_s/P) + t^1_{A^l}(m_s)/P \approx (t^P_{read}(m_s) + t^1_{A^l}(m_s))/P \quad (2)$$

We will verify our model (2) with experiments in Section 3.2.

On q cores per node, the ideal parallel efficiency with our OpenMP parallelization would be $\beta(q) = t^1_{A_i}/(t^q_{A_i} \cdot q) = 1$. However, with an increasing number of threads sharing the cache, cache misses increase considerably and thus our OpenMP implementation scales worse than ideally on multi-core machines. For example, on a quad-core processor, efficiency is $\beta(4) = 0.5$.

Note that on systems where file I/O and MPI communication share the same network, the overlapping of reading of data and communication might be limited due to the network's bandwidth. Hence, in the worst case, with the network fully loaded by MPI communication, $t^P_{CompOver} = t^1_{A^l}(m_s)/P$.

3 Optimization of Non-blocking Collectives

In this section, we explain the optimized implementation of non-blocking collective operations for the needs of the parallel list-mode OSEM reconstruction algorithm.

3.1 Implementation with LibNBC

We used the InfiniBand optimized version of LibNBC for this work. This version uses an overlap-optimized InfiniBand transport layer which achieves better computation/communication overlap than open source MPI implementations [9]. The algorithm that is used to all-reduce large messages in LibNBC uses a pipelined communication scheme to maximize overlap and to use the network bandwidth as efficiently as possible. On P processes, it divides the data into P

chunks. Every process receives a chunk from its left neighbor, computes it and passes it on to the next neighbor in a ring-like fashion. This algorithm finishes the reduction in $2 \cdot P - 2$ communication/computation cycles.

Fig. 4(a) shows a comparison of the "blocking performance"[1] of LibNBC 0.9.3 with the "tuned" collective module of Open MPI 1.2.6rc2. The measurements were done with NBCBench [10] on the *odin* cluster at Indiana University. *Odin* consists of 128 dual core dual socket 2 GHz AMD Opteron 270 processors connected with SDR InfiniBand, uses NFSv3 over Gigabit Ethernet as file system and the Intel compiler suite version 9.1. LibNBC's allreduce uses multiple communication rounds (cf. [6]). This requires the user to ensure progress manually by calling NBC_Test or run a separate thread that manages the progression of LibNBC (i.e., progress thread). Fig. 4(b) shows the communication overhead with and without a progress thread under the assumption that the whole communication latency can be overlapped with computation (i.e., the overhead is a lower bound) and the progress thread runs on a spare CPU core (the overhead with a progress thread is constantly $3\mu s$, due to the fully asynchronous processing, and thus at the very bottom of Fig. 4(b)).

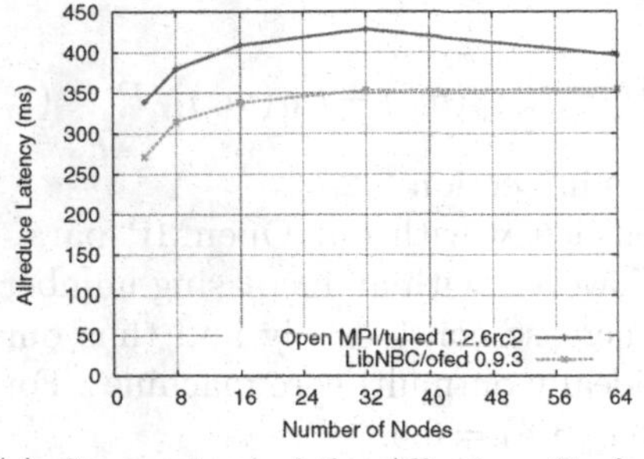

(a) Comparison of the Allreduce Performance of LibNBC and Open MPI

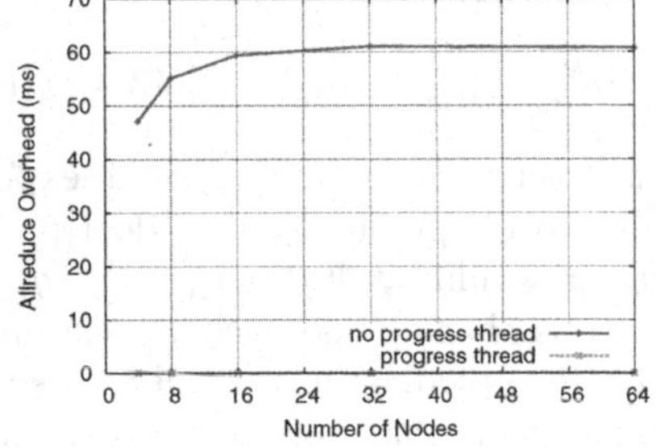

(b) Comparison of the Allreduce Overheads of different LibNBC Options

Fig. 4. Allreduce Performance Results for a 48MiB Summation of Doubles

3.2 Benchmark Results

In our benchmarks, we study the reconstruction of data collected by the *quadHIDAC* small-animal PET scanner [8]. We used 10^7 events divided into 10 subsets and performed one iteration over all events. The reconstruction image has the size $N = (150 \times 150 \times 280)$ voxels. We ran a set of different benchmarks on the *odin* system. We compared the non-threaded and threaded versions of LibNBC using the InfiniBand optimized transport. We progressed the non-threaded version with $4 \times P$ calls to NBC_Test that are equally distributed over the overlapped time. The threaded version of LibNBC is implemented by using InfiniBand's blocking semantics and the application did not call NBC_Test at all. We benchmarked all configurations of LibNBC and the original MPI implementation on

[1] NBC_Iallreduce immediately followed by NBC_Wait.

8, 16 and 32 nodes with 1, 2, 3 or 4 OpenMP threads per node three times and report the average times across all runs and processes (the variance between the runs was very low).

Computational Overlap. The computational overlap time per subset $t^P_{CompOver}$ decreases—as expected from our model—linearly with increasing number of processes P. The average time was $833.5\,ms$ on 8, $469.9\,ms$ on 16 and $241.8\,ms$ on 32 nodes. With reading time t^P_{read} ranging from $55.4\,ms$ on 8 to $11.2\,ms$ on 32 nodes and computation time $t^P_{A_l}$ ranging from $778.1\,ms$ to $230.6\,ms$ on 8 and 32 nodes, respectively, we are able to verify our model (2) with an error of about 6%.

Fig. 5(a) shows the application running time on 32 nodes with different numbers of OpenMP threads per node. We see that the non-threaded version of LibNBC is able to improve the running time in every configuration. However, the threaded version is only able to improve the performance if it has a spare core available because of scheduler congestion on the fully loaded system. The performance gain also decreases with the number of OpenMP threads. This is because we studied a strong scaling problem and the overlappable computation time gets shorter with more threads computing the static workload and is eventually not enough to overlap the full communication. Another issue for smaller node-counts is that our transformed implementation is, as described in Section 2.1, slightly less cache-friendly which limits the application speedup at smaller scale.

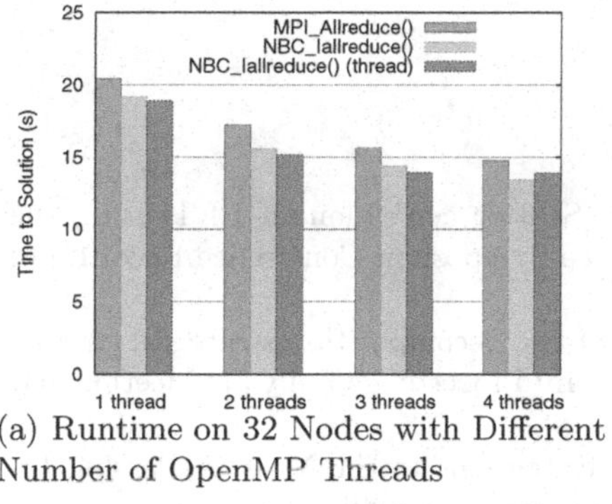

(a) Runtime on 32 Nodes with Different Number of OpenMP Threads

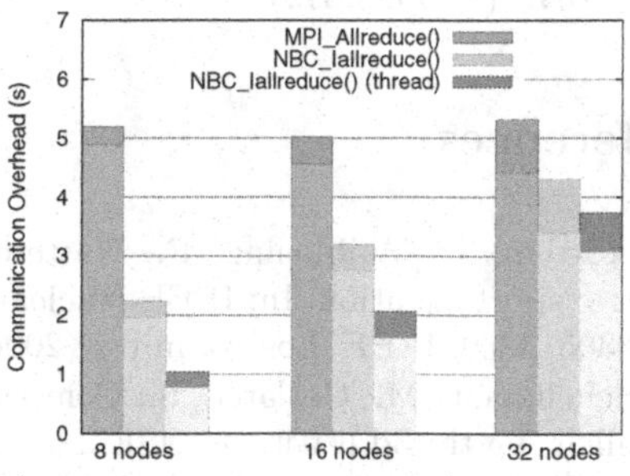

(b) Communication Overhead for Different Node Counts and a single Thread

Fig. 5. Application Benchmark for different number of OpenMP threads and nodes

Fig. 5(b) shows the communication overhead for different node counts with one thread per node[2]. Our implementation achieves significantly smaller communication overhead for all configurations. However, the workload per node that can be overlapped decreases, as described above, with the number of nodes, while the communication time of the 48 MiB Allreduce remains nearly constant. Thus, the time to overlap shrinks with the number of nodes and limits the performance gain of the non-blocking collectives.

[2] The lower part of the bars denotes the Allreduce overhead.

4 Conclusions

We applied non-blocking collective operations to the mixed MPI/OpenMP parallel implementation of the list-mode OSEM algorithm and analyzed the performance gain for a fixed problem size (strong scaling) on different setups of MPI processes and OpenMP threads.

The conducted study demonstrates that the overlap optimization potential of non-blocking collectives depends heavily on the time to overlap (amount of work to do while communicating) which usually decreases while scaling to larger process counts. However, even in the worst case (smallest workload) of our example, running 128 threads on 32 nodes, LibNBC was able to reduce the communication overhead from 40.31% to 37.3%. In the best case, with one thread on 8 nodes (highest workload per process), the communication overhead could be efficiently halved from 12.0% to 5.6%.

Acknowledgments

This work was partially supported by a grant from the Lilly Endowment, National Science Foundation grant EIA-0202048 and a gift from the Silicon Valley Community Foundation on behalf of the Cisco Collaborative Research Initiative. This work was also partly funded by the Deutsche Forschungsgemeinschaft, SFB 656 MoBil (Project B2).

References

1. Kösters, T., Wübbeling, F., Natterer, F.: Scatter correction in PET using the transport equation. In: IEEE Nuclear Science Symposium Conference Record, pp. 3305–3309. IEEE, Los Alamitos (2006)
2. Schellmann, M., Gorlatch, S.: Comparison of two decomposition strategies for parallelizing the 3d list-mode OSEM algorithm. In: Proceedings Fully 3D Meeting and HPIR Workshop, pp. 37–40 (2007)
3. Message Passing Interface Forum: MPI-2: Extensions to the Message-Passing Interface. Technical Report, University of Tennessee, Knoxville (1997)
4. Gorlatch, S.: Send-receive considered harmful: Myths and realities of message passing. ACM Trans. Program. Lang. Syst. 26(1), 47–56 (2004)
5. Brightwell, R., Riesen, R., Underwood, K.D.: Analyzing the impact of overlap, offload, and independent progress for message passing interface applications. Int. J. High Perform. Comput. Appl. 19(2), 103–117 (2005)
6. Hoefler, T., Lumsdaine, A., Rehm, W.: Implementation and Performance Analysis of Non-Blocking Collective Operations for MPI. In: Proceedings of the 2007 International Conference on High Performance Computing, Networking, Storage and Analysis, SC 2007. IEEE Computer Society/ACM (November 2007)
7. Reader, A.J., Erlandsson, K., Flower, M.A., Ott, R.J.: Fast accurate iterative reconstruction for low-statistics positron volume imaging. Phys. Med. Biol. 43(4), 823–834 (1998)

8. Schäfers, K.P., Reader, A.J., Kriens, M., Knoess, C., Schober, O., Schäfers, M.: Performance evaluation of the 32-module quadHIDAC small-animal PET scanner. Journal Nucl. Med. 46(6), 996–1004 (2005)
9. Hoefler, T., Lumsdaine, A.: Optimizing non-blocking Collective Operations for InfiniBand. In: Proceedings of the 22nd IEEE International Parallel & Distributed Processing Symposium (IPDPS) (April 2008)
10. Hoefler, T., Schneider, T., Lumsdaine, A.: Accurately Measuring Collective Operations at Massive Scale. In: Proceedings of the 22nd IEEE International Parallel & Distributed Processing Symposium (IPDPS) (April 2008)

A Simple, Pipelined Algorithm for Large, Irregular All-gather Problems*

Jesper Larsson Träff[1], Andreas Ripke[1], Christian Siebert[1], Pavan Balaji[2], Rajeev Thakur[2], and William Gropp[3]

[1] NEC Laboratories Europe, NEC Europe Ltd.
Rathausallee 10, D-53757 Sankt Augustin, Germany
[2] Mathematics and Computer Science Division
Argonne National Laboratory, Argonne, IL 60439, USA
[3] Department of Computer Science
University of Illinois, Urbana, IL 61801, USA

Abstract. We present and evaluate a new, simple, pipelined algorithm for *large, irregular* all-gather problems, useful for the implementation of the `MPI_Allgatherv` collective operation of MPI. The algorithm can be viewed as an adaptation of a linear ring algorithm for regular all-gather problems for single-ported, clustered multiprocessors to the irregular problem. Compared to the standard ring algorithm, whose performance is dominated by the largest data size broadcast by a process (times the number of processes), the performance of the new algorithm depends only on the total amount of data over all processes. The new algorithm has been implemented within different MPI libraries. Benchmark results on NEC SX-8, Linux clusters with InfiniBand and Gigabit Ethernet, Blue Gene/P, and SiCortex systems show huge performance gains in accordance with the expected behavior.

1 Introduction

The *all-gather* problem is a basic collective communication operation, in which *each* participant of a predefined group wants to broadcast personal data to all other group members. In the MPI standard, this functionality is embodied in the *regular* `MPI_Allgather` collective, in which each process contributes the same amount of data, and in the *irregular* `MPI_Allgatherv` collective, where the amount of data can be freely chosen for the different processes [8]. For both MPI collectives, all participating processes know the sizes of the data to be broadcast by all other processes. The irregular all-gather operation is used for instance in linear algebra kernels for matrix multiplication and LU factorization [1].

The regular all-gather problem has been intensively studied (theoretically under the term *gossiping*, but is also known as *broadcast-to-all*, *all-to-all-broadcast*,

* This work was supported in part by the Mathematical, Information, and Computational Sciences Division subprogram of the Office of Advanced Scientific Computing Research, Office of Science, U.S. Department of Energy, under Contract DE-AC02-06CH11357.

A. Lastovetsky et al. (Eds.): EuroPVM/MPI 2008, LNCS 5205, pp. 84–93, 2008.

as well as many other names) [5,6], and many algorithms have been proposed and/or implemented as part of MPI libraries for various systems and communication models [1,2,3,7,9,10]. The more challenging, irregular all-gather problem has received much less attention, and MPI libraries typically use the same algorithm for both `MPI_Allgather` and `MPI_Allgatherv`. For irregular problems with considerable differences between the amount of data contributed by the processes, this can have huge performance drawbacks. For extreme cases, the resulting performance loss can amount to orders of magnitude (cf. Section 3).

In this paper, we present an algorithm for large, irregular all-gather problems. The underlying idea is quite simple and can be viewed as an adaptation to the irregular problem of a ring-based algorithm for regular all-gather problems for single-ported, clustered multiprocessors. The algorithm has been implemented for several MPI libraries, and evaluated on diverse systems, namely NEC SX-8, two Linux clusters, IBM Blue Gene/P, and SiCortex 5832. We demonstrate significant performance improvements over a standard `MPI_Allgatherv` algorithm, depending on the amount of irregularity in the benchmark scenarios.

2 Algorithm and Implementation(s)

In the following, p is the number of participating (MPI) processes, numbered consecutively from 0 to $p-1$. We let m_i denote the size of the data contributed by process i, and $m = \sum_{i=0}^{p-1} m_i$ the total amount of data that eventually has to be gathered by all processes. For large data, we assume that the time for transmitting a message of size m' is simply $O(m')$. For most of the following discussion, a detailed communication cost model is unnecessary.

2.1 Standard, Linear Ring Algorithm

A basic (folklore) algorithm for large, regular all-gather problems is the *linear ring*. The algorithm performs $p-1$ communication rounds. In each round process i sends (starting with its own data) an already known block of data of size m' to process $(i+1) \bmod p$ and receives an unknown block of data from process $(i-1) \bmod p$. For regular problems where all blocks are of the same size $m_i = m'$, the completion time of the ring algorithm is $O((p-1)m') = O(m-m')$. The number of communication start-ups (latency) scales linearly with p. This is unproblematic for large m', but for small problems, an algorithm with a logarithmic number of start-ups is clearly preferable [1,3,10]. The linear ring algorithm is straightforward to implement. For systems with single-ported, bidirectional communication capabilities (where each process can at the same time send data to another process and receive data from a possibly different process) it can use the system communication bandwidth to full capacity. For irregular all-gather problems, where the data sizes m_i can vary arbitrarily over the processes, the algorithm can however perform poorly. The running time is determined by the largest amount of data $m' = \max_{i=0}^{p-1} m_i$, which has to be sent along the ring in each round, and is therefore $O((p-1)m')$. In particular, $(p-1)m'$ can be much larger than the total amount of data m.

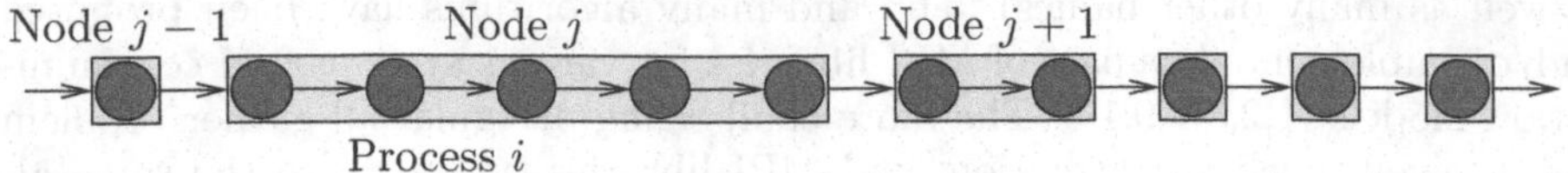

Fig. 1. The linear ring algorithm on a cluster of SMP nodes with different number of MPI processes per node. The processes are (virtually) ranked such that one process at each node receives data from another node, and one process sends data to another node in each round.

2.2 Pipelined (Blocked) Ring Algorithm

We first observe that the linear ring algorithm can also be used for the regular all-gather problems on clustered multiprocessors (like clusters of SMP nodes) with a single-ported communication network. In that case the ring is organized such that exactly one process i per SMP node has its predecessor $(i-1) \bmod p$ on another SMP node, and exactly one process j per SMP node has its successor $(j+1) \bmod p$ on another node. To accomplish this, a (virtual) reranking of the MPI processes might be necessary. The clustered, linear ring algorithm is now communication-bandwidth optimal, because in each round one process on each node receives a block of data and one process sends a block of data. This holds also for the case where the number of MPI processes per cluster node is not identical, and is illustrated in Figure 1.

In [11] it is observed that regular collective communication problems like the all-gather problem induce corresponding irregular problems over a set of nodes in a clustered system. Therefore, if the communication capabilities of processors and nodes in a cluster are similar (for instance, single ported), an algorithm for solving a regular problem on a clustered system (with possibly different number of processes per cluster node) can be used to solve its irregular counterpart over a set of processors. This observation can be exploited to convert the clustered linear ring algorithm into an algorithm for the irregular all-gather problem.

To accomplish this the data of process i of size m_i is associated with a virtual cluster node, and divided into $b_i = \max(1, \lceil m_i/B \rceil)$ blocks of size at most B. Each block is associated with a virtual processor in the node. The total number of blocks is $b = \sum_{i=0}^{p-1} b_i$ (note that $b \geq p$). Every actual process with data size m_i will play the role of a cluster node with b_i virtual processors. The linear ring algorithm with regular blocks of size (at most) B now solves the problem in $b-1$ instead of $p-1$ communication rounds. The resulting, *pipelined* (or *blocked*) *ring* algorithm is illustrated in Figure 2. Compared to the linear ring, the advantage of the pipelined ring algorithm is that (more) regular blocks are sent and received in each round, for a total time of $O((b-1)B)$. A small value for B increases the number of start-ups, and a large value increases the possible round up error. Therefore a proper balancing needs to be applied to find an optimal value for the block size parameter. We note that for extremely irregular all-gather problems where only one process has all the data, the pipelined ring algorithm is equivalent to a linear broadcast pipeline. For regular problems where $m_i = m'$ for all i, the

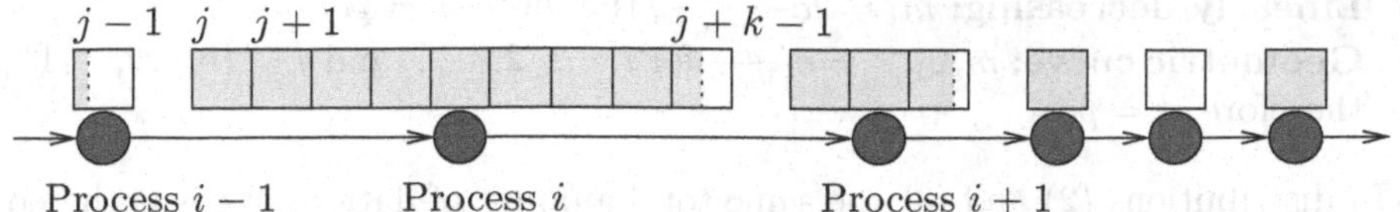

Fig. 2. The clustered, linear ring algorithm viewed as a pipelined (blocked) algorithm for solving the irregular all-gather problem. For each process, the data m_i is divided into blocks of some maximum block size B (partially full blocks are partially colored). Process i starts sending block $j+k-1$ and receiving block $j-1$. After $b-1$ rounds, where b represents the total number of blocks, all processes have gathered all the data.

block size B can be set to m', in which case the algorithm is identical to the standard, linear ring. Thus, by choosing B properly, the pipelined ring algorithm should never perform worse than the linear ring algorithm.

2.3 Determining an Optimal Block Size

We note that for partially full blocks, only the actual data are sent and received (see again Figure 2). In particular, the empty blocks which arise for processes with $m_i = 0$ are neither sent nor received. Nevertheless, they contribute to the total number of communication rounds. We estimate the optimal block size B as follows, assuming that z denotes the number of processes with $m_i = 0$:

- If $z = 0$ we take $B = \min_{i=0}^{p-1} m_i$ (as long as this is not too small). This ensures that all processes are both sending and receiving blocks in (almost) all rounds.
- If $z \neq 0$ we try to minimize the time needed for $b-1$ communication rounds. Assuming that the remainders in the m_i/B terms are equally distributed, we get an average padding of $B/2$ for all partially full blocks. We can therefore simplify $b = \sum_{i=0}^{p-1} \max(1, \lceil m_i/B \rceil)$ to $b = \frac{m}{B} + \frac{p+z}{2}$. Assuming linear communication costs, where sending and receiving messages of size m' takes time $\alpha + \beta m'$, the estimated total running time is $(b-1)(\alpha + \beta B)$. Minimizing this term gives an (approximated) optimal block size of $B = \sqrt{\frac{2\alpha m}{\beta(p+z-2)}}$.

3 Experimental Evaluation

We have benchmarked the new `MPI_Allgatherv` implementations with the following distributions of *contiguous data* over the p MPI processes. A base count c (which is varied over some interval) is used as seed for the following distributions:

1. **Regular:** all $m_i = c$ are identical, therefore $m = pc$.
2. **Broadcast:** $m_0 = c$, all other $m_i = 0$, therefore $m = c$.
3. **Spike:** similar to broadcast but all processes contribute some data, $m_0 = c/2$ and $m_i = c\frac{1}{2(p-1)}$, therefore $m = c$.
4. **Half full:** $m_{2\lfloor i/2 \rfloor} = 2c$, and $m_{2\lfloor i/2 \rfloor + 1} = 0$, therefore $m = pc$.

5. **Linearly decreasing:** $m_i = 2c\frac{(p-1-i)}{p-1}$, therefore $m = pc$.
6. **Geometric curve:** $m_{i-1+j} = c\frac{p}{i \log p}$ for $i = 1, 2, 4, \ldots$ and $j = \{0, \ldots, i-1\}$, therefore $m = pc$.

In distributions (2) and (3) the same total amount of data $m = c$ is gathered by all processes, so similar running times can be expected (comparable to the regular distribution with p times smaller data size). The case for distributions (1), (4), (5) and (6) is analogous, where the total amount of data is $m = pc$.

We compare our implementations of the new `MPI_Allgatherv` algorithm with implementations of the standard linear ring algorithm that is still used in many MPI libraries [9]. The reported running times are minimum times for the last process to finish over a (small) number of iterations [4].

3.1 Results on an NEC SX-8 Vector System

The pipelined ring has been implemented for MPI/SX for the NEC SX-series of parallel vector computers. It has been benchmarked with the distributions described above on 30 SX-8 nodes at HLRS in Stuttgart, with 1 and 8 MPI processes per node, respectively. Selected results are shown in Figure 3.

For the extreme broadcast distribution (2) the pipelined ring outperforms the standard linear ring by more than a factor of 10 on 30 SX-8 nodes. For 32 MBytes with a fixed block size B of 1 MByte an improvement of a factor $\frac{32\times 29}{29+31} \approx 15$ would have been best possible. Significant improvements can also be observed for the other distributions. The performance of the standard ring and the pipelined ring are similar for the regular (1) and the half full (4) distributions. Running on a randomly permuted communicator instead of `MPI_COMM_WORLD` gives almost identical results. This is a desirable property of an algorithm for a symmetric (i.e. non-rooted) collective operation like `MPI_Allgatherv` [12].

3.2 Results on a Linux Cluster with InfiniBand

To show the effect of the block size B, the algorithm has also been integrated into NEC's MPI/PC version and evaluated on an Intel Xeon based SMP cluster with InfiniBand interconnect. The running time is compared to the standard, non-pipelined algorithm for $B = 32K, 64K, 128K, 512K, 1024K$. Results are shown in Figure 4. For the spike distribution (3) the pipelined algorithm is faster for all block sizes. However, the best block size depends not only on the size of the problem but also on the distribution of data over the processes. This can be seen in the case of the decreasing distribution (5) where a too small block size makes the pipelined algorithm perform worse than the standard ring.

3.3 Results on a Linux Cluster with Gigabit Ethernet

We ran the benchmarks on a Linux cluster at Argonne National Laboratory with 24 nodes, each with two dual-core 2.8 GHz AMD Opteron CPUs (total of 4 cores per node or 96 cores in the system), and Gigabit Ethernet. We used

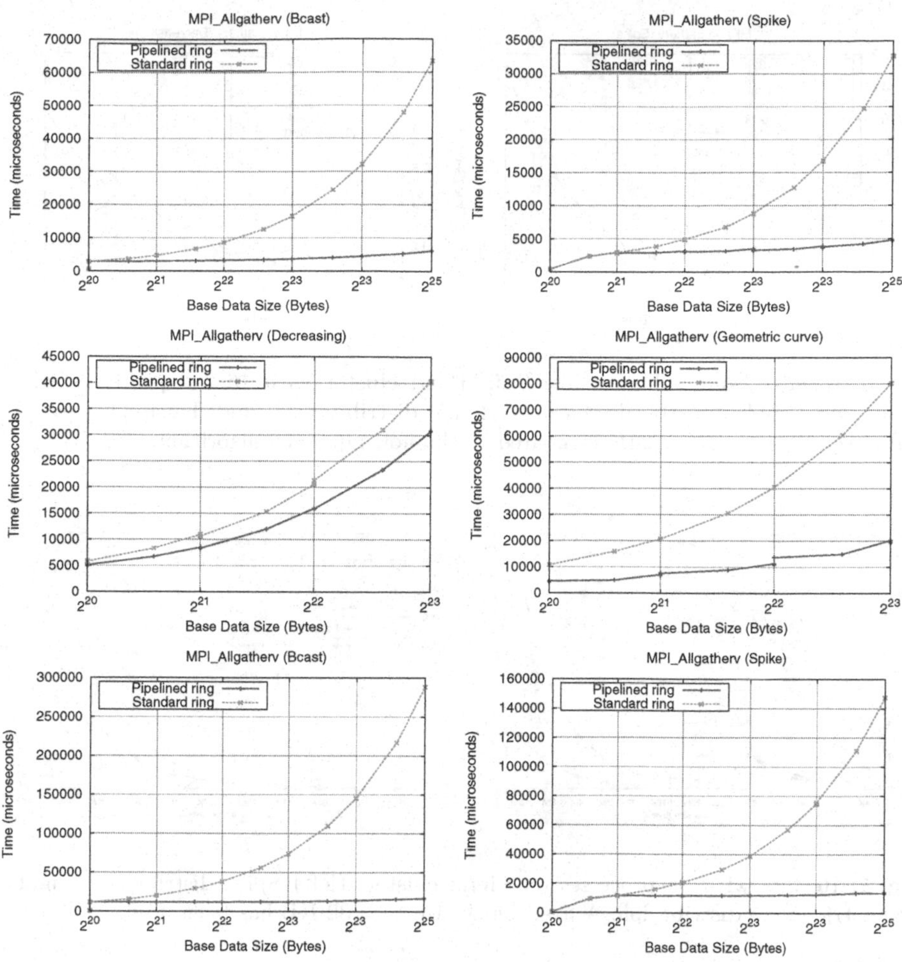

Fig. 3. Results (left to right, top to bottom) for distributions (2), (3), (5) and (6) on an NEC SX-8 with 30 nodes and 1 MPI process per node, and distributions (2) and (3) with 8 MPI processes per node. A fixed block size $B = 1$ MByte has been used. The base data size is the base count c multiplied by the size of an `MPI_INT`.

MPICH2 1.0.7 as the MPI implementation. Selected results are shown in Figures 5 and 6. For small problem sizes, the pipelined algorithm performs only slightly better than the standard algorithm, but as problem size increases, the difference in performance becomes considerable. Figure 6(right) shows the distribution of communication and idle times for the two algorithms. As expected, the standard algorithm suffers because many processes remain idle for a long time, whereas in the pipelined algorithm, communication is more balanced. We also collected traces of the program execution and plotted them using the Jumpshot tool, as shown in Figure 7. The penalty due to idle time incurred by the standard algorithm is clearly visible as the yellow bars.

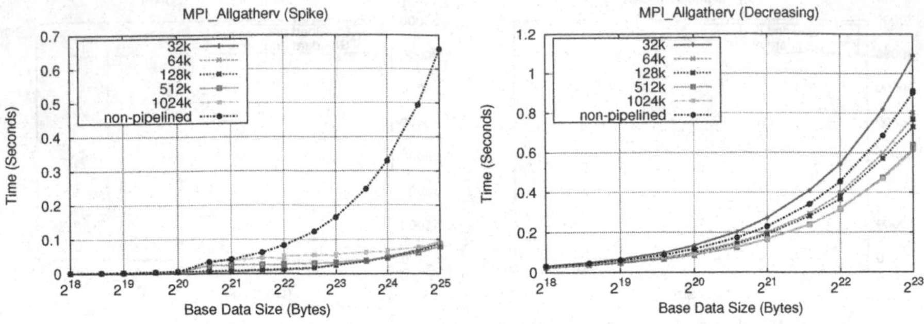

Fig. 4. Results from a Linux Xeon/InfiniBand cluster with 16×2 processes with spike (left) and linearly decreasing (right) distributions, and block size $B = 32K, 64K, 128K, 512K, 1024K$ compared to the non-pipelined algorithm

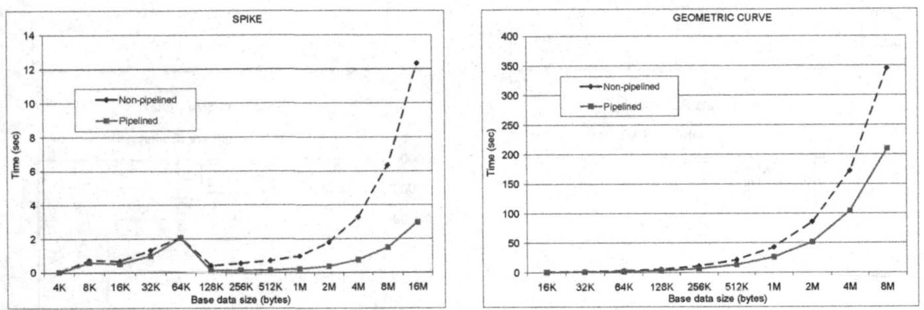

Fig. 5. Results with 96 processes on Linux cluster: (left) Spike distribution (right) Geometric curve distribution. A fixed block size B = 32 KB has been used.

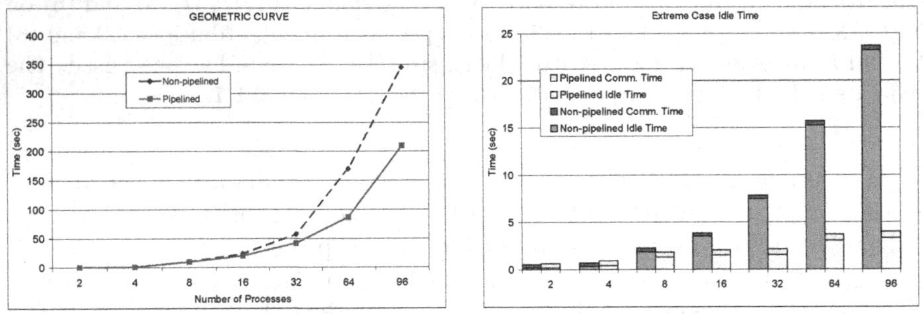

Fig. 6. Linux cluster: (left) Geometric curve distribution with varying number of processes, (right) Communication versus idle time in the extreme case of broadcast distribution

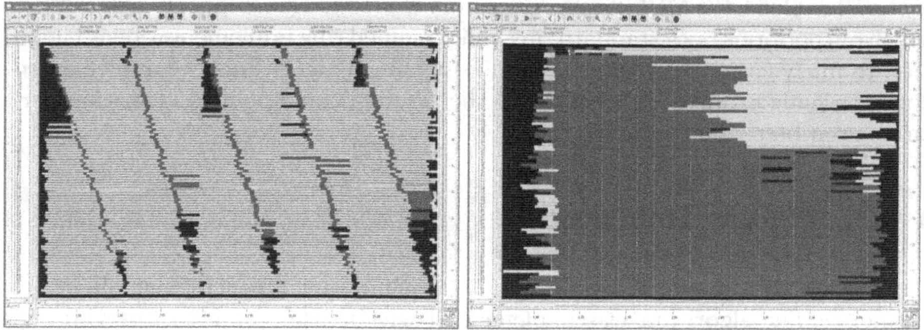

Fig. 7. Jumpshot plot of program trace on Linux cluster for several iterations of allgatherv with broadcast distribution: (left) Non-pipelined algorithm, (right) Pipelined algorithm. Yellow (light) is idle time, purple (dark) is communication time.

3.4 Results on SiCortex

Benchmarks were also performed on the SiCortex 5832 system at Argonne. This machine has 972 nodes, each with 6 cores, for a total of 5832 processors. The nodes are connected by a Kautz graph network. In some of our experiments the native SiCortex MPI implementation failed. We therefore implemented the standard linear ring algorithm ourselves and compared it with the pipelined algorithm. Figure 8 shows the results for a test run with a geometric curve distribution on 5784 processors. The pipelined algorithm significantly outperforms the standard algorithm as the message size increases.

3.5 Results on IBM Blue Gene/P

Finally, we performed the tests on one rack of the IBM Blue Gene/P at Argonne National Laboratory (4096 cores). The native implementation of `MPI_Allgatherv`

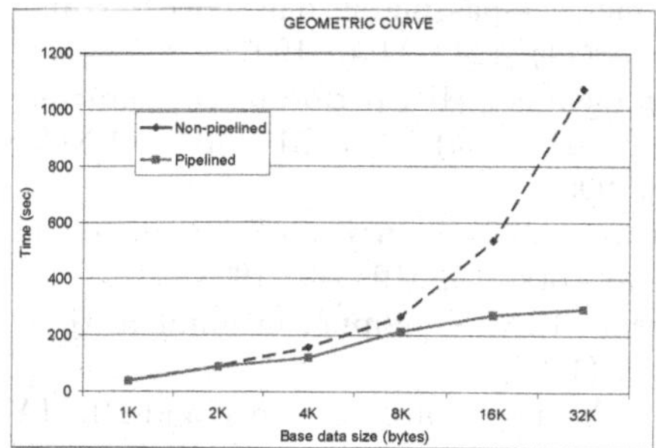

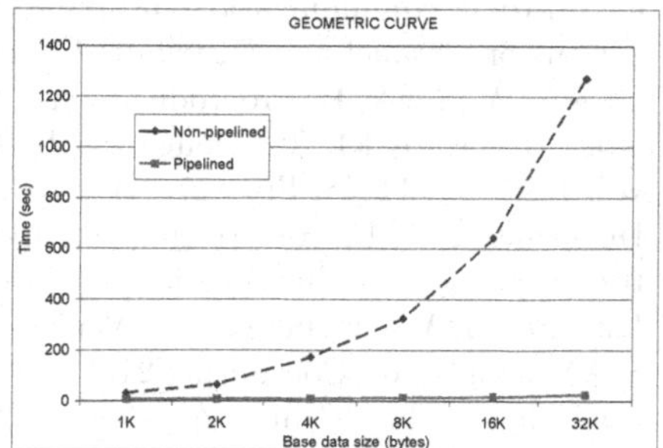

Fig. 8. Results for the geometric curve distribution: (left) with 5784 processes on the SiCortex machine and a fixed block size of $B = 1$ MB, (right) with 4096 processes on 1 rack of the Blue Gene/P and a fixed block size of $B = 64$ KB

in the Blue Gene/P's MPI library uses a very fast hardware-supported algorithm, which outperforms both standard ring and pipelined ring implementations. Therefore, to fairly compare pipelined and non-pipelined algorithms, we implemented both these algorithms. Figure 8 shows the results. The pipelined algorithm performs even better on this machine.

4 Concluding Remarks

We described a simple, pipelined ring algorithm for large, irregular all-gather problems. The algorithm was implemented within different MPI libraries and benchmarked on various systems, and in all cases showed considerable improvements over a commonly used linear ring algorithm for problems with significant irregularity in the individual message sizes. Determining the best possible pipeline block size for all distributions of input data still requires more (experimental) work. On regular problem instances the pipelined algorithm performs similarly to the linear ring, which is bandwidth optimal for that case. Ring algorithms can likewise be implemented to be largely independent on process placement in an SMP system. This is an important property for users expecting (self-) consistent performance of their MPI library [12].

References

1. Balaji, P., Buntinas, D., Balay, S., Smith, B.F., Thakur, R., Gropp, W.: Nonuniformly communicating noncontiguous data: A case study with PETSc and MPI. In: 21st International Parallel and Distributed Processing Symposium (IPDPS 2007), pp. 1–10 (2007)
2. Benson, G.D., Chu, C.-W., Huang, Q., Caglar, S.G.: A comparison of MPICH allgather algorithms on switched networks. In: Dongarra, J., Laforenza, D., Orlando, S. (eds.) EuroPVM/MPI 2003. LNCS, vol. 2840, pp. 335–343. Springer, Heidelberg (2003)
3. Bruck, J., Ho, C.-T., Kipnis, S., Upfal, E., Weathersby, D.: Efficient algorithms for all-to-all communications in multiport message-passing systems. IEEE Transactions on Parallel and Distributed Systems 8(11), 1143–1156 (1997)
4. Gropp, W., Lusk, E.: Reproducible measurements of MPI performance characteristics. In: Margalef, T., Dongarra, J., Luque, E. (eds.) PVM/MPI 1999. LNCS, vol. 1697, pp. 11–18. Springer, Heidelberg (1999)
5. Hedetniemi, S.M., Hedetniemi, T., Liestman, A.L.: A survey of gossiping and broadcasting in communication networks. Networks 18, 319–349 (1988)
6. Krumme, D.W., Cybenko, G., Venkataraman, K.N.: Gossiping in minimal time. SIAM Journal on Computing 21(1), 111–139 (1992)
7. Mamidala, A.R., Vishnu, A., Panda, D.K.: Efficient shared memory and RDMA based design for mpi_allgather over InfiniBand. In: Mohr, B., Träff, J.L., Worringen, J., Dongarra, J. (eds.) PVM/MPI 2006. LNCS, vol. 4192, pp. 66–75. Springer, Heidelberg (2006)
8. Snir, M., Otto, S., Huss-Lederman, S., Walker, D., Dongarra, J.: MPI – The Complete Reference, 2nd edn., vol. 1, The MPI Core. MIT Press, Cambridge (1998)

9. Thakur, R., Gropp, W.D., Rabenseifner, R.: Improving the performance of collective operations in MPICH. International Journal on High Performance Computing Applications 19, 49–66 (2004)
10. Träff, J.L.: Efficient allgather for regular SMP-clusters. In: Mohr, B., Träff, J.L., Worringen, J., Dongarra, J. (eds.) PVM/MPI 2006. LNCS, vol. 4192, pp. 58–65. Springer, Heidelberg (2006)
11. Träff, J.L.: Relationships between regular and irregular collective communication operations on clustered multiprocessors. Parallel Processing Letters (2008) (Forthcoming)
12. Träff, J.L., Gropp, W., Thakur, R.: Self-consistent MPI performance requirements. In: Cappello, F., Herault, T., Dongarra, J. (eds.) PVM/MPI 2007. LNCS, vol. 4757, pp. 36–45. Springer, Heidelberg (2007)

MPI Reduction Operations for Sparse Floating-point Data

Michael Hofmann* and Gudula Rünger

Department of Computer Science
Chemnitz University of Technology, Germany
{mhofma,ruenger}@cs.tu-chemnitz.de

Abstract. This paper presents a pipeline algorithm for MPI_Reduce that uses a *Run Length Encoding* (RLE) scheme to improve the global reduction of sparse floating-point data. The RLE scheme is directly incorporated into the reduction process and causes only low overheads in the worst case. The high throughput of the RLE scheme allows performance improvements when using high performance interconnects, too. Random sample data and sparse vector data from a parallel FEM application is used to demonstrate the performance of the new reduction algorithm for an HPC Cluster with InfiniBand interconnects.

Keywords: MPI, performance optimization, pipelining, reduction operation, run length encoding.

1 Introduction

The *Message Passing Interface* (MPI) is the de facto standard for distributed memory parallel programming in the area of scientific high performance computing and the optimization of MPI libraries and communication operations is still an active field of research. Emerging high performance interconnects such as Quadrics, Myrinet, SCI, or InfiniBand have led to continuing efforts for improving the performance of MPI implementations, too. Especially for collective MPI operations, there exists a variety of different algorithms. Automatic tuning as well as static and dynamic optimizations are used to adapt to specific system architectures and applications by selecting appropriate algorithms or algorithmic parameters [1,2]. Good overall performance of communication operations requires a transition from latency-optimal algorithms for small messages to bandwidth-optimal algorithms for large messages. Pipelining techniques are used for achieving high bandwidth, especially with high performance interconnects [3,4]. Improved algorithms for global reduction operations (e.g., MPI_Allreduce) are presented in [5].

The contribution of this paper is to apply the well known *Run Length Encoding* (RLE) to floating-point data and to incorporate it into a pipeline algorithm for MPI_Reduce. The RLE scheme reduces the amount of communication and increases the performance, especially for sparse vector data. General purpose compression algorithms and specific algorithms for floating-point data are used

* Supported by Deutsche Forschungsgemeinschaft (DFG).

A. Lastovetsky et al. (Eds.): EuroPVM/MPI 2008, LNCS 5205, pp. 94–101, 2008.

to improve the performance of MPI communication operations on clusters with Fast Ethernet interconnects [6,7]. However, the throughputs of these algorithms are unstable and too small for high performance interconnects. Our target platform is the HPC Cluster CHiC [8] consisting of 530 compute nodes with InfiniBand interconnects. MPI reduction operations are used, for instance in parallel numerical methods for implementing global error control. We use random sample data as well as sparse vector data from a parallel FEM application to investigate the performance of the new reduction algorithm with RLE.

The rest of this paper is organized as follows. Section 2 investigates the feasibility of applying data compression for optimizing communication operations and introduces the RLE scheme for floating-point data. Section 3 describes a pipeline algorithm for `MPI_Reduce` and the application of the RLE scheme. Section 4 presents performance results and Section 5 concludes the paper.

2 Compression of Floating-Point Data

The communication time for large messages is mainly determined by the bandwidth of the communication network. To achieve a benefit from transferring compressed data instead of uncompressed data, the additional computational time of the compression algorithm has to be lower than the time saved during the communication. Under optimal conditions, the compression and decompression operations perfectly overlap and the message size is reduced so that the communication time can be neglected. To benefit from the compression in that case, the throughput of the compression/decompression operation has to be at least as high as the bandwidth of the communication network.

A general purpose data compression library like zlib [9] achieves throughputs of 0.5-22 MB/s (depending on the specified compression level) using a 2.6 GHz AMD Opteron processor. The algorithm of Ratanaworabhan et al. for compressing scientific floating-point data achieves throughputs of about 22-47 MB/s using a 3.0 GHz Pentium 4 processor [10]. When using high performance interconnects such as InfiniBand, the performance of these algorithms is insufficient. The HPC cluster CHiC reaches bandwidths of about 970 MB/s for unidirectional point-to-point communication with `MPI_Send`/`MPI_Recv`.

This estimation about the required throughput assumes that compression and decompression occur as additional tasks before and after the data transmission. Nevertheless, it is also possible to incorporate the compression algorithm into operations that are already existing. For example, the compression can be done when the message is copied to communication buffers or when the reduction operation of `MPI_Reduce` is applied. In that case, a compression/decompression throughput equal to the bandwidth of the communication network helps to prevent a loss of performance even if the size of the message can not be reduced.

2.1 Run Length Encoding for Floating-Point Data

Run Length Encoding is a well known compression scheme that works by replacing repetitions of equal values with the information about the number of repetitions. Non-repeating values remain unchanged. The RLE scheme is useful for data that contains long sequences of equal values, e.g. sparse vector data with many zero values. The encoded repetition of a value requires a marker that is distinguishable from the not-encoded values. We use *Not a Number* (NaN) values in the IEEE 754 representation of floating point numbers as markers. Considering 64-Bit floating-point numbers, NaN values have an arbitrary sign bit, all 11 bits of the exponent set to one and a non-zero mantissa (52 bits). We use the non-zero mantissa to save the number of repetitions as a 52-Bit integer.

This RLE scheme for floating-point data can be adapted to different use cases. If only repetitions of one fixed value (e.g., zero) are considered, then every NaN in the encoded data represents a sequence of at least two of these values. If repetitions of arbitrary values are considered, then the specific value that is repeated has to be saved together with the NaN. In this case, it is appropriate to skip the encoding of sequences of size two, since their encoded size is the same as their original size. This RLE scheme does not increase the size of the encoded data. If NaN values are included in the original data, a distinction between original and encoded NaNs is required. To preserves most of the information of the original NaN, the arbitrary sign bit can be used for this distinction.

2.2 Throughputs of RLE for Sparse Floating-Point Data

The throughput of the RLE scheme is evaluated using an implementation for repetitions of zeros. Compressible random sample data is used that consists of floating-point vectors with randomly placed non-zero values. All operations are written in C and compiled with the PathScale 3.1 compiler (optimization `-O3`). The throughput values are calculated according to the size of *one* vector. Throughputs with respect to the total amount of data read and written by the operations (without RLE) can be obtained, by applying a factor of two for the copy operation and a factor of three for the vector addition operations.

Figure 1 (left) shows throughputs of the compression and decompression operation depending on the amount of non-zero values in the input data. The throughputs of both operations strongly depend on the amount of non-zero values. The compression operation shows a significant loss of performance when having more than 20 % non-zero values. With 100 % non-zero values the performance of both operations is comparable to the copy operation.

Figure 1 (right) shows throughputs of vector addition operations with and without RLE depending on the amount of non-zero values in the input data. Operations $\boldsymbol{a}_C+\boldsymbol{b} \rightarrow \boldsymbol{a}_C$ and $\boldsymbol{a}_C+\boldsymbol{b} \rightarrow \boldsymbol{c}_C$ represent the addition of a compressed vector $\boldsymbol{a}_C$ and an uncompressed vector $\boldsymbol{b}$ where the result (in $\boldsymbol{a}_C$ or $\boldsymbol{c}_C$) is compressed, too. The compressed version using only two vector arrays ($\boldsymbol{a}$ and $\boldsymbol{b}$) reaches about 70-95 % of the performance of the uncompressed version. Similar to the compression operation, the results show a loss of performance if the amount of non-zero values increases, but still good results with 100 % non-zero values.

In comparison to these results, the `memcpy` operation of the PathScale compiler achieved throughputs of about about 3 GB/s and the vector addition operation (`daxpy`) of the AMD Core Math Library achieved throughputs of about 2 GB/s.

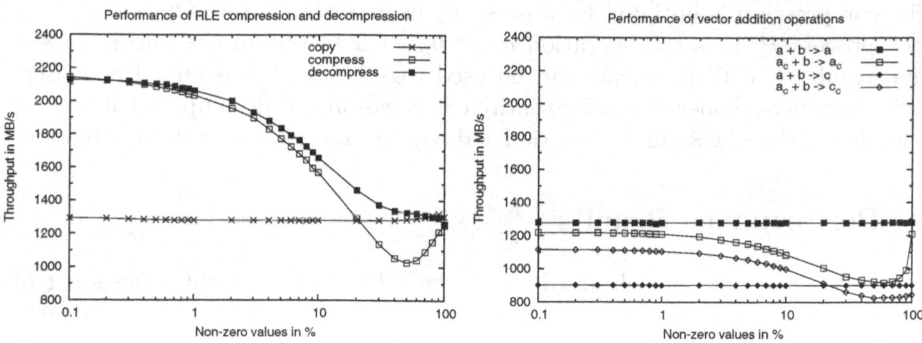

Fig. 1. Throughputs of RLE compression and decompression operations (left) and vector addition operations with and without RLE (right)

The performance of these highly optimized operations shows, that there is still room for improving the compiler optimized implementations. The addition operation in conjunction with repetitions of zeros provides several characteristics that ease the implementation of the corresponding vector operations. Nevertheless, the RLE scheme is not limited to repetitions of zeros (see Section 2.1) and applicable to other operations, too. A general vector operation with RLE can be achieved, by incorporating the corresponding compression operation into the process of writing the resulting vector.

3 `MPI_Reduce` with Run Length Encoding

As shown in Section 2, the throughput of data compression algorithms is insufficient in comparison to the bandwidth of high speed interconnects like InfiniBand. Therefore, we incorporate the RLE scheme into the already existing process of applying the reduction operation of `MPI_Reduce`. We start with a pipeline algorithm for `MPI_Reduce` that is appropriate to achieve high bandwidths for large messages. Each process P_i sends data only to process P_{i+1}. The last process P_p is the root process of `MPI_Reduce`. Process P_i performs operation $\boldsymbol{a} \otimes \boldsymbol{b_i} \rightarrow \boldsymbol{a}$ to apply the reduction operation $\otimes$ to the incoming data $\boldsymbol{a}$ and its local data $\boldsymbol{b_i}$. The result is placed in $\boldsymbol{a}$ and send to process P_{i+1}. The data is divided into equal blocks and the sending and receiving of blocks is overlapped using `MPI_Sendrecv`.

We incorporate the compression and the decompression with the RLE scheme into the process of performing the reduction operation for floating-point data. The regular operation $\boldsymbol{a} \otimes \boldsymbol{b_i} \rightarrow \boldsymbol{a}$ is replaced by the compressed version $\boldsymbol{a_C} \otimes \boldsymbol{b_i} \rightarrow \boldsymbol{a_C}$. The reduction operation $\otimes$ is applied to the compressed incoming data $\boldsymbol{a_C}$ and the uncompressed data $\boldsymbol{b_i}$ of process P_i. The compressed result is placed in $\boldsymbol{a_C}$ and is sent to process P_{i+1}. The first process P_1 has no incoming data and therefore no reduction operation to perform. We avoid the overhead of an additional compression operation by sending uncompressed data from P_1 to P_2.

The compression is initiated by process P_2 using operation $\boldsymbol{a} \otimes \boldsymbol{b_i} \rightarrow \boldsymbol{a_C}$. The root process P_p uses the operation $\boldsymbol{a_C} \otimes \boldsymbol{b_p} \rightarrow \boldsymbol{a}$ to obtain the uncompressed final result. The RLE scheme can be used together with predefined and user-defined reduction operations. The entire compression and decompression process is hidden in the `MPI_Reduce` operation and requires no changes to the application.

4 Performance Results

Performance results are obtained using the HPC Cluster CHiC consisting of 530 compute nodes each with two 2.6 GHz Dual-Core AMD Opteron processors, 4 GiB main memory and InfiniBand interconnect. One process is used per node.

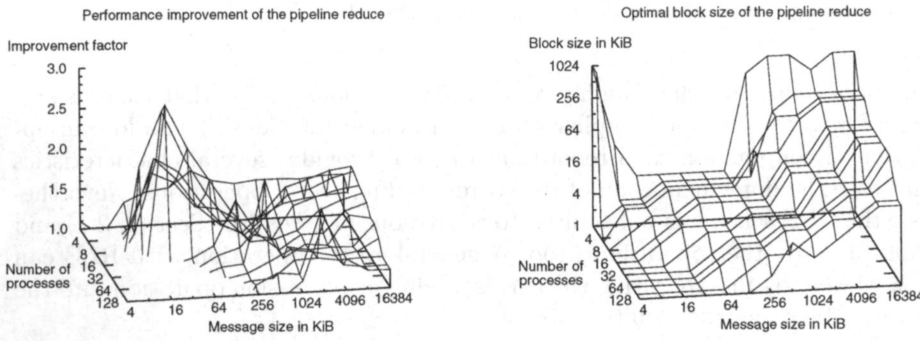

Fig. 2. Performance improvement of the pipeline reduce algorithm in comparison to the native `MPI_Reduce` (left) and optimal block size (right)

The impact of the pipeline reduce algorithm is demonstrated first. Figure 2 (left) shows the relative improvement of the pipeline reduce algorithm in comparison to the native `MPI_Reduce` of OpenMPI (version 1.2.4) depending on the number of processes and the message size. The results show a decrease in performance for message sizes up to 8 KiB in general and up to 256 KiB for large numbers of processes. For large message sizes, the pipeline reduce algorithm achieves improvements up to a factor of about 1.8 (single peaks show improvements up to about 2.9). Figure 2 (right) shows the corresponding block sizes of the pipeline algorithm that achieve the best performance. The results show that the block size increases with increasing message sizes and decreasing numbers of processes.

Figure 3 (left) shows latency results for `MPI_Reduce` and the pipeline reduce with and without RLE depending on the amount of non-zero values in the input data with 16 MiB messages, and 128 processes. The RLE scheme is incorporated into the vector addition operation to compress repetitions of zero values. Results are shown for identical and non-identical random distributions of non-zero values in the different messages of the processes. In contrast to the constant performance of `MPI_Reduce` and the pipeline reduce, RLE pipeline reduce shows a dependence

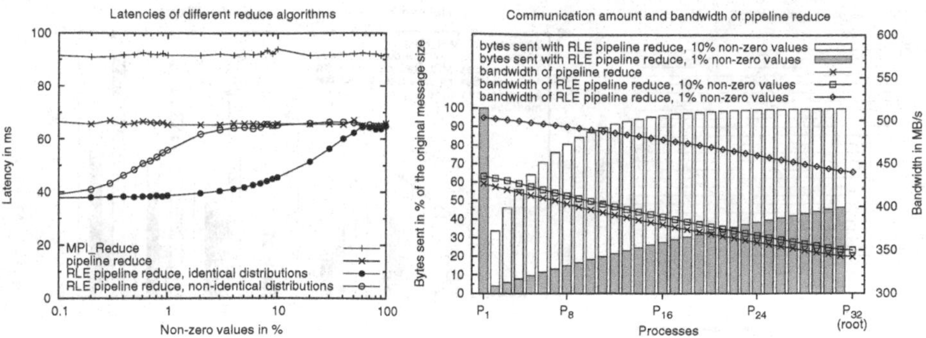

Fig. 3. Latency of the native MPI_Reduce and the pipeline reduce with and without RLE using different distributions of non-zero values (left). Communication amount and bandwidth of the different processes participating in the pipeline reduce (right).

on the amount of non-zero values. With 0.1 % non-zero values the latency of RLE pipeline reduce falls below 40 ms. With 100 % non-zero values, using the RLE scheme causes no overhead and is as fast as pipeline reduce without RLE (about 65 ms). The distribution of the non-zero values has a significant influence on the performance of RLE pipeline reduce, too. With identical distributions on all processes, improvements are achieved with up to 50 % non-zero values while with non-identical distributions the improvements vanish when using more than 10 % non-zero values.

Figure 3 (right) shows the amount of data sent by the individual processes using 1 % and 10 % non-identically distributed non-zero values, 16 MiB messages, and 32 processes. Bandwidths calculated from the latencies of the individual processes are also shown. Since process P_1 sends uncompressed data to P_2, a decreased amount of transferred data is first observed for P_2. Because of the non-identical distributions of the non-zero values, the number of non-zero values increases when the data approaches the root process. This reduces the efficiency of the RLE scheme and increases the amount of data transferred by the latter processes in the pipeline. With 10 % non-zero values, starting at process P_{11} over 90 % of the original data is transferred. With 1 % non-zero values, the amount of transferred data increases more slowly resulting in a higher improvement.

Next, we use data from a parallel adaptive FEM application. The elements of the floating-point vectors correspond to the nodes of the mesh used by the FEM application. Adaptive mesh refinement increases the number of mesh nodes and therefore the size of the vectors, too. According to the distribution of the mesh nodes to the different processes, each process contributes only to a subset of elements of the solution vector. The complete solution vector is obtained by a summation of all local contributions using `MPI_Reduce`. Figure 4 (left) shows an example for the sparse structure of the vectors supplied to `MPI_Reduce`. The vectors consist of 564 elements with about 6-11 % non-identically distributed non-zero values.

Figure 4 (right) shows bandwidths of the root process for `MPI_Reduce` and the pipeline reduce with and without RLE using the data of the FEM application

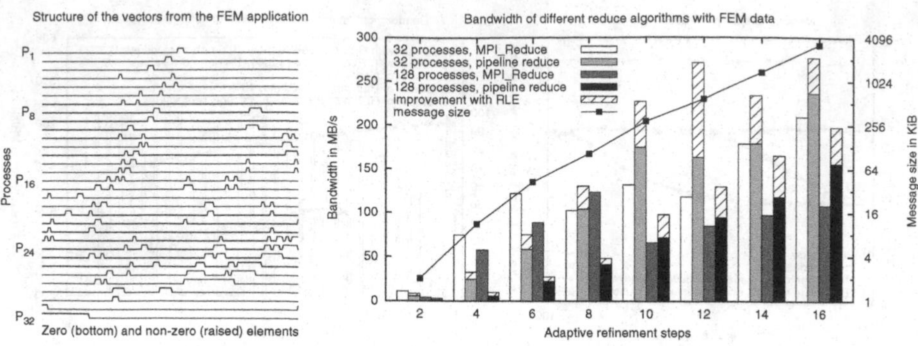

Fig. 4. Zero and non-zero elements of vectors from the FEM application (left). Bandwidth of the native `MPI_Reduce` and the pipeline reduce with and without RLE using data of the FEM application (right).

after different adaptive refinement steps. As previously seen in Figure 2, pipeline reduce achieves performance improvements only for large messages while the native `MPI_Reduce` is better for small messages. With 32 processes, improvements are achieved after more than six refinement steps (≈ 42 KiB messages) and with 128 processes after more than eight refinement steps (≈ 72 KiB messages). Pipeline reduce with RLE has always a higher performance than without RLE. However, the improvements are most significant for large messages. In comparison to the native `MPI_Reduce`, the bandwidth of RLE pipeline reduce increases up to 228 % with 32 processes and up to 182 % with 128 processes.

Instead of integrating this kind of optimization into the MPI operations, it is also possible to utilize an appropriate sparse vector format inside the application. However, these unconventional formats prevent the usage of operations like `MPI_Reduce` and require that optimized communication algorithms are implemented on the application level, too. The RLE scheme is rather simple and the high throughputs of the RLE operations prevent a loss of performance when using incompressible input data. Integrated into an MPI library, the RLE compression scheme could be enabled by default or optionally used with a new special MPI datatype.

5 Conclusion

In this paper, we have shown that a fast RLE scheme can be used to improve the performance of `MPI_Reduce` even with high performance interconnects such as InfiniBand. We have introduced an RLE scheme for floating-point data and incorporated the compression and decompression process into the reduction operation of `MPI_Reduce`. Performance results show that the pipeline reduce algorithm and the RLE scheme lead to significant performance improvements for large messages. The improvements due to the RLE scheme strongly depend on the input data. However, the marginal overhead of the RLE scheme prevents a decrease in performance when using incompressible input data. Results with sparse floating-point data from a parallel FEM application show improvements in bandwidth up to a factor of two.

Acknowledgment

We thank Arnd Meyer and his group from the Department of Mathematics, Chemnitz University of Technology, for providing the data of the parallel FEM application.

References

1. Faraj, A., Yuan, X., Lowenthal, D.: STAR-MPI: Self Tuned Adaptive Routines for MPI Collective Operations. In: ICS 2006: Proc. of the 20th annual international conference on Supercomputing, pp. 199–208. ACM Press, New York (2006)
2. Pješivac-Grbović, J., Bosilca, G., Fagg, G.E., Angskun, T., Dongarra, J.J.: MPI collective algorithm selection and quadtree encoding. Parallel Computing 33(9), 613–623 (2007)
3. Worringen, J.: Pipelining and Overlapping for MPI Collective Operations. In: LCN 2003: Proc. of the 28th Annual IEEE International Conference on Local Computer Networks, pp. 548–557. IEEE Computer Soceity, Los Alamitos (2003)
4. Almási, G., et al.: Optimization of MPI Collective Communication on BlueGene/L Systems. In: ICS 2005: Proc. of the 19th annual international conference on Supercomputing, pp. 253–262 (2005)
5. Rabenseifner, R., Träff, J.L.: More Efficient Reduction Algorithms for Non-Power-of-Two Number of Processors in Message-Passing Parallel Systems. In: Kranzlmüller, D., Kacsuk, P., Dongarra, J. (eds.) EuroPVM/MPI 2004. LNCS, vol. 3241, pp. 36–46. Springer, Heidelberg (2004)
6. Calderón, A., García, F., Carretero, J., Fernández, J., Pérez, O.: New Techniques for Collective Communications in Clusters: A Case Study with MPI. In: ICPP 2001: Proc. of the Int. Conf. on Parallel Processing, pp. 185–194. IEEE Computer Society Press, Los Alamitos (2001)
7. Ke, J., Burtscher, M., Speight, E.: Runtime Compression of MPI Messages to Improve the Performance and Scalability of Parallel Applications. In: SC 2004: Proc. of the ACM/IEEE Conf. on Supercomputing, p. 59. IEEE Computer Society Press, Los Alamitos (2004)
8. `http://www.tu-chemnitz.de/chic/`
9. `http://www.zlib.net/`
10. Ratanaworabhan, P., Ke, J., Burtscher, M.: Fast Lossless Compression of Scientific Floating-Point Data. In: DCC 2006: Proceedings of the Data Compression Conference, pp. 133–142. IEEE Computer Society Press, Los Alamitos (2006)

A Prototype Implementation of MPI for SMARTMAP

Ron Brightwell

Sandia National Laboratories*
Scalable System Software Department
Albuquerque, NM USA
rbbrigh@sandia.gov

Abstract. Recently the Catamount lightweight kernel was extended to support direct access shared memory between processes running on the same compute node. This extension, called SMARTMAP, allows each process read/write access to another process' memory by extending the virtual address mapping. Simple virtual address bit manipulation can be used to access the same virtual address in a different process' address space. This paper describes a prototype implementation of MPI that uses SMARTMAP for intra-node message passing. SMARTMAP has several advantages over POSIX shared memory techniques for implementing MPI. We present performance results comparing MPI using SMARTMAP to the existing MPI transport layer on a quad-core Cray XT platform.

1 Introduction

Catamount [1] is a third-generation lightweight kernel developed by Sandia National Laboratories and Cray, Inc., as part of the Sandia/Cray Red Storm project [2]. Red Storm is the prototype of the Cray XT series of massively parallel machines. Recently, Catamount was enhanced using a technique called SMARTMAP – Simple Memory of Address Region Tables for Multi-core Aware Programming. SMARTMAP allows the processes running on a compute node as part of the same parallel job to efficiently read and write each other's memory. Unlike POSIX shared memory, SMARTMAP allows a process to access another process' memory by simply manipulating a few bits in a virtual address. This mechanism has several advantages for efficiently implementing MPI for intra-node communication.

We have developed a prototype MPI implementation using Open MPI that is able to use SMARTMAP for intra-node communication. Initial performance results show that SMARTMAP is able to achieve significant improvement for

* Sandia is a multiprogram laboratory operated by Sandia Corporation, a Lockheed Martin Company, for the United States Department of Energy's National Nuclear Security Administration under contract DE-AC04-94AL85000.

A. Lastovetsky et al. (Eds.): EuroPVM/MPI 2008, LNCS 5205, pp. 102–110, 2008.

intra-node point-to-point and collective communication operations. The following section describes the advantages of SMARTMAP compared to existing approaches for intra-node MPI. Section 3 provides a detailed description of the implementation of MPI communication using SMARTMAP, which is followed by a performance comparison between SMARTMAP and the existing transport for Red Storm. Relevant conclusions and an outline of future work are presented in Section 5.

2 Background

SMARTMAP takes advantage of the fact that Catamount only uses a single entry in the top-level page table mapping structure (PML4) on each core of a multi-core AMD Opteron processor. Each PML4 slot covers 39 bits of address space, or 512 GB of memory. Normally, Catamount only uses the first entry covering physical addresses in the range `0x0` to `0x007FFFFFFFFF`. The Opteron supports a 48-bit address space, so there are 512 entries in the PML4.

Each core writes the pointer to its PML4 table into an array at core 0 startup. Each time the kernel enters the routine to start a new context, the kernel copies all of the PML4 entries from every core into every other core. This allows every process on a node to see every other process' view of the virtual memory at an easily computed offset in its own virtual address space. The following routine can be used by a process to manipulate a "local" virtual address into a "remote" virtual address on a different core:

```
static inline void *remote_address( unsigned core, void *vaddr )
{
  uintptr_t addr = ((uintptr_t) vaddr) & ( (1UL<<39) - 1);
  addr |= ((uintptr_t) (core+1)) << 39;
  return (void*) addr;
}
```

SMARTMAP also takes advantage of Catamount's physically contiguous address space mapping and the fact that the address mappings are static. Unlike traditional UNIX-based operating systems, Catamount determines the mapping from virtual to physical addresses when a process is created and the mapping is never changed. Because each process from the same executable will have the same virtual address mapping, the location of variables with global scope will be identical across all of the processes – both on node and off node.

There is much previous work on using shared memory for intra-node MPI communications [3,4,5]. The traditional approach is to use a POSIX shared memory to allocate a region of memory that is shared between communication processes on a node. This memory is divided up among the processes and message queues are built inside the region. In order to send a message, the sender copies data into the shared region and the receiver copies it out. Other approaches that use only a single copy have been implemented and studied. One such implementation is a Linux kernel module that handles re-mapping of user memory pages into kernel space so that the operating system can do a single memory copy between processes [6]. Another approach is to have an intelligent or programmable network interface perform a single copy between processes on the same node. A comprehensive analysis of the different approaches for intra-node MPI communication was presented in [7].

Both of these approaches are currently used for the MPI implementation on Red Storm using the Portals data movement layer [8]. There are two different implementations of Portals available on Red Storm. The default implementation interrupts the operating system to service the network. For intra-node transfers, the operating system simply copies data between the processes. This approach is much like the Linux kernel module, except that Catamount's static memory mapping avoid having to do any re-mapping of pages. For larger messages, it is more efficient for the OS to use the SeaStar [9] network interface to perform the copy. In the second implementation of Portals, all network processing is performed on the SeaStar. Therefore, all intra-node transfers must go through the network interface.

The simplified memory model of Catamount means that there is no registration overhead, no system call, and no setup or teardown time necessary for one process to directly move data into another process. In addition, there is no serialization of processes through the operating system. Processes are free to move data without any OS involvement. Relative to using an intelligent network adapter, there is no need to have data traverse a I/O bus and there is no serialization of requests through the network interface. There is no synchronization mechanism needed to transfer large messages. Our current implementation has a single protocol for moving data. Since data only moves when both sender and receiver have initiated communication, there are no unexpected messages and no protocols needed to distinguish between the various MPI point-to-point send modes. We are able to achieve low latency, high throughput, significantly increased small message rate, overlap of computation and communication, and we are able to support both the active and passive MPI-2 one-sided functionality. Our approach is much simpler in term of resource allocation and management. For example, we are not constrained by a shared resource, such as how to best divide up shared memory regions between control and data.

In addition to these benefits, SMARTMAP allows for additional capabilities that existing approaches do not. For example, non-contiguous data transfers can simply be copied directly from sender to receiver with no intermediate copies or packing/unpacking. Collective operations can operate directly on the buffers involved in the communication. In particular, reduction operations can operate directly on the buffer in-place at the root of the operation. Using MPI_IN_PLACE, copying can be avoided altogether.

3 Prototype Implementation

Catamount supports multiple cores by running in virtual node mode, where each core is treated as a node that runs a process in the parallel job. The memory on a single physical node is divided evenly among the available cores. When a process is started, the parallel runtime system is responsible for setting the following values in the process' address space:

- number of processes in the job (_my_nnodes)
- global rank of the process (_my_rank)
- number of active cores on the node (_my_vnm_degree)
- core rank on the node (_my_core)

We then use these values to determine a global rank to core rank mapping and a core rank to global rank mapping. First, we allocate an array for the global rank to core rank mapping and initialize all entries to -1. We then allocate an array for the core rank to global rank mapping. We take advantage of the SMARTMAP capability to determine the core rank to global rank mapping. Each core loops from 0 to _my_vnm_degree. If the loop variable is equal to the local core's rank, it fills in the core rank to global rank mapping with its global rank. If the loop variable is not the process' core rank, it accesses the _my_rank value on the other core and fills in the array with this value. At this point, we can use this array to set the corresponding values in the global rank to core rank array to the appropriate values.

Each process has a single queue for posted receives and a queue per core for posted sends. These queues are doubly-linked lists. Each queue element contains: buffer address, local source rank, buffer length, context id, tag, request address, and completion flag.

In order to send a message to another core on the same node, we first check the global rank to core rank mapping to discover whether the destination rank is a local process. If it is, we check the destination core to see if the message is being sent to the sending process. If so, we traverse the posted receive queue looking for a match. If the message is destined for a different core on the same node, we simply pop a queue element off of a stack, fill in the contents based on the send request, and enqueue it on the send queue for that particular core. If the send is blocking, we then call into the progress routine, which is described in detail below.

When posting a receive, we check to see if the posted receive queue is currently empty. If it is, we can check for a match right away. In order to check for a match, we get the address of our core's send queue in our address space. Because this virtual address is identical across all of the processes on the node, we can easily use SMARTMAP to get the address of this queue in the other process' address space. If the source of the receive is specified, we get the address of our send queue in the other process' address space and proceed to traverse this queue looking for a matching send queue entry. Because the pointers are local to the destination process, every next pointer needs to be converted to a remote pointer to traverse the queue. When a match is found, the start address of the send buffer is converted to a remote pointer and the send buffer is copied from the sending process' memory into the local receiving process' memory. We then set the completion flag of the send queue element in the sending process' memory. If the source of the receive is MPI_ANY_SOURCE, we simply traverse our send queue in all of the other process' address space looking for match. If our posted receive queue is not empty, we simply pop a queue element off of a stack, fill in the contents, and enqueue it. We then call into the progress function to see if any outstanding requests have been or can be completed. In order to probe for an incoming messages, we do the same steps as posting a receive; however, when a match is found, we simply fill in the appropriate status information.

The progress function first checks to see if any outstanding posted receives can be completed. It does this in the same way as was just described in the previous paragraph – traversing our send queue in the other process' address space looking for match. After traversing the posted receive queue, the progress function traverses all of the send queues to see if any of the outstanding sends have been marked as completed by the other cores. Any send queue elements that have been marked as completed by the receiving core are dequeued and any cleanup routine for the request is run.

In this strategy, all queues are managed by the local core, so there are no race conditions with enqueueing and dequeueing elments. Receives are completed by the local core, copying any data from the remote core. Sends are completed by the remote core and dequeued by the local core.

4 Performance Evaluation

The platform used to gather our performance results is a Red Storm development system that contains 2.2 GHz quad-core Opterons. Our prototype implementation was done using Open MPI. Open MPI already had support for Portals on the Cray XT, using either a path that does matching inside the MPI library (BTL) or one that does matching inside Portals (MTL). See [10] for a complete discussion of these two approaches. We also compare results to the Cray MPI implementation, which is an modified version of MPICH2.

We used the Intel MPI Benchmark Suite (IMB) version 2.3 to measure point-to-point and collective communication performance. In order to characterize small message rate, we used the Ohio State message rate benchmark that has been modified by PathScale (now Qlogic).

Since we are interested in intra-node communication, all of our results are from a single quad-core node. We limited our measurements to the interrupt-driven version of Portals because it is more efficient at intra-node transfers. The ability to have the operating system perform a copy between processes is more efficient than having the SeaStar adapter perform the copy. Due to limitations of the SeaStar, send operations must go through the OS, so in addition to serializing requests through a slower network interface, requests must also be serialized through the OS.

Figures 1(a) and 1(b) show ping-pong latency and bandwidth respectively. SMARTMAP is able to achieve a zero-byte latency of 520 ns, while Cray's MPI achieves 2.98 μs. SMARTMAP's bandwidth peaks at more than 9.3 GB/s, while the MTL in Open MPI is able to achieve nearly 5.8 GB/s. This difference is

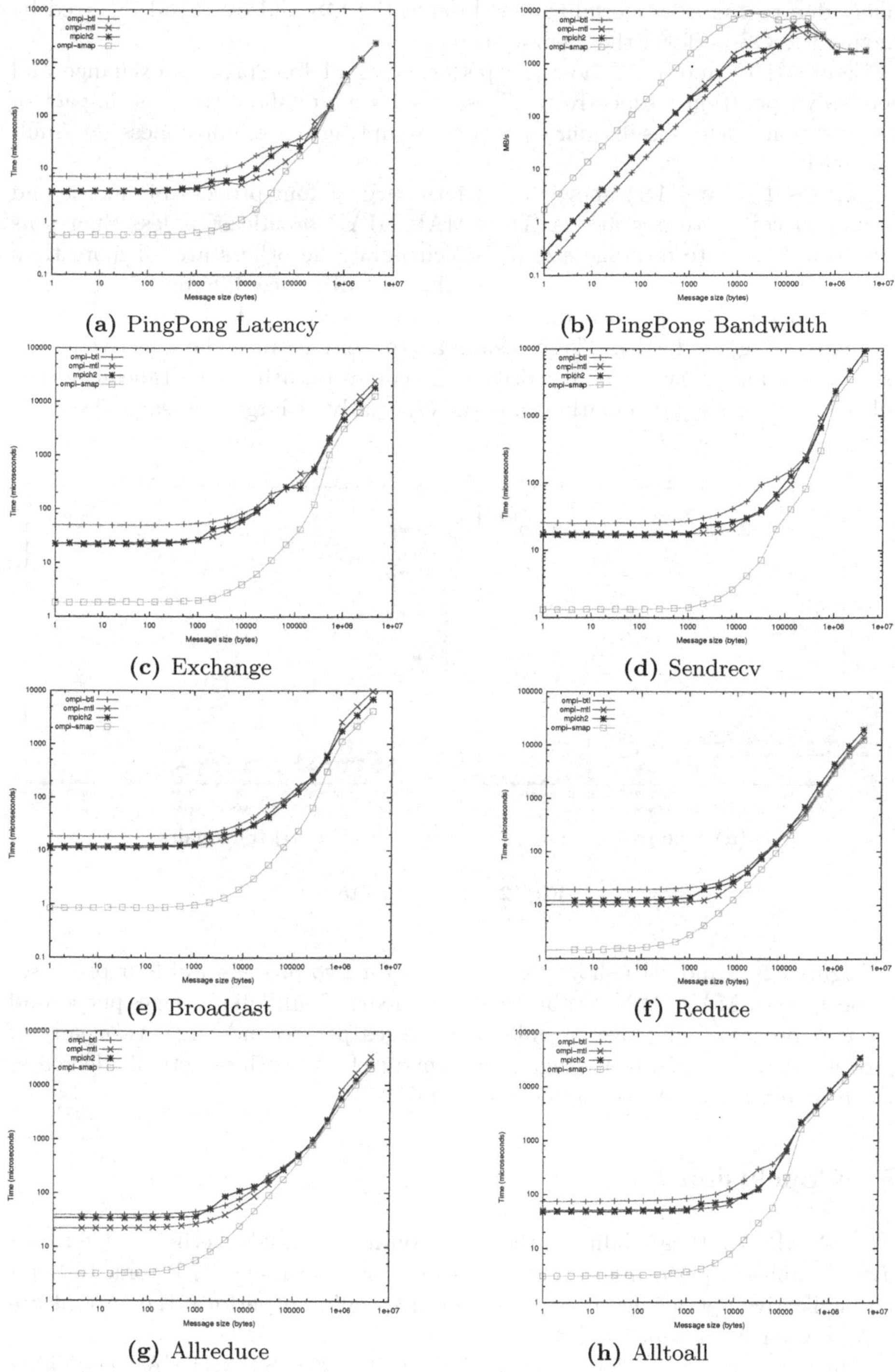

(a) PingPong Latency

(b) PingPong Bandwidth

(c) Exchange

(d) Sendrecv

(e) Broadcast

(f) Reduce

(g) Allreduce

(h) Alltoall

Fig. 1. IMB Results

likely due to the extra serialization through the OS and overhead incurred by having the OS perform the memory copies.

Figures 1(c) and 1(d) show the performance of four-process exchange and sendrecv operations respectively. These results again show that the impact of serialization worsens as all four cores are attempting to exchange message simultaneously.

Figures 1(e) and 1(f) show the performance of four-process broadcast and reduction collective operations. The SMARTMAP broadcast is less than 1 μs out to a 1024-byte message size, at which point the others are all more than 12 μs. The reduce operation shows a similar performance differential.

We conclude IMB performance results with allreduce and alltoall performance in Figures 1(g) and 1(h). These results again demonstrate the effectiveness of using shared memory when all processes are communicating simultaneously. The alltoall results are particularly dramatic, especially at larger message sizes.

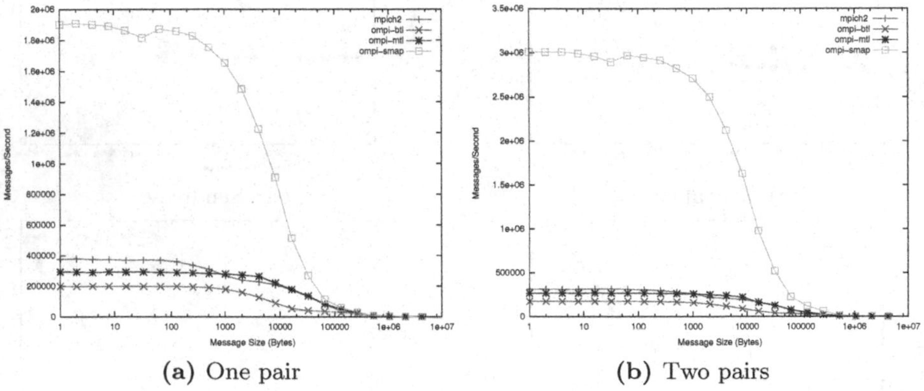

(a) One pair (b) Two pairs

Fig. 2. Message Rate

Figures 2(a) and 2(b) show message rates for two process and four processes respectively. SMARTMAP is able to achieve nearly 2 million messages per second for two processes and over 3 million messages per second when two pairs of processes are exchanging messages. Performance for the others actually decreases when increasing from two to four processes.

5 Conclusion

The SMARTMAP capability in the Catamount lightweight kernel is able to deliver significant performance improvements for intra-node MPI point-to-point and collective operations. It is able to dramatically outperform the current approaches for intra-node MPI data movement.

There is much work left to do to fully utilize the SMARTMAP capability for MPI. First, because the Portals data movement layer encapsulates the MPI posted receive queue, the complexity of handling MPI_ANY_SOURCE receives is significantly increased. In most other networks, the MPI posted receive queue exists inside the MPI library, so the atomicity required to handle a non-specific receive is straightforward. Because Portals contains the posted receive queue for network transfers (either in the OS or in the network interface) and the MPI library contains the posted receive queue for intra-node transfers, there is no mechanism by which atomicity can be enforced. There are two possible choices for handling MPI_ANY_SOURCE receives for communicators that are not either completely on-node or completely off-node. If the request cannot be satisfied by any current in-progress communication, all current shared memory transfers need to queued as unexpected messages and the peer processes need to be informed that all subsequent messages should use Portals rather than shared memory. Alternatively, the implementation could modify the Portals implementation so that all matching occurs inside the MPI library, rather than inside Portals. Cray's MPI and the Open MPI BTL already support this mode of operation. Either way – by putting the MPI posted receive queue completely in the network or completely inside MPI – great care must be taken to avoid race conditions and to maintain MPI ordering semantics, both in order to satisfy the non-specific receive and to possibly switch back after it has been completed. We are currently adapting both the Open MPI Portals MTL and the shared memory BTL to be able to use SMARTMAP. This will allow us to support both intra- and inter-node communication and will allow for direct comparisons between SMARTMAP and the POSIX shared memory approach.

From a complexity standpoint, it is much easier to enhance MPI collective operations to use the SMARTMAP capability. It is not necessary to handle non-specific receive operations or non-blocking operations. We can also make use of direct shared memory for certain operations, like barrier, when incrementing a shared counter may be more efficient than exchanging messages. We are currently implementing a collective module in Open MPI to use SMARTMAP directly and hope to take advantage of the existing hierarchical collective module to make use of it.

We would also like to use SMARTMAP to handle on-node non-contiguous data transfers with no intermediate buffering, and there is an opportunity to enhance the on-node MPI-2 one-sided operations using SMARTMAP as well.

Acknowledgments

Trammell Hudson is responsible for the implementation of SMARTMAP in Catamount, which was an outcome of discussions between the author and Kevin Pedretti. This work would also not have been possible without John Van Dyke, who is responsible for implementing virtual node mode support in Catamount. The author would also like to thank Sue Kelly and the Cray support staff at Sandia for assistance with the Red Storm development systems.

References

1. Kelly, S.M., Brightwell, R.: Software architecture of the light weight kernel, Catamount. In: Proceedings of the 2005 Cray User Group Annual Technical Conference (2005)
2. Camp, W.J., Tomkins, J.L.: Thor's Hammer: The First Version of the Red Storm MPP architecture. In: Proceedings of the SC 2002 Conference on High Performance Networking and Computing, Baltimore, MD (2002)
3. Buntinas, D., Mercier, G., Gropp, W.: Implementation and evaluation of shared-memory communication and synchronization operations in MPICH2 using the Nemesis communication subsystem. Parallel Computing 33(9), 634–644 (2007)
4. Buntinas, D., Mercier, G., Gropp, W.: Implementation and shared-memory evaluation of MPICH2 over the Nemesis communication subsystem. In: Proceedings of the 2006 European PVM/MPI Users' Group Meeting (2006)
5. Buntinas, D., Mercier, G., Gropp, W.: Design and evaluation of Nemesis, a scalable, low-latency, message-passing communication subsystem. In: Proceedings of the 2006 International Symposium on Cluster Computing and the Grid (2006)
6. Jin, H.W., Sur, S., Chai, L., Panda, D.K.: Limic: Support for high-performance MPI intra-node communication on Linux. In: Proceedings of the 2005 Cluster International Conference on Parallel Processing (2005)
7. Buntinas, D., Mercier, G., Gropp, W.: Data transfers between processes in an SMP system: Performance study and application to MPI. In: Proceedings of the 2006 International Conference on Parallel Processing (2006)
8. Brightwell, R., Hudson, T., Pedretti, K., Riesen, R., Underwood, K.: Implementation and performance of Portals 3.3 on the Cray XT3. In: Proceedings of the 2005 IEEE International Conference on Cluster Computing (2005)
9. Brightwell, R., Hudson, T., Pedretti, K.T., Underwood, K.D.: SeaStar interconnect: Balanced bandwidth for scalable performance. IEEE Micro. 26(3) (2006)
10. Graham, R.L., Brightwell, R., Barrett, B., Bosilca, G., Pjesivac-Grbovic, J.: An evaluation of Open MPI's matching transport layer on the Cray XT. In: Proceedings of the 14th European PVM/MPI Users' Group Conference (2007)

Gravel: A Communication Library to Fast Path MPI

Anthony Danalis, Aaron Brown, Lori Pollock, Martin Swany, and John Cavazos

Department of Computer and Information Sciences
University of Delaware, Newark, DE 19716
{danalis,brown,pollock,swany,cavazos}@cis.udel.edu

Abstract. Remote Direct Memory Access (RDMA) technology allows data to move from the memory of one system into another system's memory without involving either one's CPU. This capability enables communication-computation overlapping, which is highly desirable for addressing the costly communication overhead in cluster computing. This paper describes the consumer-initiated and producer-initiated protocols of a companion library for MPI called Gravel. Gravel works in concert with MPI to achieve increased communication-computation overlap by separating the meta-data exchange from the application data exchange, thus allowing different communication protocols to be implemented at the application layer. We demonstrate performance improvements using Gravel for a set of communication patterns commonly found in MPI scientific applications.

1 Introduction

The communication overhead of cluster computing continues to challenge MPI programmers trying to maximize the performance of their applications. Remote Direct Memory Access (RDMA) technology holds the promise of hiding these overheads by facilitating the overlap of communication operations with computation. To exploit the RDMA for communication-computation overlap, the communication library must provide support for one-sided communication and two-sided communication with low-overhead rendezvous protocols, and the application must contain communication and computation patterns that are amenable to overlap.

There already exist communication libraries that provide for asynchronous communication and have the goal of exploiting RDMA support; however, none provides the set of features that the proposed library, *Gravel*, provides. MPI provides asynchronous communication operations (e.g., Isend, Irecv, and Wait) and even one-sided communication support, although it enforces strict rules regarding the use of the latter. The User Direct Access Programming Library, uDAPL [1], provides functionality necessary to enable RDMA from applications, with support for memory registration and connection establishment, but does not provide a "message" abstraction nor does it provide a high level and intuitive interface for domain scientists and engineers to embrace. ARMCI [3] aims to be a portable library for RDMA communication. However, it requires the use of a custom memory allocator, which makes it unsuitable for substituting arbitrary MPI operations in Fortran applications without major restructuring of the application buffers. GASNet [4] provides similar capabilities and is intended to be used internally by a compiler or transformation system, but as its name implies, is targeted to Global Address Space parallel languages. Other communication libraries are provided by the hardware vendor as a

A. Lastovetsky et al. (Eds.): EuroPVM/MPI 2008, LNCS 5205, pp. 111–119, 2008.

means of implementing higher-level libraries, such as MPI, for the given interconnect (e.g., VAPI [5], GM [2]), or are usable only on specific hardware interconnects (e.g., MX [6]). Additionally, most of these libraries require either C pointer manipulation, or the use of memory returned by C library functions (i.e., `lib_specific_malloc()`), both of which are not possible directly from within FORTRAN programs.

In contrast to these approaches, our goal was to design a communication library that provided minimal messaging protocol, maximal overlap potential, support for Fortran and support for further communication optimization through code motion and transformation. Fortran support implies several requirements, most notably explicit memory registration functionality rather than special memory allocation routines. This paper describes a portable communication library, *Gravel*, which works in conjunction with MPI and is designed to (1) replace only key data exchange calls in MPI programs (2) separate the meta-data exchange from the application data exchange, and (3) improve the potential of code motion to increase the communication-computation overlapping and hide communication latency.

2 Communication Library

Gravel is a minimal library designed to be used in conjunction with MPI by replacing only key data exchange calls in MPI programs to exploit the potential communication-computation overlap in applications. It implements a simple messaging abstraction designed for RDMA-capable networks. The Gravel system is currently built on top the uDAPL [1] in the OpenFabrics software suite. However, it is not dependent on uDAPL, therefore can be ported to additional network interconnects.

Gravel is neither aimed to be a replacement for MPI, nor a low level library that one should use for implementing MPI. Rather, Gravel is designed to be used in MPI applications to improve performance by replacing selected MPI calls. While systems like ARMCI provide synchronization mechanisms, Gravel relies on those provided by MPI. Gravel's API is designed to be similar to that of MPI to facilitate automatic replacement (e.g., within a compiler) of performance-critical MPI calls with Gravel calls. Gravel places a number of restrictions on when it can be used. Gravel never copies messages, instead all message exchanges are "rendezvous" style, in MPI parlance. Further, as is the case with all RDMA layers, memory must be registered to be a source or destination for messages.

Gravel provides minimal RDMA-based messaging functionality that is decomposed to maximize potential for overlapping communication with computation. In general, a message can be seen as the exchange of message metadata (e.g., a message header) and the exchange of the data itself, followed by some indication of completion. Gravel separates this functionality into independent parts by providing distinct functions for implementing explicit registration of memory, communication rendezvous (exchanging message metadata), and performing the actual data transfer.

2.1 Rendezvous Protocols

Sur et al. [7] discuss the behavior and performance of the simple rendezvous found in a typical MPI implementation. They present a more advanced alternative which they implemented within *MVAPICH*, the Ohio State University implementation of MPI. Both rendezvous protocols are "sender initiated". The legacy protocol uses RDMA-write for the transfer, and the advanced protocol uses RDMA-read to eliminate unnecessary round-trip delays. Shipman et al. [8] also discuss advanced protocols used in MPI, but for the OpenMPI implementation of the MPI standard.

Gravel provides distinct functions for transferring data and metadata in order to allow an automatic program transformation tool (e.g., a compiler) or programmer to use the most efficient model given the constraints of the application. Efficiency in this case comes from appropriate overlapping of orthogonal computation.

Unlike highly abstracted libraries such as MPI, Gravel does not provide fixed data exchange protocols hidden inside calls such as MPI_Isend. It rather provides the necessary API and infrastructure for implementing any protocol at the application layer. Details are hidden behind the library API, so that the application does not need to specify anything other than which task should do what and when. By doing so, Gravel enables "function separation", i.e., the handshake is separated from the data transfer. As a result, each application can implement the appropriate exchange protocols that best fit the structure of each data exchange in that application. At the same time, the high-level, abstract API enables programmers to easily implement highly efficient exchange protocols, even in languages such as FORTRAN that does not support C pointers (pointers to arbitrary memory locations, or pointers returned by C library functions). Four protocols that can be implemented with Gravel are presented in Figure 1 and described below. In the following text, we use the term "producer" for the node that will send the **application data** message, and "consumer" for the node that will receive the **application data** message.

Consumer Initiated RDMA-Write Protocol. In this case (shown in Figure 1(a)), the consumer initiates the handshake (post_receive_buffer_rdma()) by sending a

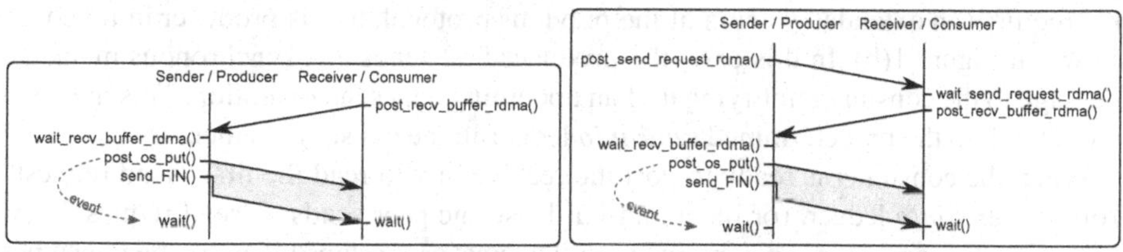

(a) Consumer Initiated RDMA write protocol (b) Producer Initiated RDMA write protocol

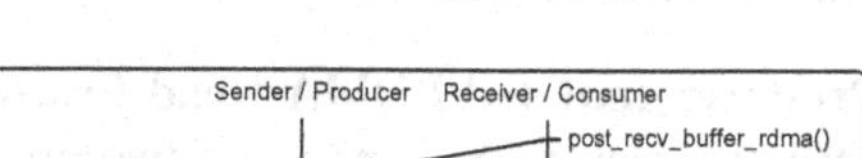

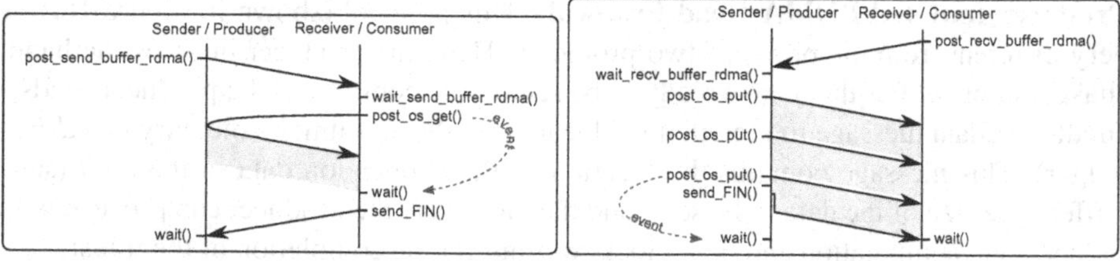

(c) Producer Initiated RDMA read protocol (d) Advanced RDMA write based protocol

Fig. 1. Timing schematics of RDMA_write and RDMA_read based rendezvous protocols

metadata message to a predefined location in the producer's memory (initialized by `gravel_init()`) referred to as the *receive-info* ledger. The details concerning the ledger are hidden from the application and are automatically handled by Gravel. Thus, the application does not need to allocate the ledger, register it, exchange its address between all the peers, etc. Furthermore, an application programmer, or compiler that uses Gravel does not even need to know that there is such an entity as a ledger. With this handshake message, the consumer passes to the producer the start of the local memory where the data will be received (receive buffer), the size of the expected data, an application defined tag, and a request handle to the call. After initiating the non-blocking transfer of the handshake, the consumer can proceed with independent computation, but it must assume that the receive buffer can be altered at any time between this call and the return of the corresponding wait operation.

When the producer is ready to post a send, it reads the next metadata message from its *receive-info* ledger (or blocks until the metadata message from the consumer arrives). The producer then initiates a non-blocking `RDMA-write` (`post_os_put()`) operation to transfer the message data to the consumer, followed by an additional non-blocking `RDMA-write` to send a small metadata message (into the *RDMA-FIN* ledger)[1]. Then, the producer can proceed with independent computation. Finally, both sides will wait for the completion of the transfer.

This protocol is well-suited for cases when the receive operation can be posted early because the computation does not have data dependencies with the receive buffer. In this scenario, the handshake overhead will be entirely overlapped with the independent computation and the actual data transfer will be performed by an RDMA-write operation without any delays or copying, regardless of the size of the transfer, leading to an efficient data exchange.

Producer Initiated RDMA-Write Protocol. Many MPI programs use the wildcard `MPI_ANY_SOURCE` such that any peer could be the producer of the data. Supporting this requires an extended version of the previous protocol, that is producer initiated, as shown in Figure 1(b). In this case, the producer first sends an asynchronous metadata message to the consumer notifying it of an upcoming data send operation. This message is written into the predetermined *send-info* ledger in the consumer's memory.

When the consumer is ready to post the receive, it will read the first "send request" from its *send-info* ledger, (or block until at least one peer sends a "send request"). At this point, the consumer can continue with the exact same steps taken by the consumer initiated RDMA-write protocol.

Producer Initiated RDMA-Read Protocol. This protocol (shown in Figure 1(c)) is very different from the previous two protocols. Here, the producer initiates the handshake, but only after the data is ready to be sent to the consumer. The producer sends a small metadata message to a predefined location in the consumer's memory (*send-info* ledger). This message contains the location of the application data in memory (send buffer), the size of the data to be sent, and the tag. Then, the producer can proceed with independent computation and then block, waiting for the completion of the transfer.

---

[1] If the underlying network does not support message ordering, appropriate measures need to be taken for the FIN to arrive at the consumer after the application data.

On the other side, the consumer reads the next entry from the *send-info* ledger, or blocks waiting for the arrival of a metadata message from the producer. At this point, the consumer initiates the transfer with a non-blocking `RDMA-read` operation. This provides the RDMA engine with the necessary information about the transfer and returns immediately. Thus, the consumer can execute independent computation during the data transfer. Before the consumer can notify the producer about the completion of the transfer, the consumer must wait for the `RDMA-read` to complete. When the transfer is completed, the consumer writes to the *RDMA-FIN* ledger of the producer, signaling the completion of the data transfer. Clearly, the producer must assume that the data is being used at any point between the initiation of the rendezvous and the corresponding `wait()` function, and cannot alter the data.

This protocol is expected to perform less efficiently than the "consumer initiated" protocol described earlier, but is necessary for cases that meet all the following criteria: 1) the communication is symmetric and every node is both consumer and producer, 2) the send operation takes place before the receive operation, and 3) the data or control dependencies prevent the receive operation from being hoisted above the send operation.

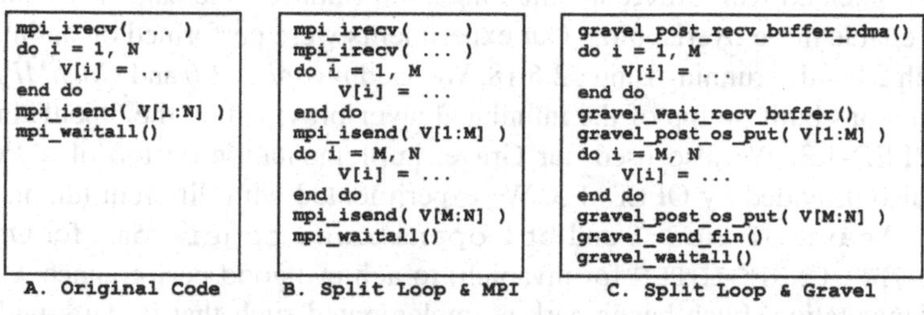

Fig. 2. Communication overlapping before and after splitting computation

Advanced RDMA-Write Based Protocol. Figure 2(a) shows an example of a parallel program where every task computes, stores into an array `V` and transfers some data to one or more neighbors. This simple case is common and does not provide an opportunity for overlapping the communication generated by the `send` operation with useful computation. However, if there are no dependencies on array `V` throughout the iteration space of the loop, it can be transformed to enable overlapping. In Figure 2(b), we see that the loop is split and therefore the `mpi_isend(V[1:M])` call that transfers the first part of the array can be overlapped with the computation of the second part of the array. Although this could lead to performance benefits, due to overlapping, it also has two major drawbacks. Namely, smaller messages experience lower throughput and every additional message adds contention and overhead through multiple handshakes. The latter concern can be alleviated with Gravel through a more advanced rendezvous scheme, specialized for pipelined transfers. In particular, only one handshake message is necessary, even if the transfer takes place in multiple segments. Indeed as is shown in Figure 2(c) only one call to the metadata transfer functions (`post_recv_buffer()`, `wait_recv_buffer()`, `send_fin()`) is performed, but the data is transferred in two steps by calling `post_os_put()` twice, achieving overlap without the overhead of additional control messages. Clearly, when `post_os_put()` is called multiple times, each call needs to be given an "offset" equal to the cumulative amount of data transferred by the previous calls. Figure 1(d) shows a schematic of a similar communication pattern, where the data transfer takes place in three steps.

This example of an advanced protocol demonstrates the difference between MPI and Gravel. MPI provides implicit protocols hidden behind calls such as `MPI_Isend` and `MPI_Irecv`, designed without any knowledge of a particular application. In contrast, Gravel provides the appropriate API and infrastructure for implementing the exchange protocols at the application layer. This way, the particular characteristics of each application can be exploited for maximizing communication-computation overlap.

3 Experimental Study

Experiment Design. For evaluation, we designed our experiments to explore how a program's performance is affected when key MPI calls are replaced by the equivalent Gravel library calls and how the previously mentioned different Gravel protocols affect performance.

We experimented with different communication libraries, message sizes, and strip-mining to enable more overlapping. Our experiments were performed on an infiniband cluster with 24 nodes running Linux 2.6.18. We used *mvapich-1.0* and *OpenMPI-1.2.5*[2] implementations built on top of the infiniband layer provided by the OpenFabrics Alliance's OFED-1.3. We also used our Gravel implementation on top of uDAPL-2.0 which is also provided by OFED-1.3. We experimented with different tuning parameters (`mpi_leave_pinned=1` and `btl_openib_use_eager_rdma` for OpenMPI and `LAZY_MEM_UNREGISTER` for mvapich) to achieve good performance with each MPI implementation. Each benchmark is implemented such that it starts with 1,000 cold runs that are not timed, so that the MPI engine is warmed up and exhibits "steady state" performance before the timing begins. After the timer is started the code segment that performs the computation and communication is also executed 1,000 times so that the timing errors are amortized and anomalous behavior is averaged out.

```
mpi_irecv(rBuf(1), size, prev, rreq(1), ... )
do i=1, size
   indx = 1+MOD(i-1,128)
   sBuf(i) = temp(indx)
enddo
mpi_isend(sBuf(1), size, next, sreq(1), ... )
mpi_wait(rreq(1), ...)
do i=1, size
   indx = 1+MOD(i-1,128)
   temp2(indx) = rBuf(i)
enddo
mpi_wait(sreq(1), ...)
```

Fig. 3. 1-D Ring micro-benchmark critical part

[2] We also experimented with the trunk version of OpenMPI-1.3 and found that it performs significantly faster than OpenMPI-1.2.5, but compares with Gravel the same way. Since the trunk does not constitute a stable version of the library that others can use to replicate our experiments, we do not report those results.

We compared the runtime performance of a set of MPI Fortran micro-benchmarks which represent communication-computation patterns found in many scientific applications. We used two micro-benchmarks: (1) a *1D-ring* pattern where every task receives data from the previous one and sends data to the next one and (2) a *2D-wavefront* where the tasks are organized in a 2-D grid and every task receives data from "north" and "west" and sends data to "south" and "east". The *1D-ring* pattern appears in well known codes including *SP*, *BT* and *LU* in the *NAS* suite; the *2D-wavefront* pattern appears in *LU* of the *NAS* suite and *Sweep3D*. The critical part of each micro-benchmark performs some minimal computation (array to array copy) that fills the message buffer to be sent, transfers the message and then performs some computation (copy) with the message it received from its peer. Figure 3 demonstrates the critical part of the *1-D Ring* micro-benchmark in pseudocode.

Results. Figure 4(a) and Figure 4(b) show the performance results for the *1D-ring* and *2D-wavefront* micro-benchmarks, respectively. The Y axis of both graphs presents execution time normalized to the ideal case, where communication is infinitely fast and

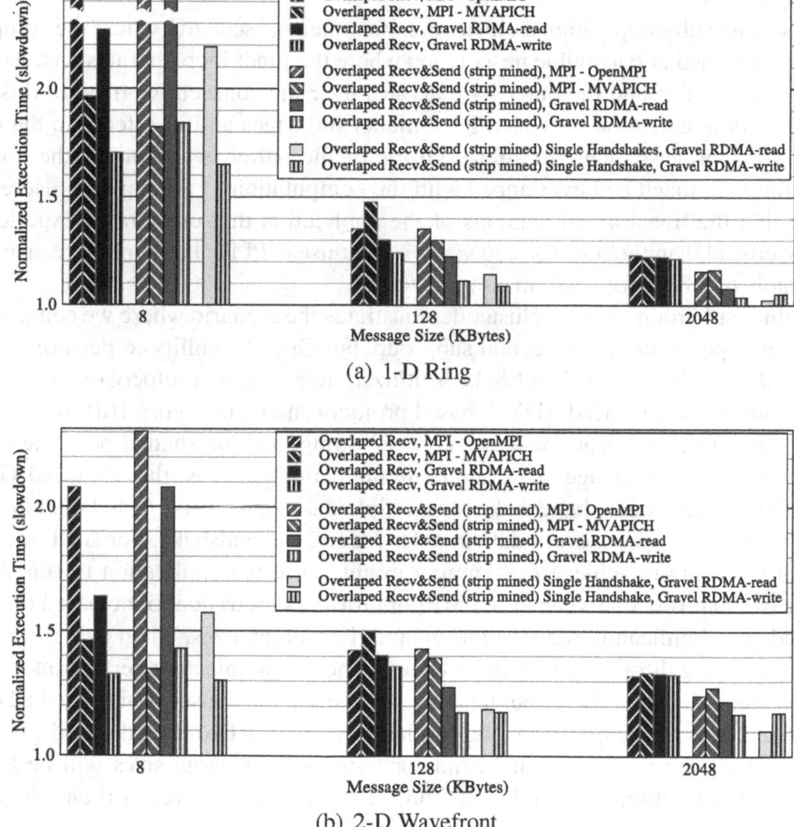

(a) 1-D Ring

(b) 2-D Wavefront

Fig. 4. Experimental evaluation of Gravel and MPI for 1-D Ring and 2-D Wavefront data exchanges

causes no overhead. This case is simulated by a version of the benchmark that does not perform any communication, just the computation loops. All values above 1 designate the slowdown factor caused by the communication. Each micro-benchmark was run several times, and the graph plots the minimum execution time for each scenario. We found very little variation across several runs of the same micro-benchmark using the same configuration. Each graph shows three different clusters of bars corresponding to the three different message sizes we evaluated. Within each cluster, there are three subgroups using dark, medium and light shadings, respectively. For all subgroups, the "computation" is a simple buffer to buffer copy.

The first subgroup to the left end of each bar cluster demonstrates the scenario where the transmission of data is not overlapped with computation and only the `recv` operation has potential for overlapping. Here, we attempt to answer the question of how the program's performance is affected when MPI calls are simply replaced by equivalent Gravel library calls. By looking at the bars within this subgroup and in particular the bar with the vertical stripes representing the non-transformed code using Gravel's consumer initiated RDMA-write protocol, one can see that the execution time of this version is lower than that achieved with either MPI version across sizes and data exchange patterns.

The second subgroup within a cluster demonstrates the scenario where the computation is strip-mined into a double nested loop where the inner loop operates on a section, or tile, of the buffer and the outer loop iterates over the consecutive tiles. In this scenario, the communication is broken into smaller messages and inserted into the outer loop of the loop nest such that after each tile of the buffer is computed, the transfer of that tile is initiated and overlapped with the computation of the next tile. Here, one can see that the transformed versions of the application that use Gravel experienced lower overhead than the transformed versions that use MPI for large message sizes and comparable overhead for small message sizes.

The third subgroup within a cluster demonstrates the scenario where we compare the same overlapped code as the second subgroup, but Gravel's ability to perform several data transfers with only one handshake is utilized, to minimize protocol overhead. This corresponds to the advanced RDMA based protocol shown in Figure 1(d). By comparing the bars of this subgroup with the corresponding medium shaded bars, the reader can see that across message sizes and communication patterns, the advanced Gravel protocols that use a single handshake for multiple data transfers perform better than the MPI-like protocols where every data transfer requires a handshake. For small message sizes, strip-mining to achieve overlapping might cause the application to run slower than the original version (when either MPI or Gravel is used) due to increased protocol overhead and significantly reduced throughput. For larger sizes, when either library is used, overlapping through strip-mining benefits the communication performance.

The results show that for all but the very large sizes, the consumer-initiated RDMA-write based protocol outperforms the producer-initiated RDMA-read based protocol. The reason for the reversal of this behavior witnessed for large sizes will be further investigated in the future. Also, by studying the graphs of Figures 4(a) and 4(b), one can see that for every bar cluster, there is a trend going from more to less overhead as we move from the left-most bar to the right-most bar. This is due to moving from a simpler form of the code to an optimized form as well as moving from pure MPI code, to code that combines MPI and Gravel, and finally to code that combines MPI and an advanced use of Gravel.

4 Conclusions and Future Work

In this paper we presented Gravel, a communication library designed to inter-operate with MPI to fast-path key data transfers of parallel applications. We have described rendezvous protocols that can be implemented at the application layer when using Gravel as opposed to MPI's exchange protocols that are fixed and do not exploit the structure of each particular application. In addition, we have demonstrated the performance improvements that a parallel application can achieve with Gravel through communication-computation overlapping. Currently, we are working on using Gravel with our compiler transformation tool [9], to enable communication optimization of parallel applications without the need for user intervention.

References

1. uDAPL: User Direct Access Programming Library, `http://www.datcollaborative.org/uDAPL_doc_062102.pdf`
2. GM reference manual, `http://www.myri.com/scs/GM/doc/refman.pdf`
3. Nieplocha, J., Carpenter, B.: ARMCI: A portable remote memory copy library for distributed array libraries and compiler run-time systems. In: RTSPP IPPS/SDP 1999 (1999)
4. Bonachea, D.: GASNet specification. Technical Report CSD-02-1207, University of California, Berkeley (October 2002)
5. Mellanox Technologies Inc.: Mellanox IB-Verbs API (VAPI) (2001)
6. Myricom Inc.: Myrinet EXpress (MX): A High Performance, Low-level, Message-Passing Interface for Myrinet (2003), `http://www.myri.com/scs/`
7. Sur, S., Jin, H.W., Chai, L., Panda, D.K.: RDMA read based rendezvous protocol for MPI over InfiniBand: design alternatives and benefits. In: PPoPP 2006: Proceedings of the eleventh ACM SIGPLAN symposium on Principles and practice of parallel programming, pp. 32–39 (2006)
8. Shipman, G.M., Woodall, T.S., Bosilca, G., Graham, R.L., Maccabe, A.B.: High performance RDMA protocols in HPC. In: Proceedings, 13th European PVM/MPI Users' Group Meeting, Bonn, Germany. LNCS. Springer, Heidelberg (2006)
9. Danalis, A., Pollock, L., Swany, M., Cavazos, J.: Implementing an open64-based tool for improving the performance of mpi programs. In: Open64 Workshop in conjunction with IEEE/ACM International Symposium on Code Generation and Optimization (CGO), Boston, MA (April 2008)

Toward Efficient Support for Multithreaded MPI Communication

Pavan Balaji[1], Darius Buntinas[1], David Goodell[1], William Gropp[2], and Rajeev Thakur[1]

[1] Mathematics and Computer Science Division, Argonne National Laboratory, Argonne, IL 60439, USA
[2] Department of Computer Science, University of Illinois, Urbana, IL, 61801, USA

Abstract. To make the most effective use of parallel machines that are being built out of increasingly large multicore chips, researchers are exploring the use of programming models comprising a mixture of MPI and threads. Such hybrid models require efficient support from an MPI implementation for MPI messages sent from multiple threads simultaneously. In this paper, we explore the issues involved in designing such an implementation. We present four approaches to building a fully thread-safe MPI implementation, with decreasing levels of critical-section granularity (from coarse-grain locks to fine-grain locks to lock-free operations) and correspondingly increasing levels of complexity. We describe how we have structured our implementation to support all four approaches and enable one to be selected at build time. We present performance results with a message-rate benchmark to demonstrate the performance implications of the different approaches.

1 Introduction

Processor development is clearly heading to an era where chips comprising multiple processor cores (tens or even hundreds) are ubiquitous. As a result, parallel systems are increasingly being built with multiple CPU cores on a single node, all sharing memory, and the nodes themselves are connected by some kind of interconnection network. On such systems, it is of course possible to run applications as pure MPI processes, one per core. However, as the total number of processes gets very large, the local problem size per process in some applications may decrease to a level where the program does not scale any further. Also, on some systems, running multiple MPI processes per node may restrict the amount of resources, such as TLB space or memory, available to each process. To alleviate these problems, researchers are evaluating other programming models that involve fewer MPI processes per node and use threads to exploit loop-level and other parallelism. Such a hybrid model can be achieved by either explicitly writing a multithreaded MPI program, using say POSIX threads (Pthreads), or by augmenting an MPI program with OpenMP directives [15]. In either case, MPI functions could be called from multiple threads of a process.

A. Lastovetsky et al. (Eds.): EuroPVM/MPI 2008, LNCS 5205, pp. 120–129, 2008.

MPI implementations have traditionally not provided highly tuned support for multithreaded MPI communication. In fact, many implementations do not even support thread safety. For example, the versions of the following MPI implementations available at the time of this writing do not support thread safety: Microsoft MPI, SiCortex MPI, NEC MPI, IBM MPI for Blue Gene/L, Cray MPI for XT4, and Myricom's MPICH2-MX. Other MPI implementations, such as MPICH2, Open MPI, MVAPICH2, IBM MPI for Blue Gene/P and Power systems, and Intel, HP, SGI, and SUN MPIs do support thread safety. With the increasing use of threads, just supporting thread safety is not sufficient—*efficient* support for multithreaded MPI is needed. Designing an efficient, thread-safe MPI implementation is a non-trivial task. Several issues must be considered, as outlined in [6]. In this paper, we describe our efforts at improving the multithreaded support in our MPI implementation, MPICH2 [10]. We present four approaches to building a fully thread-safe MPI implementation, with decreasing levels of critical-section granularity and correspondingly increasing levels of complexity. We describe how we have structured our implementation to support all four approaches and enable one to be selected at build time. We present performance results with a message-rate benchmark to demonstrate the performance implications of the different approaches.

Related Work. The issue of efficiently supporting multithreaded MPI communication has received only limited attention in the literature. In [6], we described and analyzed what the MPI Standard says about thread safety and what it implies for an implementation. We also presented an efficient multithreaded algorithm for generating new context ids, which is required for creating new communicators. Protopopov and Skjellum discuss a number of issues related to threads and MPI, including a design for a thread-safe version of MPICH-1 [12,13]. Plachetka describes a mechanism for making a thread-unsafe PVM or MPI implementation quasi-thread-safe by adding an interrupt mechanism and two functions to the implementation [11]. García et al. present MiMPI, a thread-safe implementation of MPI [5]. TOMPI [3] and TMPI [14] are thread-based MPI implementations, where each MPI rank is actually a thread. (Our paper focuses on conventional MPI implementations where each MPI rank is a process that itself may have multiple threads, all having the same rank.) USFMPI is a multithreaded implementation of MPI that internally uses a separate thread for communication [2]. A good discussion of the difficulty of programming with threads in general is given in [8].

2 Thread Safety in MPI

For performance reasons, MPI defines four "levels" of thread safety [9] and allows the user to indicate the level desired—the idea being that the implementation need not incur the cost for a higher level of thread safety than the user needs. The four levels of thread safety are as follows:

1. `MPI_THREAD_SINGLE`. Each process has a single thread of execution.
2. `MPI_THREAD_FUNNELED`. A process may be multithreaded, but only the thread that initialized MPI may make MPI calls.

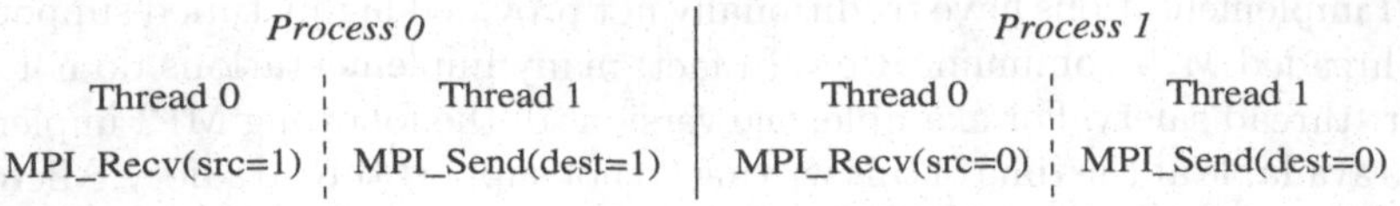

Process 0		*Process 1*	
Thread 0	Thread 1	Thread 0	Thread 1
MPI_Recv(src=1)	MPI_Send(dest=1)	MPI_Recv(src=0)	MPI_Send(dest=0)

Fig. 1. An implementation must ensure that this example never deadlocks for any ordering of thread execution

3. `MPI_THREAD_SERIALIZED`. A process may be multithreaded, but only one thread at a time may make MPI calls.
4. `MPI_THREAD_MULTIPLE`. A process may be multithreaded, and multiple threads may simultaneously call MPI functions (with a few restrictions mentioned below).

MPI provides a function, `MPI_Init_thread`, by which the user can indicate the level of thread support desired, and the implementation will return the level supported. A portable program that does not call `MPI_Init_thread` should assume that only `MPI_THREAD_SINGLE` is supported. This paper focuses on the `MPI_THREAD_MULTIPLE` (fully multithreaded) case.

For `MPI_THREAD_MULTIPLE`, the MPI Standard specifies that when multiple threads make MPI calls concurrently, the outcome will be as if the calls executed sequentially in some (any) order. Also, blocking MPI calls will block only the calling thread and will not prevent other threads from running or executing MPI functions. (The example in Figure 1 must not deadlock for any ordering of thread execution.) MPI also says that it is the user's responsibility to prevent races when threads in the same application post conflicting MPI calls. For example, the user cannot call `MPI_Info_set` and `MPI_Info_free` on the same `info` object concurrently from two threads of the same process; the user must ensure that the `MPI_Info_free` is called only after `MPI_Info_set` returns on the other thread. Similarly, the user must ensure that collective operations on the same communicator, window, or file handle are correctly ordered among threads.

3 Choices of Critical-Section Granularity

To support multithreaded MPI communication, the implementation must protect certain data structures and portions of code from being accessed by multiple threads simultaneously in order to avoid race conditions. A portion of code so protected is called a *critical section* [4]. The granularity (size) of the critical section and the exact mechanism used to implement the critical section can have a significant impact on performance. In general, having smaller critical sections allows more concurrency among threads but incurs the cost of frequently acquiring and releasing the critical section. A critical section can be implemented either by using mutex locks or in a lock-free manner by using assembly-level atomic operations such as compare-and-swap or fetch-and-add [7]. Mutex locks are comparatively expensive, whereas atomic operations are non-portable and can make the code more complex.

We describe four approaches to the selection of critical-section granularity in a thread-safe MPI implementation.

Global. There is a single, global[1] critical section that is held for the duration of most MPI functions, except if the function is going to block on a network operation. In that case, the critical section is released before blocking and then reacquired after the network operation returns. A few MPI functions have no thread-safety implications and hence have no critical section (e.g., `MPI_Wtime`) [1,6]. This is the simplest approach and is used in the past few releases of MPICH2.

Brief Global. There is a single, global critical section, but it is held only when required. This approach permits concurrency between threads making MPI calls, except when common internal data structures are being accessed. However, it is more difficult to implement than Global, because determining where a critical section is needed, and where not, requires care.

Per Object. There are separate critical sections for different objects and classes of objects. For example, there may be a separate critical section for communication to a particular process. This approach permits even more concurrency between threads making MPI calls, particularly if the underlying communication system supports concurrent communication to different processes. Correspondingly, it requires even more care in implementing.

Lock Free. Instead of critical sections, lock-free (or wait-free) synchronization methods [7] are implemented by using atomic operations that exploit processor-specific features. This approach offers the potential for improved performance and greater concurrency. Complexity-wise, it is the hardest of the four.

To manage building and experimenting with these four options in MPICH2, we have developed a set of abstractions built around named critical sections and related concepts. These are implemented as compile-time macros, ensuring that there is no extra overhead. Each section of code that requires atomic access to shared data structures is enclosed in a begin/end of a named critical section. In addition, the particular object (if relevant) is passed to the critical section. For example,

```
MPIU_THREAD_CS_BEGIN(COMM,vc)
... code to access a virtual communication channel vc
MPIU_THREAD_CS_END(COMM,vc)
```

In the Global mode, there is an "`ALLFUNC`" (all functions) critical section, and the other macros, such as the `COMM` one above, are defined to be empty so that there is no extra overhead. In the Brief-Global mode, the `ALLFUNC` critical section is defined to be empty, and others, such as the above `COMM` critical section, are defined to acquire and release a common, global mutex. The `vc` argument to the macro is ignored in that case. In the Per-Object mode, the situation is similar to that in Brief Global, except that instead of using a common, global mutex,

[1] Global here means global to all threads of a process.

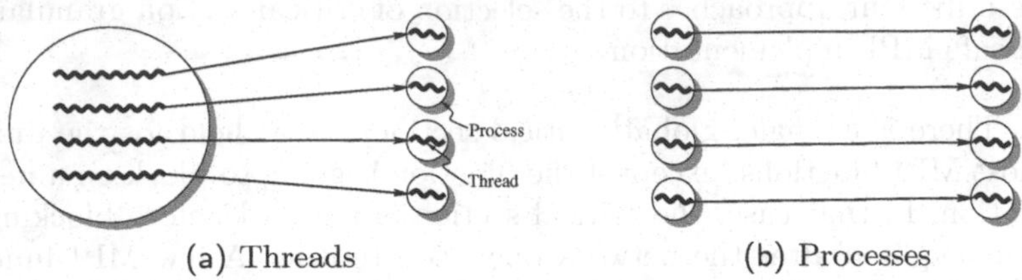

Fig. 2. Illustration of test programs. Multiple threads or processes send messages to a different single-threaded receiving process.

the critical-section macro uses a mutex that is part of the object passed as the second argument of the macro.

For Lock Free, a different code path must be followed. To help with this case, we have implemented a portable library of atomic operations (such as compare-and-swap, test-and-set, fetch-and-add) that are implemented separately for different architectures by using assembly-language instructions. By using these atomic operations, we can replace many of the critical sections with lock-free code. (This part of the code is still under development.)

4 Performance Experiments

To assess the performance of each granularity option, we wrote a test that measures the message rate achieved by n threads of a process sending to n single-threaded receiving processes, as shown in Figure 2(a). The receiving processes prepost 128 receives using `MPI_Irecv`, send an acknowledgment to the sending threads, and then wait for the receives to complete. After receiving the acknowledgment, the threads of the sending process send 128 messages using `MPI_Send`. This process is repeated for 100,000 iterations. The acknowledgment message in each iteration ensures that the receives are posted before the sends arrive, so that there are no unexpected messages. The sending process calls `MPI_Init_thread` with the thread level set to `MPI_THREAD_MULTIPLE` (even for runs with only one thread, in order to show the overhead of providing thread safety). The message rate is calculated as $n/avg\_latency$, where n is the number of sending threads or processes, and $avg\_latency$ is $avg\_looptime/(niters * 128)$, where $avg\_looptime$ is the execution time of the entire iteration loop averaged over the sending threads.

To provide a baseline message rate, we also measured the message rate achieved with separate processes (instead of threads) for sending. For this purpose, we used a modified version of the test that uses multiple single-threaded sending processes, as shown in Figure 2(b). The sending processes simply call `MPI_Init`, which sets the thread level to `MPI_THREAD_SINGLE`.

We performed three sets of experiments to measure the impact of critical-section granularity. The first set does not perform any actual communication (send to `MPI_PROC_NULL`), the second does blocking communication, and the third does nonblocking communication.

The tests were run on a single Linux machine with two 2.6 GHz, quad-core Intel Clovertown chips (total 8 cores), with our development version of MPICH2 in which the ch3:sock (TCP) channel was modified to incorporate the thread-safety approaches described in this paper.

4.1 Performance with MPI_PROC_NULL

This test is intended to measure the threading overhead in the MPICH2 code in the absence of any network communication. For this purpose, we use `MPI_PROC_NULL` as the destination in `MPI_Send` and as a source in `MPI_Irecv`. In MPICH2, an `MPI_Send` to `MPI_PROC_NULL` is handled at a layer above the device-specific code and does not involve manipulation of any shared data structures.

Figure 3 shows the aggregate message rate of the sending threads or processes as a function of the number of threads or processes. In the multiple-process case, the message rate increases with the number of senders because there is no contention for critical sections. In the multithreaded case with Brief Global, the performance is almost identical to multiple processes because Brief Global acquires critical sections only as needed, and in this case no critical section is needed as there is no communication. With the Global mode, however, there is a significant decline in message rate because, in this mode, a critical section is acquired on entry to an MPI function, which serializes the accesses by different threads.

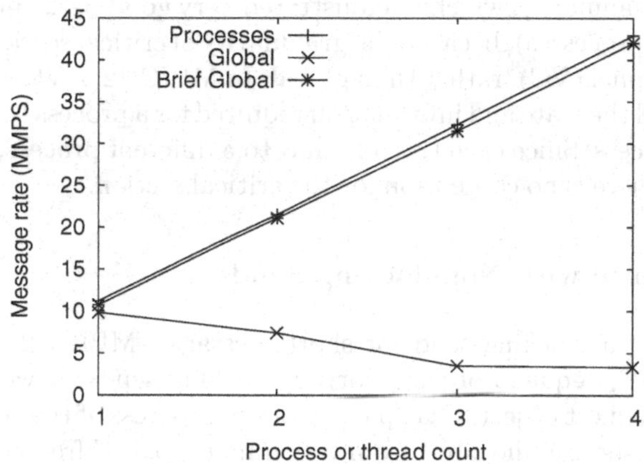

Fig. 3. Message rate (in million messages per sec.) for a multithreaded process sending to `MPI_PROC_NULL` with Global and Brief-Global granularities, compared to that with multiple processes

4.2 Performance with Blocking Sends

This test measures the performance when the communication path is exercised, which requires critical sections to be acquired. The test measures the message rate for zero-byte blocking sends. (Even for zero-byte sends, the implementation must send the message envelope to the destination because the receives could have been posted for a larger size.)

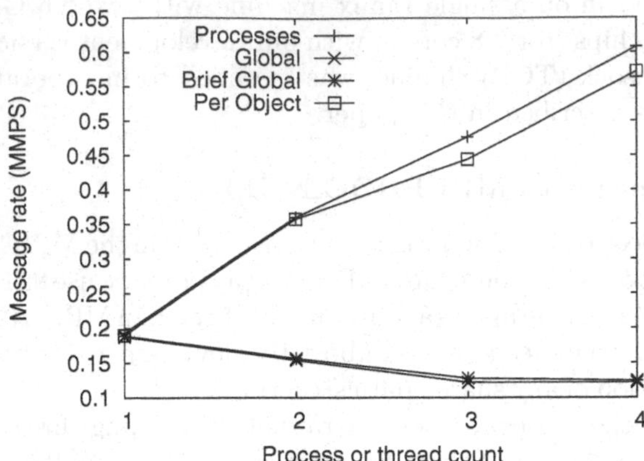

Fig. 4. Message rates with blocking sends for Global, Brief-Global, and Per-Object granularities

Figure 4 shows the results. Notice that because of the cost of communication, the overall message rate is considerably lower than with MPI_PROC_NULL. In this test, even Brief Global performs as poorly as Global, because it acquires a large critical section during communication, which dominates the overall time. We then tried the Per-Object granularity, which demonstrated very good performance (comparable to multiple processes), because the granularity of critical sections in this case is per virtual channel (VC), rather than global. In MPICH2, a VC is a data structure that holds all the state and information required for a process to communicate with another process. Since each thread sends to a different process, they use separate VCs, and there is no contention for the critical section.

4.3 Performance with Nonblocking Sends

When performing a blocking send for short messages, MPICH2 does not need to allocate an `MPI_Request` object. For nonblocking sends, however, MPICH2 must allocate a request object to keep track of the progress of the communication operation. Requests are allocated from a common pool of free requests, which must be protected by a critical section. When a request is completed, it is freed and returned to the common pool. As a result, the common request pool becomes a source of critical-section contention.

Each request object also uses a reference count to determine when the operation is complete and when it is safe to free the object. Since any thread can cause progress on communication, any thread can increment or decrement the reference count. A critical section is therefore needed, which can become another source of contention. All this makes it more difficult to minimize threading overhead in nonblocking sends than blocking sends.

We modified the test program to use nonblocking sends and measured the message rates. Figure 5 shows the results. Notice that the performance of Per-Object

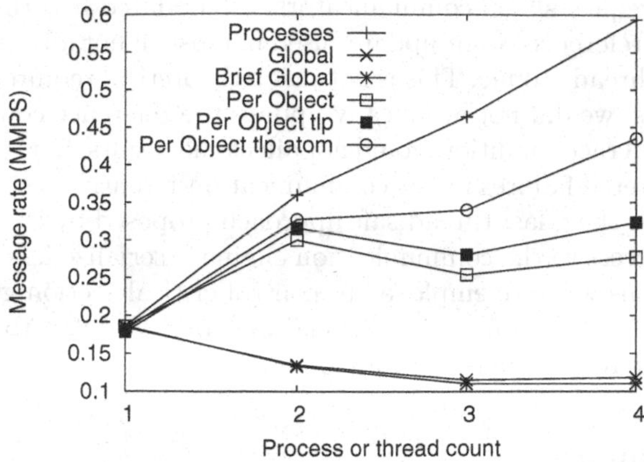

Fig. 5. Message rates with nonblocking sends. "Per Object tlp" is the thread-local request-pool optimization and "Per Object tlp atom" is the update of reference counts using atomic assembly instructions.

granularity is significantly affected by the contention on the request pool, and the message rate does not increase beyond more than two threads.

To reduce the contention on the common request pool, we experimented with providing a local free pool for each thread. These thread-local pools are initially empty. When a thread needs to allocate a request and its local pool is empty, it will get it from the common pool. But when a request is freed, it is returned to the thread's local pool. The next time the thread needs a request, it will allocate it from its local pool and avoid acquiring the critical section for the common request pool. The graph labeled "Per Object tlp" in Figure 5 shows that by adding the thread-local request pool, the message rate improves, but only slightly. The contention for the reference-count updates still hurts.

To alleviate the reference-count contention, we modified MPICH2 to use atomic assembly instructions for updating reference counts (instead of using a mutex). The graph labeled "Per Object tlp atom" in Figure 5 shows that the message rate improves even further with this optimization, and increases with the number of threads. It is still less than in the multiple-process case, but some performance degradation is to be expected with multithreading because critical sections cannot be completely avoided.

5 Conclusions and Future Work

We have studied the problem of improving the multithreaded performance of MPI implementations and presented several approaches to reduce the critical-section granularity, which can impact performance significantly. Such optimizations, however, require careful implementation.

While it is clear that atomic use and update of the communication engine is essential, it is equally important to ensure that all shared data structures, including

MPI datatypes, requests, and communicators, are updated in a thread-safe way. For example, the reference-count updates used in most (if not all) MPI implementations must be thread atomic. This is not just a theoretical requirement: In some early experiments, we did not atomically update the reference counts, assuming that the very small race condition would not affect the results; but, by doing so, we regularly encountered failures in our communication-intensive tests. This experience suggests that the quasi-thread-safe approach proposed by Plachetka [11], in which only the access to the communication engine is serialized, is not sufficient.

The abstractions we have employed to control critical-section granularity are similar to what is required for transactional memory. We plan to use these abstractions to explore the use of transactional memory.

Acknowledgments

This work was supported by the Mathematical, Information, and Computational Sciences Division subprogram of the Office of Advanced Scientific Computing Research, Office of Science, U.S. Department of Energy, under Contract DE-AC02-06CH11357. We thank Sameer Kumar and others in the MPI group at IBM Research and IBM Rochester for discussions about efficient support for thread safety in MPI.

References

1. Analysis of thread safety needs of MPI routines, `http://www.mcs.anl.gov/research/projects/mpich2/design/threadlist.htm`
2. Caglar, S.G., Benson, G.D., Huang, Q., Chu, C.-W.: USFMPI: A multi-threaded implementation of MPI for Linux clusters. In: Proceedings of the IASTED Conference on Parallel and Distributed Computing and Systems (2003)
3. Demaine, E.D.: A threads-only MPI implementation for the development of parallel programs. In: Proceedings of the 11th International Symposium on High Performance Computing Systems, July 1997, pp. 153–163 (1997)
4. Dijkstra, E.W.: Solution of a problem in concurrent programming control. Communications of the ACM 8(9), 569 (1965)
5. García, F., Calderón, A., Carretero, J.: MiMPI: A multithread-safe implementation of MPI. In: Margalef, T., Dongarra, J., Luque, E. (eds.) PVM/MPI 1999. LNCS, vol. 1697, pp. 207–214. Springer, Heidelberg (1999)
6. Gropp, W., Thakur, R.: Thread safety in an MPI implementation: Requirements and analysis. Parallel Computing 33(9), 595–604 (2007)
7. Herlihy, M.: Wait-free synchronization. ACM Transactions on Programming Languages and Systems 11(1), 124–149 (1991)
8. Lee, E.A.: The problem with threads. Computer 39(5), 33–42 (2006)
9. Message Passing Interface Forum. MPI-2: Extensions to the Message-Passing Interface, July (1997), `http://www.mpi-forum.org/docs/docs.html`
10. MPICH2, `http://www.mcs.anl.gov/mpi/mpich2`
11. Plachetka, T. (Quasi-) thread-safe PVM and (quasi-) thread-safe MPI without active polling. In: Kranzlmüller, D., Kacsuk, P., Dongarra, J., Volkert, J. (eds.) PVM/MPI 2002. LNCS, vol. 2474, pp. 296–305. Springer, Heidelberg (2002)

12. Protopopov, B.V., Skjellum, A.: A multithreaded message passing interface (MPI) architecture: Performance and program issues. Journal of Parallel and Distributed Computing 61(4), 449–466 (2001)
13. Skjellum, A., Protopopov, B., Hebert, S.: A thread taxonomy for MPI. In: Proceedings of the 2nd MPI Developers Conference, pp. 50–57 (June 1996)
14. Tang, H., Yang, T.: Optimizing threaded MPI execution on SMP clusters. In: Proceedings of the 15th ACM International Conference on Supercomputing, pp.381–392 (June 2001)
15. Wu, M.-S., Aluru, S., Kendall, R.A.: Mixed mode matrix multiplication. In: Proceedings of the IEEE International Conference on Cluster Computing (Cluster 2002), September 2002, pp. 195–203 (2002)

MPI Support for Multi-core Architectures: Optimized Shared Memory Collectives*

Richard L. Graham and Galen Shipman

Oak Ridge National Laboratory
Oak Ridge, TN USA
rlgraham@ornl.gov, gshipman@ornl.gov

Abstract. With local core counts on the rise, taking advantage of shared-memory to optimize collective operations can improve performance. We study several on-host shared memory optimized algorithms for MPI_Bcast, MPI_Reduce, and MPI_Allreduce, using tree-based, and reduce-scatter algorithms. For small data operations with relatively large synchronization costs fan-in/fan-out algorithms generally perform best. For large messages data manipulation constitute the largest cost and reduce-scatter algorithms are best for reductions. These optimization improve performance by up to a factor of three. Memory and cache sharing effect require deliberate process layout and careful radix selection for tree-based methods.

Keywords: Collectives, Shared-Memory, MPI_Bcast, MPI_Reduce, MPI_Allreduce.

1 Introduction

As HPC systems continue to grow rapidly the scalability of many scientific applications are limited by the scalability of collective communication. These systems are growing in both node counts and core counts per node. These multi-core nodes provide a way to increase the scalability of collective communication for applications which use more than a single MPI task per node. Implementing these algorithms in terms of on-host and off-host phases reduces network traffic improves their overall scalability. This paper will study in detail the options for implementing the on-host, shared-memory collective algorithms and compare these with Open MPI's point-to-point implementations using shared-memory merely as a transport layer. Memory traffic, cache conflicts and synchronization are barriers to the scalability and performance of shared memory collectives. These algorithms are aimed at reducing memory traffic by limiting the number of writers to a given memory segment and balancing synchronization and memory access costs. In addition to reducing memory traffic on socket and balancing synchronization we also aim take advantage of shared caches and reduce inter-socket memory traffic.

* Research sponsored by the Mathematical, Information, and Computational Sciences Division, Office of Advanced Scientific Computing Research, U.S. Department of Energy, under Contract No. DE-AC05-00OR22725 with UT-Battelle, LLC.

A. Lastovetsky et al. (Eds.): EuroPVM/MPI 2008, LNCS 5205, pp. 130–140, 2008.

This paper describes the shared memory implementation of MPI_Reduce, MPI_Allreduce, and MPI_Bcast, and provides the results of benchmark studies of these collective algorithms. These are used by a large number of scientific simulation codes with reductions frequently being the collective communications pattern that hinders scalability the most. The range of reduction sizes used by scientific codes varies from single word sizes used to determine the convergence of iterative algorithms to many megabytes used to aggregate simulation results. Broadcast collective communication is often used to initialize application data structures.

The remainder of this paper is organized as follows; Section 2 provides an overview of previous work on shared memory optimizations of collective operations. Section 3 describes the shared-memory MPI_Reduce, MPI_Allreduce, and MPI_Bcast algorithms implemented within the framework of Open MPI. Results of numerical experiments are discussed in Section 5. Conclusions and future work are then discussed in Section 6.

2 Background

The recent interest in shared memory communications optimizations has come primarily from the desire to take advantage of the performance gain opportunities this gives in the context of hierarchical collectives on systems with a non-uniform memory hierarchy, such as clusters of Shared Memory nodes. As such, the main goal of these studies has been to show collective communications performance improvements of these hierarchical collectives, over standard implementations. Typical collective cost models [1, 2] include a network communications term and, for operations such as reductions, a local processing term. On shared memory systems where one MPI process can directly access another's memory or a copy of this memory reduction operations can avoid the data transfer step by directly operating on another processes data. Operations such as broadcast can reduce the number of memory transfers with multiple consumers accessing a shared buffer. This paper studies the benefits of such optimizations over traditional point-to-point based collective communications with shared-memory as a mere transport mechanism.

Several MPI implementations have provided support for shared memory optimized collectives. These include but are not limited to LA-MPI [3], Sun MPI [4], and NEC's MPI [5].

Mamidala et al. [6] have studied the performance of shared-memory MPI_Bcast, MPI_Allgather, MPI_Allreduce, and MPI_Alltoall on a four core Intel Clovertown system. Their focus was on the interaction with the underlying hardware rather than the characteristics of the algorithms. Mamidala et al. [7], also developed a shared memory MPI_Allgather algorithm for use in a hierarchical implementation of this algorithm.

Tipparaju et al. [8] studied the effect of taking the shared memory hierarchies for several collective algorithms. They focused on the hierarchical collectives, and the performance gains from exploiting the memory hierarchies.

Wu et al. [9] proposed a general approach for optimizing shared memory collective operations using several communications primitives. They implemented MPI_Bcast and MPI_Scatter using these primitives showing performance improvements for these collective algorithms over point-to-point based mechanisms.

Unpublished work on shared-memory collective operations in the context of LA-MPI [10] on the 128 processor SGI Origin2000 machines showed that as the process count increases, read and write contention for the shared memory segments can greatly affect the performance of these operations. As a result, in this paper we describe collective algorithms designed to take advantage of the ability to directly access another processes shared memory segment with carefully controlled memory access. We study these algorithms to examine the characteristics of MPI collective operations on the emerging multi-core systems.

3 Algorithms

The ability of multiple processes to directly access common memory brings with it several opportunities for collective communications optimization with memory operations being the means of inter-process communications. However, factors such as process affinity and cache and memory access have a substantial impact on the performance of these algorithms. Cache thrashing causes severe performance penalties, shared caches reduce the access times to volatile memory, and memory bus contention reduces the memory bandwidth available to each MPI process. We proceed to develop algorithms that selectively eliminate extra memory traffic by taking into account cache and memory characteristics.

The approach we use in all these algorithms is to assign a fixed size segment of shared memory to each MPI communicator for use by all the MPI collective algorithms. This segment is contiguous in virtual memory, is memory-mapped at communicator creation, and is freed at communicator destruction. This provides a means of allocating the resources only when they are needed and for adjusting to the changing needs of an application. This shared memory has a control region for managing the working memory and a scratch (or working) memory region used by the collective communication routines. The scratch region is divided into two banks of memory with each bank having several segments of memory. The number of banks and segments is determined at run-time with the default values being two and eight. A bank is either available for use, or not, and once available, the buffers are used in-order without any additional availability checks. The control region is used to manage the availability of these memory banks and a non-blocking barrier structure is associated with each memory bank. When the last segment in the bank is used, a non-blocking barrier is initiated, and if not complete when an attempt to re-use the first segment is made then the process blocks completing this barrier. Multiple banks are used to allow the non-blocking barrier to complete while another block is in use reducing the synchronization costs.

The memory segments used by the collective algorithms are also divided into a control region and data regions. The control region consists of a flag the algorithms use for signaling other processes and the data region is where each process puts its data. There is one control region and one data region per process in the communicator. Each of these regions is page aligned with the size of the control region being fixed and the size of the data region being set at run-time. By default a single page is allocated for each process's data region. First touch is used to ensure memory locality (if process affinity has been enabled) and a given process only ever writes to it's own control and data region within the memory segment but may read other processes' control and data regions.

The memory cost is constant on a per-process basis with the overall segment sizes scaling linearly with the number of processes in the shared memory communicator. Memory costs also scale linearly with the number of pages used in the data segment. The overall cost of the memory bank control region per process and per bank is the cost for the data structure (a 64 bit field) for the non-blocking barrier. For two memory banks this amounts to one page per process as these data structures share pages on a per process basis.

The shared memory scratch space provides all processes in the communicator access to common data. Since the size of these data segments is fixed and the number of data segments is limited all algorithms process the user data a segment at a time. The reduction routines described in this paper can only be applied to data types that fit within the per-process data segment.

Communication patterns are pre-computed and cached at communicator creation time. The nodes in partial leaf levels of a tree are assigned a parent by distributing these uniformly across the parent layer. Process affinity is used to control process locality taking into account memory and cache hierarchies.

3.1 MPI_Allreduce

Three different algorithms are implemented for MPI_Allreduce, recursive doubling, reduce-scatter followed by an all-gather and fan-in/fan-out.

The recursive-doubling algorithm is useful for large data reductions in which data manipulation tends to dominate reduction time. Each process uses it's data-control flag to signal when it's data is ready for use. To allow both processes involved in a pair to process their data simultaneously the data segment is divided into to two sections. One section is the read section for both processes and other is the write section for the process owning that memory. The roles of these regions are reversed at each step in the reduction so that data can be used in place. For non-power of two communicators with M processes, if N is the largest power of two less than M, rank $N+K$ is paired with rank K, where $K = 0, 1, .., M - N$. Before the recursive doubling algorithm is used each pair reduces it's data, with each process reducing half the data, and the lower rank copying the higher rank's reduced data. After the recursive doubling phase ranks N and higher copy the results from their partner directly into the user's buffer. A single segment is used for all sections of a large single reduction. The only parameter that can be varied is the size of the data segment.

The reduce-scatter algorithm followed by an all-gather is efficient for large data reductions. It typically performs better than the recursive-doubling algorithm. At each step of the reduce scatter each process in the pair reduces it's portion of the data into it's temporary buffer and then reads the data directly from its partner's shared-memory buffer. The all-gather step is a simple data read from the scratch space of the other processes. Data readiness is signaled using the data-control flags and for non-powers of two are handled in a manner similar to that used in the recursive doubling algorithm. A single segment is used for all sections of a single reduction. The only parameter that can be varied is the size of the data segment.

The fan-in/fan-out algorithm is aimed at minimizing synchronization between processes participating in the reduction operation and is suited well for small data reductions where synchronization costs are prominent. We implement the fan-in and fan-out with different radixes. The fan-in phase involves synchronization of n+1 process in a tree of radix-n, as a single process, the parent process, serially reduces the data from the other n processes onto it's own data. In the fan-out phase the parent process signals M processes in a tree of radix-m that the data is ready to be read and these m processes can attempt to read this data simultaneously. To keep the synchronization cost down we use a new shared memory segment for each section of the user data to amortize the cost of "freeing" these buffers. The size of the data segments can also be varied.

3.2 MPI_Reduce

Two different algorithms are implemented for MPI_Reduce a fan-in algorithm and a reduce-scatter followed by a gather to the root. These are implemented in a manner similar to that of the MPI_Allreduce algorithms. The fan-in algorithm is just the first half of the MPI_Allreduce. For a given communicator we cache the fan-out tree for rank zero as the root and translate the nodes of the tree by n (with wrap around) for root n. The reduce-scatter algorithm differs from the MPI_Allreduce in that the results of the reduce-scatter are gathered back only to the root of the operation.

3.3 MPI_Bcast

The MPI_Bcast is implemented as a fan-out tree of radix-m which can be specified at run-time. For a given communicator we cache the fan-out tree for rank zero as the root and translate the nodes of the tree by n (with wrap around) for root n.

4 Experimental Setup

The shared-memory collective routines are implemented within the Open MPI code base [11] as a separate collective component. The working code is revision 18489 of the trunk. We take advantage of Open MPI's process affinity to control process locality and control memory locality using a first-touch approach.

The performance measurements were all taken on a 16 processor quad-socket, quad-core, 2 Gigahertz (GHz) Barcelona Opteron system with 512 kilobytes

(KB) secondary cache and 2 megabytes (MB) shared tertiary cache per socket. The system is running Linux version 2.6.18-53.1.13ccs.el5.

The performance measurements were taken using simple benchmark codes with an outer loop wrapped around an inner loop of calls to the collective routine. A barrier is called right before and right after the inner loop with the time being measured between the ends of both barrier calls. For rooted collectives an inner-most loop is added which rotates the root of the operation with all processes being the root an equal number of times in a particular measurement. We use integer data in all the measurements and we use the MPI_SUM reduction operations in the reduction operations. We report the latency of the collective operation as the average time per call.

5 Results and Discussion

Experiments were performed to study the shared-memory optimized MPI_Bcast, MPI_Reduce, and MPI_Allreduce. The following sections summarize the results of these experiments. To help keep the discussion brief we will discuss the results in two classes; short data in the range of eight to 1024 bytes; and large data in the range of 1 to 16 MB. These ranges of data sizes were selected as they are the most relevant to applications with which we are familiar. We also restrict the discussion to sixteen process MPI jobs as our experimental hardware was limited to 16 cores per node.

5.1 Memory Hierarchies

The memory layout of the quad-core Barcelona shared-memory nodes offers several opportunities. The tertiary cache shared between cores on a single socket offer an improved multi-process memory access for volatile data. When going off socket, multi-process volatile data access must go to main memory, with the cores on a given socket all sharing that socket's bandwidth. To get an idea on the order of magnitude of these effects we measured the latency of an eight byte and sixteen MB MPI_Allreduce operation laying out the MPI processes in several different configurations. For two process reductions we found that sharing the socket improved the small data operation by about 15% and the large data operation by 10%. For an eight process reduction, using all the cores on two sockets improved the small data operation by about 10% over spreading the eight processes across all four sockets. However the increased memory contention with only two sockets in the large data case reduces it's performance by about 28% relative to the four socket case. While these affects are not as large as those going between hosts, they are significant, and need to be taken into account. In the current set of experiments we used this information as a guide to setting up the communication patterns by using the Linux process affinity capability to lay out the MPI processes in a manner that results in the desired memory traffic patterns. Different layouts are used for different collective routines and different data sizes. Planned future work will explicitly include these hierarchies in algorithm implementation.

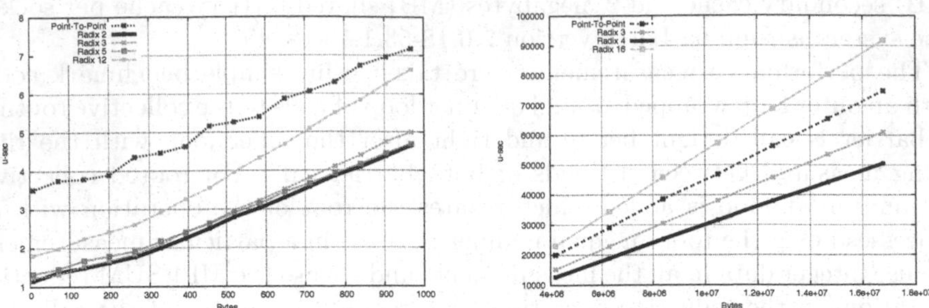

Fig. 1. MPI_Bcast as a function of fan-in radix for sixteen process communicators

5.2 MPI_Bcast

Experiments were carried out on small and large data while varying the radix of the fan-out tree. We choose radix values of 2–8,12,16 and a fragment size of $32KB$. Figure 1 shows the best set of results from these experiments carried out with 16 processes and compares this data with the point-to-point based MPI_bcast algorithm on the same machine configuration.

For small data the shared memory broadcast using a radix-2 broadcast tree gives the best performance over the range of 8 bytes to 1KB, but the data values using radix-3 and 5 have similar performance. The radix-4 tree also gives similar performance but the data is omitted so as not to clutter the figure. The worst performance is obtained using a radix-12 tree. In all cases the shared memory broadcast performs better than the point-to-point algorithm using shared-memory communications with the radix-2 shared memory optimized broadcast algorithm being about two microseconds more efficient over the range of small data; a factor of three faster at eight bytes. Synchronization for the shared memory optimized routine is far simpler and amounts to reading a flag at a pre-computed address whereas the synchronization in the point-to-point based method occurs via the MPI general-purpose send/receive matching logic.

The large data shared memory broadcast using a radix-4 broadcast tree performs noticeably better than with other reduction trees with a tree of radix-3 being about 25% slower at a message size of four MB. It should come as no surprise that the worst performance is obtained when using a tree of radix-16, with all processes trying to read one buffer thereby creating a large amount of memory contention. The radix-16 algorithm performs even worse than the point-to-point based method. The latter performs about 60% worse than radix-3 broadcast algorithm which is not surprising, given the reduced number of memory copies in the shared-memory optimized algorithm.

5.3 MPI_Reduce

Experiments were carried out using both small and large data varying the radix of the fan-in tree. We chose radix values of 2–8,12,16. As expected, for small

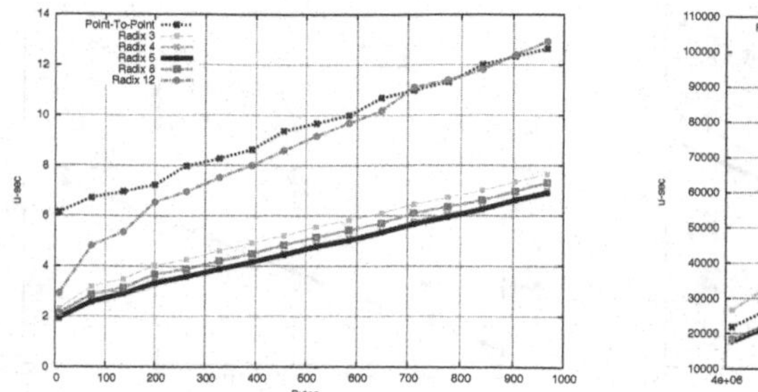

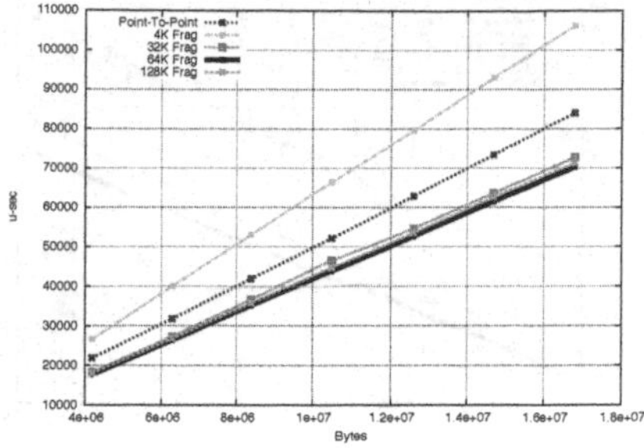

Fig. 2. Left: MPI_Reduce as a function of Fragment size for a sixteen process communicator

data fan-in reduction tree algorithms are the algorithm of choice. For large data a reduce-scatter followed by a gather to the root is the most suitable algorithm of those used. We limit our discussion to these combinations of data-size and algorithm and compare them with point-to-point based implementations of the MPI_Reduce algorithm. Figure 2 shows a select set of results from these experiments carried out at a sixteen process count.

For small data the best results are obtained with a radix 5 fan-in tree which gives virtually identical performance to that of a radix-4 tree. The results using a tree of radix-8 and radix-3 are slightly worse. Using a tree of radix-12 results in much worse performance and at the upper end of the size range this performance is even worse than that of the point-to-point based method. The point-to-point reduction routine performs quite a bit worse than the radix-5 fan-in tree based shared memory reduction. At eight bytes the shared memory version is about a factor of three faster than the point-to-point based method and at 968 bytes it is about a factor of 1.8 times faster with the slope of the point-to-point based method being steeper than that of the shared-memory based method.

For large data, communications time is reported as a function of shared memory segment data size. The best performance in this case is obtained with segments of size 64KB, with comparable performance using fragments of size 128KB and 32KB. The best shared memory results are about 23% better than the point-to-point based method for a 4 MB transfer size. However, if a segment size of 4KB is used, the shared memory implementation is slower than the point-to-point based method. At this message size the dominant factor in the overall time are the memory operations and the shared memory algorithm with 4KB segments can not use the memory subsystem as well as the 32KB sized point-to-point shared memory segments used by the point-to-point communications layer.

5.4 MPI_Allreduce

The experiments performed for the MPI_Allreduce function are very similar to those of the MPI_Reduce. These results are reported in Figure 3. In addition, a recursive doubling algorithm was implemented, but this is not competitive with

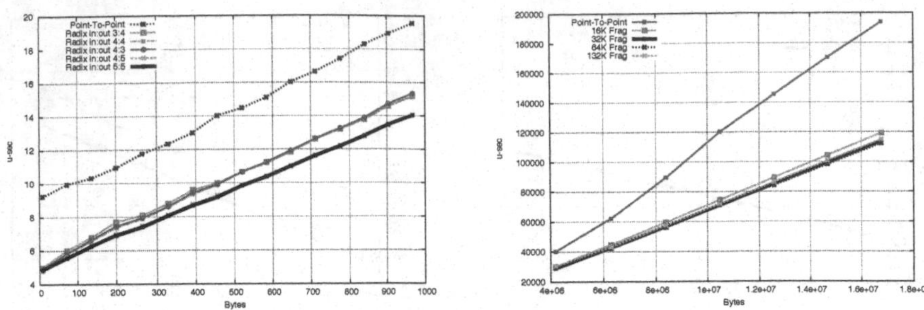

Fig. 3. MPI_Allreduce for a sixteen process communicator. Left: short message fanin/-fanout. Right: Reduce-scatter, followed by Allgather.

the reduce-scatter/allgather algorithm used for large data and therefore we do not report these results.

For small data, the results from a small set of algorithms using fan-in/fan-out tree algorithms are reported. Overall, a fan-in and fan-out radix of five and five, respectively, give the best performance, but values of four and five, respectively, give virtually identical results. The all-reduce algorithms with fan-in/fan-out radix pairs of 4:3, 4:4, and 4:3 have slightly worse performance at this range of sizes. At eight bytes the performance of all these methods is virtually identical as they are dominated by synchronization but at 968 bytes where memory operations take a larger portion of time the first set of methods are about 8% more efficient than the second group of shared memory optimized methods. Similar to the MPI_Reduce case, the shared memory algorithms perform better than the point-to-point based algorithms. The method using a fan-in radix of five and a fan-out radix of five performs about 89% better than the point-to-point based algorithm at eight bytes and about 29% better at 968 bytes. In addition to improved latency, the slope of the point-to-point based method is higher than that of the shared-memory optimized method.

For large data, the best performance is obtained when using memory segments of 32KB with the reduce-scatter/allgather algorithm which is only marginally better performance than that obtained using both 64KB and 128KB segments and slightly worse at 16KB segment size. With the reduced memory traffic of the shared-memory optimized algorithms relative to the point-to-point based algorithm it is not surprising that these perform better than the point-to-point based method. At 4MB, the 32KB reduce-scatter/allgather method is about 38% more efficient than the point-to-point based method and at 16MB it is about 71% more efficient.

6 Conclusions

Taking advantage of memory hierarchies in the implementation of hierarchical collective operations is a good way to improve overall collectives performance.

In this paper we have examined the benefits of creating shared memory optimized collectives for on-node operations. We have presented optimizations of the MP_Bcast, MPI_Reduce, and MPI_Allreduce routines aimed at taking advantage of shared memory architectures. We use instance specific control regions to set flags indicating when data is ready to be used, bypass the point-to-point matching logic, read other processes data directly out of their shared memory buffers reducing memory accesses and write only to one's "own" shared memory region. These algorithms also take into account cache and memory layout in assigning process affinity.

Shared memory optimization improve the performance of the local collective operations over point-to-point based methods. Performance improvement of 3 fold have been demonstrated for small data operations such as eight byte broadcasts and reductions across 16 processes. Improvements on the order of tens of percent were measured across a wider size of messages. In terms of absolute time per collective call the difference in performance between the shared-memory optimized and the point-to-point based algorithms increases with data size. These performance improvements require careful attention to memory layout as care must be taken not to overwhelm the memory subsystem. By varying the arity of tree based algorithms and varying the size of shared memory data regions performance can be improved dramatically.

This paper has described approaches to improve the performance of collective operations on-node. Future work will include exploring the use of these shared memory based collectives in conjunction with the hierarchical collectives framework within Open MPI or weather a tightly coupled approach in which the shared memory based collectives are integrated with point-to-point approaches for off-node communication are in order.

References

[1] Thakur, R., Gropp, W.: Improving the performance of collective operations in mpich. In: Lecture Notes In Computer Science, pp. 257–267 (2006)

[2] Rabenseifner, R.: Optimization of collective reduction operations. In: Lecture Notes In Computer Science, pp. 1–9 (2004)

[3] LA-MPI, `http://public.lanl.gov/lampi`

[4] Sistare, S., van de Vaart, R., Loh, E.: Optimization of mpi collectives on clusters of large-scale smp's. In: Proceedings of SC 1999: High Performance Networking and Computing (1999)

[5] NEC web page, `http://www.nec.de`

[6] Mamidala, A.R., et al.: Mpi collectives on modern multicore clusters: Performance optimizations and communication characteristics. In: CCGRID 2008 (accepted for publication, 2008)

[7] Mamidala, A.R., Vishnu, A., Panda, D.K.: Efficient shared memory and rdma based design for mpi_allgather over infiniband. In: Lecture Notes In Computer Science

[8] Tipparaju, V., Nieplocha, J., Panda, D.: Fast collective operations using shared and remote memory access protocols on clusters. In: Proceedings of the International Parallel and Distributed Processing Symposium (2003)

[9] Wu, M.S., Kendall, R.A., Aluru, S.: Exploring collective communications on a cluster of smps. In: Proceedings, HPCAsia2004, pp. 114–117 (2004)
[10] Graham, R.L., Choi, S.E., Daniel, D.J., Desai, N.N., Minnich, R.G., Rasmussen, C.E., Risinger, L.D., Sukalksi, M.W.: A network-failure-tolerant message-passing system for terascale clusters. International Journal of Parallel Programming 31(4) (2003)
[11] Open MPI, `http://www.open-mpi.org`

Constructing MPI Input-output Datatypes for Efficient Transpacking

Faisal Ghias Mir and Jesper Larsson Träff

NEC Laboratories Europe, NEC Europe Ltd.
Rathausallee 10, D-53757 Sankt Augustin, Germany
{mir,traff}@it.neclab.eu

Abstract. Communication and file I/O buffers in MPI can contain contiguous as well as non-contiguous, structured data. To describe non-consecutive data layouts compactly, MPI provides a powerful concept of *derived* or user-defined *datatypes*. Especially for MPI-IO, where data are transferred between file and memory buffers, the need for copying between differently typed MPI buffers arise. A straightforward solution to this *typed copy problem* consists in packing and unpacking the differently structured data via an intermediate buffer. For a maximally efficient MPI(-IO) implementation, means of copying directly (without intermediate buffers) between differently typed MPI buffers are needed.

We present a new approach to the *typed copy problem*. For any two given MPI datatypes describing the layout of input and output buffer, respectively, we show how to efficiently construct an *input-output* type that subsumes both. This type is used to copy directly from input to output buffer by means of a special *transpack* function. By completely eliminating the need for intermediate buffering, the typed copy performance can in theory be improved by up to a *factor of two*, with only a modest overhead for constructing the input-output type. An experimental evaluation shows that even more significant improvements can be achieved in practice.

1 Introduction

The MPI derived datatype mechanism [4, Chapter 3] is an extremely powerful mechanism for concise description of structured, non-contiguous data layouts, which allows for the use of structured, non-contiguous data in all communication and file I/O operations of MPI. Especially for parallel I/O, the need for copying between two structured data buffers described by different derived datatypes arises. MPI itself provides no functionality for this, and as far as we are aware, the problem of how to do this efficiently has never been seriously addressed in the MPI community. We do so in this paper.

We address the *typed copy problem* as follows. Given an input type T^i describing the data in the source buffer, and an output type T_o describing the data in the destination buffer, we first construct an *input-output type* T_o^i giving a direct map from source to target type. Known techniques for efficient packing

A. Lastovetsky et al. (Eds.): EuroPVM/MPI 2008, LNCS 5205, pp. 141–150, 2008.

and unpacking of structured data, like *flattening on the fly* [5] and related methods [1,2,3,6], can then readily be utilized to implement a *transpack* function that copies directly from input to output buffer.

Compared to the straightforward solution to the typed copy problem, in which input data are first packed into a contiguous, intermediate buffer and then unpacked into the output buffer, the use of the input-output type in combination with an efficient transpack function can in theory (leaving cache and memory effects aside) improve the performance by up to a factor of *two* by completely eliminating the intermediate copy.

In this paper we concentrate on the input-output type construction. To show the potential performance benefits of transpacking, we compare the achieved transpacking time to the time of a straightforward pack-unpack solution for a set of increasingly complex examples. Benchmarking is done on an NEC SX-8 vector processor. Even for complex examples where the construction gives rise to relatively complex input-output types, we demonstrate (more than) the expected factor of two performance improvements.

2 Problem and Properties

A derived datatype T in MPI is a compact description of a *type map* which is a sequence of basic types (like integers, bytes, floating point numbers) and relative offsets. The type map is an explicit (but inefficient, both in terms of space and handling, see eg. [7]) description of a data layout in memory. An MPI derived datatype is built from basic types (MPI_INT, MPI_FLOAT, ...) by invoking constructor functions (MPI_Type_contiguous, MPI_Type_vector, MPI_Type_indexed, MPI_Type_create_struct, ...) and can be represented as a rooted, labeled DAG (Directed Acyclic Graph). Each node of the DAG represents a datatype with child nodes as subtypes. Edges in the DAG are labeled with the number of times a subtype is to be repeated in the type map. Children are ordered from left to right, such that the type map can be constructed by ordered depth-first search traversal of the DAG. DAG nodes also have an associated offset and extent.

An *input-output type* is a DAG representation of an input-output type map: a sequence of basic types and corresponding input and output offsets. The input-output type construction problem is to construct an input-output type T^i_o from input type T^i and output type T_o (with the same *type signature*, ie. describing the same sequence of basic types), such that for each basic type in the input-output type map the input offset comes from T^i and the output offset from T_o.

A trivial solution to the problem is to construct the input and output type maps explicitly from T^i and T_o, following the rules for the MPI type constructors (amounting to DFS traversal of the ordered, labeled DAG), and use the two lists of basic types and offsets as the input-output type. Obviously, this is not the desired solution to the problem. As mentioned, the explicit list representation of the type map is not efficient for copying, and because of the repetition counts in the DAG, the type maps can be much larger than T^i and T_o. A significant improvement to the trivial solution would be to reconstruct a more compact input-output type DAG from the two type maps. This would still be more expensive than constructing the input-output type directly from T^i and T_o. Furthermore the complexity of *type reconstruction*, ie. finding the most compact DAG representation of a given type map, is not known to the authors (and constitutes an interesting, open problem).

We want to devise a function IOTYPE that takes input and output types T^i and T_o (with the same type signature) and returns an input-output type T^i_o. The input-output type should be a DAG subsuming both T^i and T_o, while retaining as much of the common structure as possible. The running time should be proportional to the size of the resulting input-output type DAG. This restricts algorithms from traversing type DAGs unboundedly.

In the following we ignore the handling of offsets and extents in both input, output and resulting input-output types. Although not trivial, the details can be filled in by the interested reader. We say that two types T^i and T_o are *structurally equivalent*, denoted $T^i \equiv T_o$, if they have the same DAG (ignoring offsets and extents).

The IOTYPE function should have the following properties.

- IOTYPE$(T^i, C_o) \equiv T^i$ where C_o is a type representing a contiguous segment of basic types in memory.
- IOTYPE$(C^i, T_o) \equiv T_o$ where C^i is a type representing a contiguous segment of basic types in memory.
- IOTYPE$(T^i, T_o) \equiv T^i \equiv T_o$ if $T^i \equiv T_o$.

This should also hold for proper subtypes, ie. subtypes $\hat{T}^i$ and $\hat{T}_o$ with the same signature describing parts of input and output type maps, respectively, with other, proper subtypes describing the parts of the type maps before and after $\hat{T}^i$ and $\hat{T}_o$.

We represent (input-output) datatype DAGs by linked structures with only two pointers per node. A type node has a child pointer to its subtype, and a repetition count for the number of times the subtype is repeated in the type map. If the subtype is a basic type, the child pointer is null, and only the repetition count is significant. For types that are part of a structured or indexed type, a sibling pointer points to the next type in the structure. The MPI type map is determined by DFS traversal of this DAG, with children traversed before siblings. Examples of MPI vector and indexed types are shown in Figure 1.

3 Constructing Input-Output Datatypes

We now give the algorithm for construction of input-output type T^i_o from input and output types T^i and T_o. The idea is to explore simultaneously paths from leaves of each of the two DAGs representing the same position in the type map toward the roots. Traversing these paths upwards toward the roots, a corresponding part of the input-output type DAG can be created that retain common parts of the two paths. We first handle the special case where the type DAGs are paths, and then briefly describe how to extend this to all MPI datatypes.

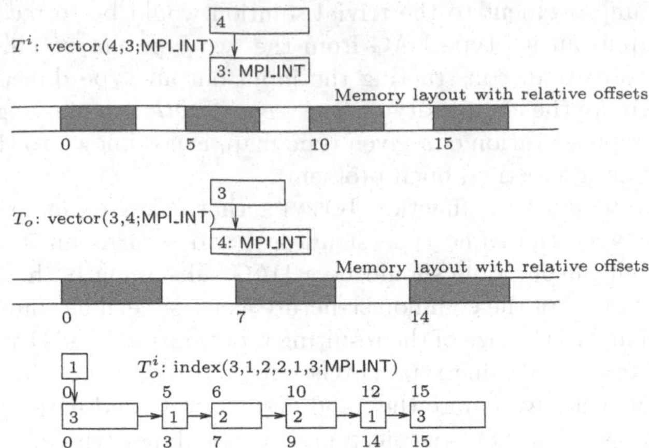

Fig. 1. Memory layout (with relative offsets) of an *input vector* T^i of 4 blocks of 3 MPI_INT against an *output vector* T_o of 3 blocks of 4 MPI_INT. The DAG representing the resulting *input-output indexed type* with blocks of lengths 3, 1, 2, 2, 1, 3 is also shown.

3.1 Types That Are Paths (MPI Vector and Contiguous Types)

We first show how to handle input and output type DAGs consisting of a single path from root to basic type. This may seem trivial, but covers already arbitrarily nested MPI vector and contiguous types. Figure 1 shows that the input-output type for two single path types (e.g. MPI vectors) with blocks of different sizes is not necessarily itself a single-path type, but can be a structured type.

We first define a function StackBuild(e, T, S) that implicitly constructs the type map of type T and locates the leaf corresponding to the eth basic type in the type map. This can be done simply by a DFS traversal of T. A stack S of the traversed type nodes is built. Each stack frame contains the repetition `count` for its child type, the total number of basic `elements` of the type node, and a number of `consumed` basic elements that will be used later in the input-output type construction (and is initialized to zero). For single-path types it holds for stack level $t, t < \text{top}(S) - 1$ that $S[t].\texttt{elements} = S[t].\texttt{count} \times S[t+1].\texttt{elements}$, ie. that the number of elements at level t is equal to the `count` times the number of `elements` on the next level. Trivially, StackBuild(e, T, S) can be implemented to take time $O(T)$, independently of e.

The input-output type construction itself is done by the Match function that simultaneously traverses the two stacks for input and output type:

IOtype(T^i, T_o):
StackBuild$(0, T^i, S^i)$
StackBuild$(0, T_o, S_o)$
return Match$(S^i, S_o, \perp)$

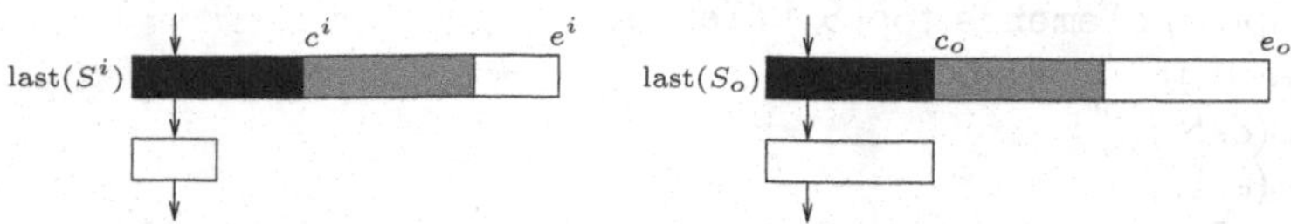

Fig. 2. Snapshot of the state of the MATCH function before an iteration of the loop. On both stacks the black portions have been consumed, and a corresponding input-output type created. Since the number of consumed elements is the same, a multiple of this type can be constructed, corresponding to the black and gray part. The remaining white parts will give rise to sibling types and are handled by the recursive MATCH call in the next iteration.

Starting from the top of the two stacks, MATCH first determines the common number of basic elements (corresponding to leaves of the type DAGs, which by assumption have the same basic type), creates a subtype input-output node for this number of elements, and consumes these elements by a call to the STACKINC function (given below). This function traverses the stack upwards and resets the `consumed` elements at each level until a level is found where the number of consumed elements is less than the total number of elements. This level is recorded as last(S). The last(S) level is initialized to 0 by STACKBUILD. After STACKINC the number of `consumed` elements on all levels above last(S) is zero.

STACKINC(e, S):
last(S) $\leftarrow$ top(S)
while S[last(S)].`consumed` $+ e =$ S[last(S)].`elements` **do**
  S[last(S)].`consumed` $\leftarrow 0$
  last(S) $\leftarrow$ last(S) $- 1$
  if last(S) < 0 **return**
  $e \leftarrow$ S[last(S)].`elements`
S[last(S)].`consumed` $\leftarrow$ S[last(S)].`consumed` $+ e$

After the number of basic elements at the leaves have been found, the MATCH function checks whether the created subtype node can be repeated. This is done by comparing the number of `consumed` elements to the total number of `elements` on the last frames of the two stacks. Repetition of the created input-output subtype node is possible if the number of `consumed` elements on the two stack frames is the same and smaller than the remaining `elements` – `consumed` elements on both frames. If there is no room for repetition, either because the number of consumed elements on the frames differ, or because the remaining elements on either frame are not sufficient, it becomes necessary to create additional input-output types by calling MATCH recursively. The situation before the next iteration is illustrated in Figure 2.

MATCH(S^i, S_o, B): /* S^i, S_o: stacks constructed by STACKBUILD, B: base type */
$l^i, l_o \leftarrow$ last(S^i), last(S_o) /* remember old state */
/* match basic elements at leaves, B pointer to element type (unless basic) */

```
e^i, e_o ← top(S^i).elements, top(S_o).elements
e ← min(e^i, e_o)
StackInc(e, S^i)
StackInc(e, S_o)
T_o^i ← MakeTypeNode(e, B) /* make leaf of input-output type */
while last(S^i) ≥ l^i ∧ last(S_o) ≥ l_o do
    t^i, t_o ← last(S^i), last(S_o)
    /* test for repetition of subtypes */
    c^i, e^i ← S^i[t^i].consumed, S^i[t^i].elements
    c_o, e_o ← S_o[t_o].consumed, S_o[t_o].elements
    if c^i = c_o ∧ c^i ≤ e^i − c^i ∧ c_o ≤ e_o − c_o then
        /* subtype can be repeated */
        e ← min(c^i⌊(e^i − c^i)/c^i⌋, c_o⌊(e_o − c_o)/c_o⌋)
        T_o^i ← MakeTypeNode(e, T_o^i)
        if e = e^i ∧ e = e_o then
            /* both input and output subtypes fully consumed, pop stacks */
            S^i, S_o ← S^i[0 ... t^i − 1], S_o[0 ... t_o − 1]
            B ← T_o^i /* input-output element type at leaves */
        StackInc(e, S^i)
        StackInc(e, S_o)
    else
        /* no room for repetition, "normalize" stack frame */
        S^i[t^i].consumed, S^i[t^i].elements ← 0, e^i − c^i
        S_o[t_o].consumed, S_o[t_o].elements ← 0, e_o − c_o
        T ← Match(S^i, S_o, B) /* recurse to find match in remainder */
        AddSibling(T_o^i, T)
        /* restore stack frame */
        S^i[t^i].consumed, S^i[t^i].elements ← S^i[t^i].consumed + c^i, e^i
        S_o[t_o].consumed, S_o[t_o].elements ← S_o[t_o].consumed + c_o, e_o
return T_o^i
```

The algorithm maintains the *invariant* that the total number of consumed elements on S^i and S_o is the same before each loop iteration. When the MATCH function terminates, all elements on either input or output stack have been fully consumed, ie. either $S^i[0]$.`elements` $=$ $S^i[0]$.`consumed` or $S_o[0]$.`elements` $=$ $S_o[0]$.`consumed` (or both). When the MATCH function is called recursively, the stack is "normalized" by subtracting the `consumed` elements at the last level. This is necessary for determining correctly in the recursive invocation whether the type node for this level can be repeated. When repetition consumes all elements at both stack frames, the created repetition type becomes the new base type for the next iteration of the loop. Both stacks are therefore popped accordingly.

Since each MATCH call goes through both its stacks (possibly causing recursive MATCH calls), the complexity is proportional to the sum of the stack depths of all recursive calls. This is also the number of nodes in the input-output DAG, and therefore $O(T_o^i)$.

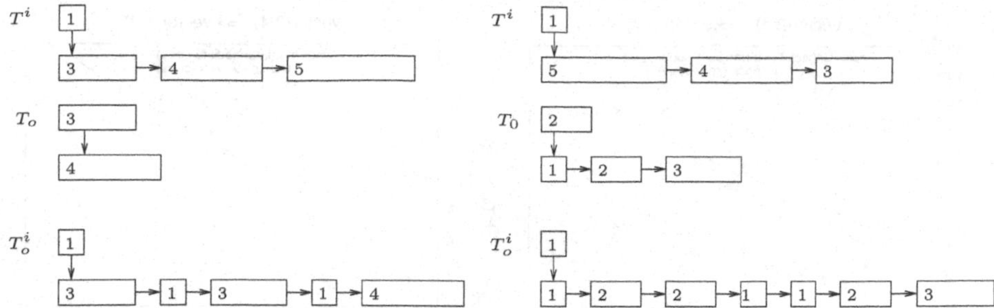

Fig. 3. Input-output types resulting from structured input and output types. Left: input index(3,4,5;MPI_DOUBLE), output vector(3,4;MPI_DOUBLE), input-output index(3,1,3,1,4;MPI_DOUBLE). Right: input index(5,4,3;MPI_DOUBLE), output contig(2,struct(1,2,3;MPI_DOUBLE)), input-output index(1,2,2,1,1,2,3;MPI_DOUBLE).

The algorithm can also be viewed as a method for finding a representation of two factorizations of the total number of elements $i_0 i_1 i_2 \cdots i_x = o_0 o_1 o_2 \cdots o_y$ of the form $c_0 c_1 \cdots (c_{z0} + c_{z1} + \ldots + c_{zz'})$ (each $c_{zz'}$ recursively of the same form).

3.2 Full Type DAGs (MPI Indexed and Structured Types)

We now extend the MATCH function to deal with structured and indexed types. The problem is that when all `elements` of a child in a structured type at level $t+1$ have been consumed, MATCHing will have to be continued with the next child of the type at level t. For structured types it generally holds that $S[t]$.`elements` $>$ $S[t]$.`count` $\times$ $S[t+1]$.`elements`.

We handle this by modifying the STACKINC function accordingly. Stack frames corresponding to structured types are extended with additional fields for efficiently finding the next child. If last(S) moves to level t, that is the `elements` of the child at level $t+1$ have been consumed, the next child of the structure has to be matched with what remains on the other stack. The part of the stack S above level t is scrapped, and replaced with the stack of types on the path to the first leaf of the next child. This is accomplished by calling STACKBUILD$(0, S[t].T[i+1], S+t)$, where $T[i+1]$ is the next child datatype and $S+t$ the point from which S has to be replaced. When a structured type has been fully consumed (that is, when last(S) moves below t), a repetition type node is created, also when the number of repetitions is only one. This prevents the MATCH function from exploring the same structures repeatedly.

Two examples of input-output type for structured input and output types are shown in Figure 3.

4 Experimental Evaluation

To assess the performance of transpacking with input-output type construction we compare our new approach to a straightforward transpack implementation

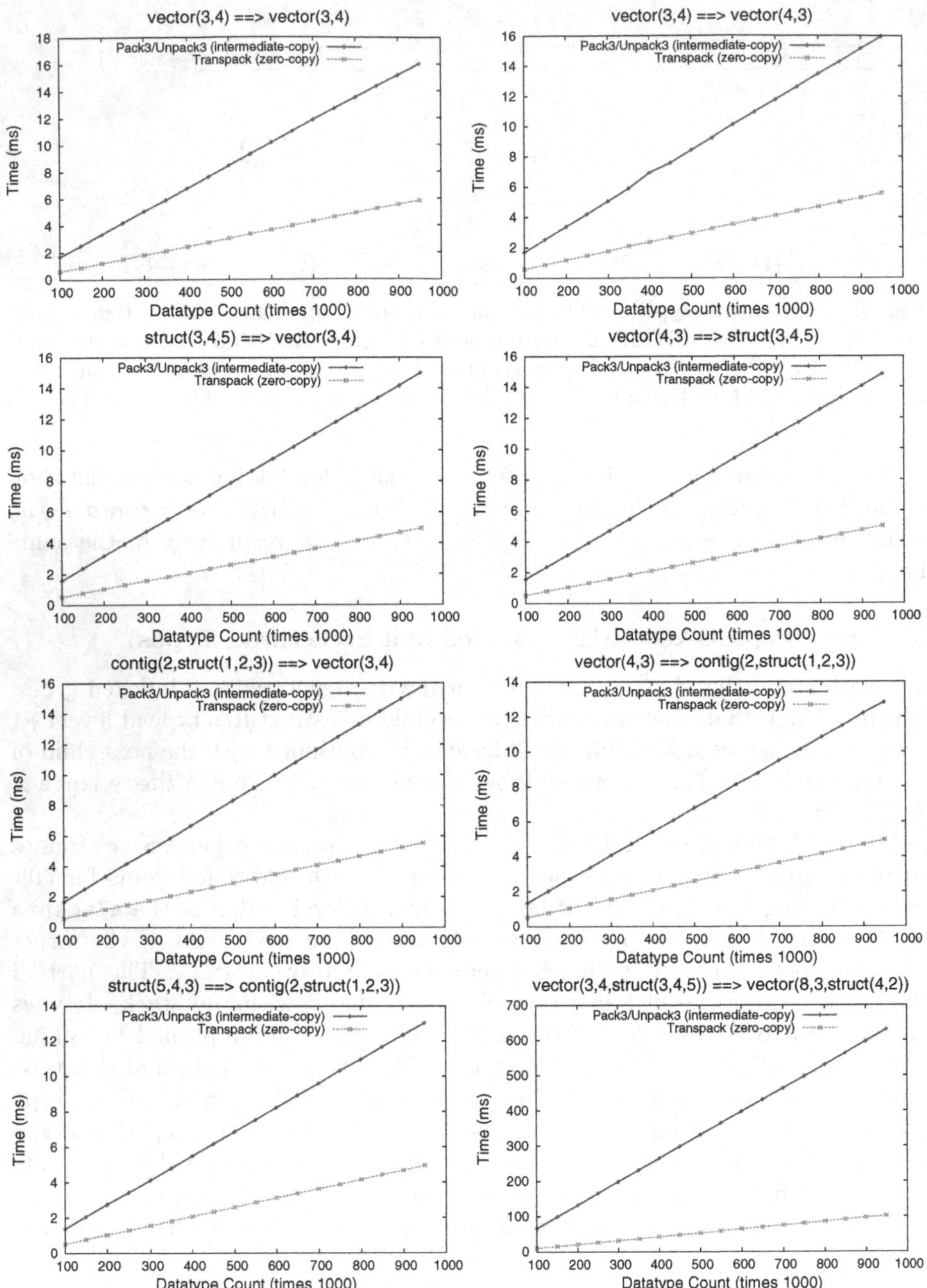

Fig. 4. Results from the test cases where both input and output types are non-contiguous. Time (in milliseconds) for input-output type based transpacking is plotted against a straightforward pack-unpack implementation for counts varying from 100,000 to 1,000,000.

Table 1. The combinations of input and output types used for performance evaluation. Construction Time for the input-output type is in microseconds. BE is the break even point in number of repetitions of the input-output type. Base type is in all cases MPI_DOUBLE. The number of leaves in the last input-output type is 71.

Input type	Output type	Input-output type	Time	BE
contig(12;)	contig(12;)	contig(12;)	24.45	
contig(12;)	vector(3,4;)	vector(3,4;)	27.77	
contig(12;)	struct(3,4,5;)	struct(3,4,5;)	54.14	
vector(3,4;)	vector(3,4;)	vector(3,4;)	29.71	2500
vector(3,4;)	vector(4,3;)	struct(3,1,2,2,1,3;)	44.17	4000
struct(3,4,5;)	vector(3,4;)	struct(3,1,3,1,4;)	46.15	4000
vector(4,3;)	struct(3,4,5;)	struct(3,3,1,2,3;)	48.05	4000
contig(2;struct(1,2,3;))	vector(3,4)	struct(1,2,1,2,1,1,1,3;)	61.02	5500
vector(4,3;)	contig(2;struct(1,2,3;))	contig(2;struct(1,2,3;))	44.07	4500
struct(5,4,3;)	contig(2,struct(1,2,3;))	struct(1,2,2,1,1,2,3;)	55.16	6000
vector(3,4;struct(3,4,5;))	vector(8,3,struct(4,2;))	struct(3,...;)	374.23	750

that packs the input into a contiguous buffer and unpacks the output from there. The straightforward solution uses direct memory copy if both types are contiguous, packs directly into the output buffer if the output type is contiguous, and unpacks directly from the input buffer if the input type is contiguous. Measurements have been performed on an NEC SX-8 vector architecture because of the high uniform bandwidth and absence of cache effects. We have used the increasingly complex combinations of input and output types shown in Table 1. In all cases the base is the 8-Byte MPI_DOUBLE entity, and structures were set up such that all subtypes are aligned on 8-Byte boundaries. This alleviates the effects of different alignment boundaries, which on the SX-architecture can be considerable. The table also gives the time in microseconds for the input-output type construction, as well as the *break even point* (in number of repetitions of input-output type) for transpacking with input-output type construction to be faster than the straightforward solution.

Results are given in Figure 4 with transpacking times in milliseconds for counts up to 1,000,000. For these large counts, input-output type construction time is completely negligible. We see that in all cases, even for the last two combinations of input and output types which give rise to quite large, structured input-output types, transpacking with input-output types achieves (often considerably more than) the theoretically best possible factor of two improvement. This still holds for even more complex type combinations that are not documented here.

5 Conclusion

We have presented a new solution to the *typed copy problem* for MPI, based on efficiently constructing input-output types from the given input and output types. We gave an algorithm for constructing input-output types that retains structure from input and output types as far as possible. The algorithm has been fully implemented and covers all MPI types and type constructors. Benchmarks shows that a performance gain of a factor of (often considerably more than) two can be achieved (bar effects of alignment changes that can be considerable on the SX-8 vector architecture, and take the results in both directions). Possible improvements to the algorithm for creating even more compact input-output types include memoization on constructed subtypes to save on recursive MATCH calls. In addition, we would like to determine the complexity of the *type reconstruction problem*.

In a follow-up paper we will detail and discuss the application of transpacking to non-contiguous memory/file I/O in MPI-IO.

References

1. Byna, S., Gropp, W.D., Sun, X.-H., Thakur, R.: Improving the performance of MPI derived datatypes by optimizing memory-access cost. In: IEEE International Conference on Cluster Computing (CLUSTER 2003), pp. 412–419 (2003)
2. Byna, S., Sun, X.-H., Thakur, R., Gropp, W.: Automatic memory optimizations for improving MPI derived datatype performance. In: Recent Advances in Parallel Virtual Machine and Message Passing Interface. 13th European PVM/MPI Users' Group Meeting. LNCS. Springer, Heidelberg (2006)
3. Ross, R., Miller, N., Gropp, W.D.: Implementing fast and reusable datatype processing. In: Dongarra, J., Laforenza, D., Orlando, S. (eds.) EuroPVM/MPI 2003. LNCS, vol. 2840, pp. 404–413. Springer, Heidelberg (2003)
4. Snir, M., Otto, S., Huss-Lederman, S., Walker, D., Dongarra, J.: MPI – The Complete Reference. The MPI Core, 2nd edn., vol. 1. MIT Press, Cambridge (1998)
5. Träff, J.L., Hempel, R., Ritzdorf, H., Zimmermann, F.: Flattening on the fly: efficient handling of MPI derived datatypes. In: Margalef, T., Dongarra, J., Luque, E. (eds.) PVM/MPI 1999. LNCS, vol. 1697, pp. 109–116. Springer, Heidelberg (1999)
6. Worringen, J., Gäer, A., Reker, F.: Exploiting transparent remote memory access for non-contiguous- and one-sided-communication. In: 16th International Parallel and Distributed Processing Symposium (IPDPS 2002), p. 163 (2002)
7. Worringen, J., Träff, J.L., Ritzdorf, H.: Fast parallel non-contiguous file access. In: Supercomputing (2003), `http://www.sc-conference.org/sc2003/tech_papers.php`

Object-Oriented Message-Passing in Heterogeneous Environments

Patrick Heckeler[1], Marcus Ritt[2], Jörg Behrend[1], and Wolfgang Rosenstiel[1]

[1] Wilhelm-Schickard-Institut für Informatik, Universität Tübingen
{heckeler,behrend,rosenstiel}@informatik.uni-tuebingen.de
[2] Instituto de Informática, Universidade Federal do Rio Grande do Sul
marcus.ritt@inf.ufrgs.br

Abstract. Heterogeneous parallel systems integrate machines with different architectural characteristics, such as endianess and word size. To use message-passing in these environments, the data must be translated by the communication layer. Message-passing standards like the Message Passing Interface (MPI) require the user to specify the type of the data sent, such that the communication layer can effect the necessary conversions.

We present an object-oriented message-passing library for C++, TPO++, which is capable to communicate in heterogeneous environments. Its functionality includes the MPI 1.2 standard, but it allows the user to communicate any data like objects or Standard Template Library (STL) containers in a type-safe way. It does not require the user to build a type representation explicitly.

We compare the performance of heterogeneous TPO++ with Boost.MPI and the C interface of OpenMPI. Our findings are that heterogeneous communication in TPO++ is possible with a very small overhead in latency compared to pure MPI. The performance of TPO++ is considerably better than that of other object-oriented communication libraries.

Keywords: Parallel computing; message-passing; object-oriented communication; heterogeneous communication.

1 Introduction

Heterogeneous architectures in parallel computing arise naturally in different contexts. Farming models, for example, allow to use existing infrastructure in organizations, which is usually heterogeneous, for parallel applications. Other contexts are meta-computing, i.e. coupling large parallel machines, and grid computing.

For a parallel application to work correctly in such environments, the message-passing layer must take care of architectural differences. The most common architectural differences are byte orders, such as little endian (e.g. x86) or big endian (e.g. PowerPC) and word size, which is usually 32 or 64 bit. To send a message from one machine to another with a different architecture, the data has to be

A. Lastovetsky et al. (Eds.): EuroPVM/MPI 2008, LNCS 5205, pp. 151–158, 2008.

translated to be interpreted correctly at receiver. To do this, the message-passing environment has to know the type of the data sent.

Message-passing standards like MPI [1,2] or PVM [3] provide mechanism to specify the type of the sent messages. For basic datatypes (e.g. integers) it is sufficient to pass the type of the data item in a send or packing primitive on the sender and receiver side. For complex datatypes, such as vectors or vectors of structures, the user has to rebuild the datatypes used in his application by calling functions from the message-passing API. This puts an extra burden on the user, and is an error-prone and tedious task. Any change of the applications' data structures must be reflected in the constructed data types of the message-passing layer. A common alternative on homogeneous architectures is to ignore this issue completely, and send untyped blocks of data. This, however, leads to non-portable applications and can cause subtle errors.

For parallel programming in C++, an object-oriented abstraction of the message-passing interface is desirable. In the last years, there have been a couple of approaches to accomplish this [4,5,6,7,8,9,10,11]. Any solution should satisfy some basic requirements. Most importantly, in parallel application, we do not want to sacrifice performance, so the overhead of the object-oriented layer should be as small as possible. At the same time, we want to make use of object-oriented features. An object-oriented message-passing library should be able to communicate objects, and in particular support inheritance. Ideally, the communication concept should be type-safe, i.e. not require the user to specify the type signature of the datatypes twice, as is done in MPI, but be able to infer it automatically. In C++, support for the communication of containers from the Standard Template Library is also indispensable.

The remainder of this article is organized as follows. In the next section, we present our approach to a type-safe, object-oriented communication in heterogeneous environments. Section 4 compares the performance of this approach with communication libraries. In section 3 we discuss some of the most recent object-oriented message-passing libraries and compare it to our approach. We conclude in section 5.

2 Type-Safe Communication in Heterogeneous Environments

TPO++ is a communication library, which provides an object-oriented interface to the functionality of the MPI 1.2 standard and the parallel I/O API from MPI 2 in C++. It supports communication of user-defined objects and STL containers, in arbitrary combinations and nesting. In a typical application, the user defines serialization methods for his objects and calls any communication method with two iterators defining the data to be sent. Similarly, the receiver provides space and defines by two iterators where he expects the data to be received. An unknown number of data can be received using network-aware inserters. Figure 1 shows how to send a vector of user-defined objects in TPO++.

```
struct Complex {
  double re, im;
  void serialize(Data& md) {
    md.insert(re);
    md.insert(im);
  }
  void deserialize(Data& md){
    md.extract(re);
    md.extract(im);
  }
}
TPO_MARSHALL(Complex)
vector<Complex> v1(10), v2;

if (comm.rank() == 0)
{
  comm.send(v1.begin(), v1.end(), 1);
  comm.send(v2.begin(), v2.end(), 1);
} else
{
  comm.recv(v1.begin(), v1.end(), 0);
  comm.recv(net_back_inserter(v2), 0);
}
```

Fig. 1. Example of two point-to-point communications. On the left, an object with its serialization methods is implemented. On the right, two point-to-point messages are sent. The first transmits a vector of 10 objects, with the receiver allocating space to hold the received elements. The second transmits a dynamic number of objects which are received using a network-aware back inserter.

More details about the functionality of TPO++ and its implementation on homogeneous architectures can be found in [8,12,13]. In the following we focus on the communication on heterogeneous architectures.

To map object-oriented data types to MPI, TPO++ uses a serialization class, which recursively decomposes complex data structures into a stream of blocks of basic datatypes. The serializer differentiates between basic datatypes, arrays, STL containers, structures and objects. It is based on generic programming and uses traits to categorize the data, with respect to the necessary operations for communicating them. For most used data structures, this categorization can be done at compile-time, which results in a low overhead of the C++ layer. For example, when sending an array of basic datatypes, the serialization class can recognize a random access iterator over a basic datatype and therefore has not to iterate over all contained elements. To communicate structures or objects, the user has to equip them with serialization and deserialization methods. (In C++ this cannot be avoided, since the language provides no means to introspect data types.) When sending objects, the serializer calls these methods to determine the members to communicate. The homogeneous version of TPO++ contains an optimization, which avoids the serialization calls for POD (plain old data) types [14]. Figure 2 shows the serialization steps starting from the send call down to the serialization class (`Message_data_service`).

To achieve type-safe heterogeneous communication, the serialization class records the type of the data to be sent along with its layout. When the message is completely decomposed, TPO++ can decide the most effective way to transmit the data. If it consists of a basic datatype or a simple sequence of basic datatypes, this information can be passed directly to the corresponding MPI call, therefore generating no communication overhead. If, on the other hand, the data has some complex structure or layout, the information of the serialization class creates a corresponding MPI datatype and passes it on to MPI.

The advantage of this approach is that TPO++ builds the MPI datatypes for the user automatically as needed. It is also not necessary to copy the data before transmission, because TPO++ passes the structural information directly

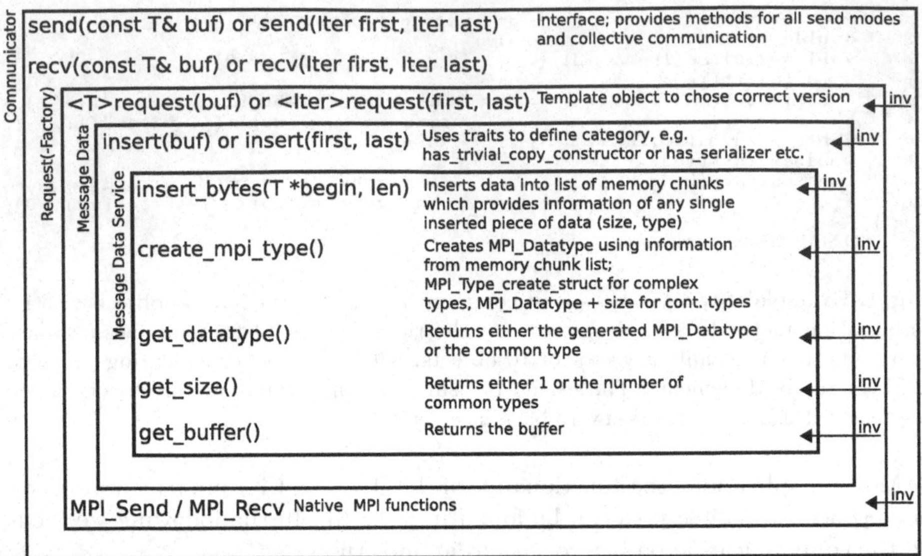

Fig. 2. Mapping of a communication call in TPO++ to MPI. The figure shows the methods invoked (inv) in different layers of TPO++.

to MPI. This allows MPI to choose the best transmission mode, and allows for zero-copy transfers. Observe that the mapping of TPO++ communication calls to MPI is one-to-one. In particular, we do not separately transmit the datatypes to the receiver. Both sides built the type information independently based on local information. To transmit data of unknown size, the receiver provides information about the underlying element data type and TPO++ determines the message size using `MPI_Probe`.

3 Related Work

Boost.MPI Boost.MPI is a recent addition to the Boost collection of C++ libraries [11], providing an object-oriented interface to message-passing. It supports the functionality of the MPI 1.2 standard, except sending modes, combined send and receive and a reduce-scatter operation. It can communicate basic and structured datatype like objects and STL containers. Sending of structured data is based on serialization, i.e. the user has to implement a method, which declares the members to send. The serialization engine, Boost.Serialize, also is part of Boost. Serialization of structured data results in a stream of pack and unpack calls for the basic datatypes contained in it, and therefore needs an extra buffer space for holding the packed data. A serialized message is transmitted with two communication calls, one for transmitting the buffer size and another for transmitting the data.

Boost. MPI provides optimized communication for special cases. On homogeneous architectures, user-defined objects can be declared "simple" or bit-wise serializable to avoid the overhead of packing and unpacking and datatype conversion. For repeatedly sending data of a fixed structure, Boost.MPI allows the user to extract the so-called "skeleton" of the data, i.e. its MPI typemap. This skeleton can be sent once to the receiver and afterwards the objects contents can be sent repeatedly without the overhead of packing and unpacking the data.

OOMPI. OOMPI introduces the concept of communication endpoints, called ports, which encapsulate the MPI communicator, and messages, which represent the data to be sent or received. OOMPI can transmit basic datatypes and user-defined objects, but has no STL-compatible interface and no support for the transmission of STL containers. User-defined objects can be transmitted only completely, as the serialization concept of OOMPI permits no selection of individual members. To make objects transmittable, they have to inherit from an user-type class, which makes it impossible to communicate existing classes.

MPI2 C++ bindings. The current MPI-2 standard contains bindings for C++. Unfortunately there is no significant gain of features achieved with this enhancement. It consists of some wrapper classes of the MPI C bindings. Especially an easy and comfortable way for sending and receiving STL containers is missing.

4 Performance Results

We made a series of performance measurements to compare TPO++ with other implementations and to determine the communication overhead of the object-oriented layer. For a comparison with another object-oriented approach, we chose Boost.MPI (Version 1.35.0), since it is one of the most recent and mature libraries available. To determine the overhead of the object-oriented layer, we compare the performance also with the C interface of MPI.

The measurements have been done on the two systems described in Table 1. We measured performance in two configurations. A loopback test on the AMD Opteron machine with two processes quantifies the pure message-passing overhead, since there is no need for a datatype conversion. To validate the heterogeneous communication we did measurements between the two machines with different architectures.

To compare the performance with different types of data layout, we sent arrays of basic datatypes, vectors of basic datatypes, and vectors of user-defined objects, all of increasing size from 1 byte to 4 MB in the heterogeneous case and to 8 MB in the homogeneous case. Each measurement consists of a warmup phase, and 20 measurement iterations. The results reported are the minimum, average and maximum transmission times for a complete message round-trip divided by 2 (ping-pong test).

Observe that we have been able to compare our approach with Boost.MPI only in the homogeneous configuration. When we performed the tests in the heterogeneous configuration, Boost.MPI failed, reporting truncated messages, when transmitting arrays, vectors or lists. Boost.MPI transmits first the size of the communication buffer in a separate message. We believe that this fails in heterogeneous configurations, since the message size on the sender and receiver side can be different.

Table 1. Architectural details of machines used in the performance measurements

	SUN X4100 Server	PowerMac G4
CPU	Dual Core AMD Opteron 2.2GHz x 2	7410 (Nitro) 533MHz
Memory	4 GB	1 GB
Network	GigaBit Ethernet	100 MBit Ethernet
OS	Debian Linux (2.6.22-14-server SMP)	Debian Linux (2.6.18-5-powerpc)
Compiler	g++ (GCC) 4.1.3	g++ (GCC) 4.1.2
MPI	OpenMPI v1.2.4	OpenMPI v1.2.4
Endianess	Big-Endian	Little-Endian
Word size	8 Byte	4 Byte
Float format	IEEE 754 32-bit	IEEE 754 32-bit
Double format	IEEE 754 64-bit	IEEE 754 64-bit
Long double format	not standard conform	IEEE 754 64-bit

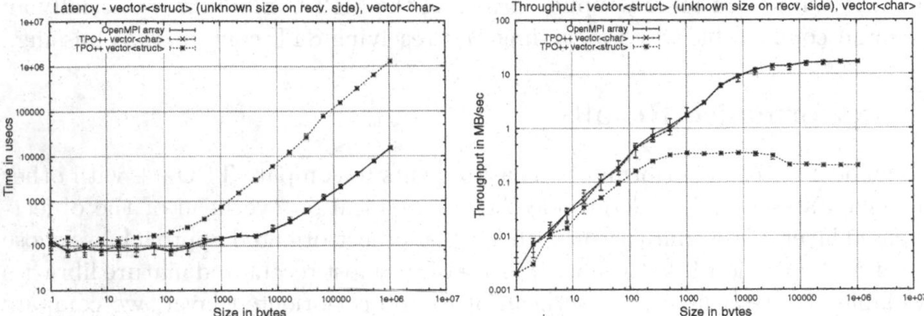

Fig. 3. Comparison of latency (left) and throughput (right) of Average latency and throughput of OpenMPI and TPO++ in the heterogeneous configuration. The error bars are indicating the maximum and minimum values.

Figure 3 shows the results of OpenMPI and TPO++ in the heterogeneous configuration. The difference between the transmission of an array of chars in MPI and a STL vector of chars in TPO++ is negligible, considering latency as well as throughput. This is a results of the optimizations, which are possible, when the data is a simple sequence of basic datatypes, such that it is not necessary to iterate over the data or to create MPI datatypes. In comparison, to send a vector of structures, both, sender and receiver have to serialize and deserialize the structures individually. Therefore the transmission takes considerably more time. For small message sizes, the latency is almost the same, but the throughput drops significantly.

Figure 4 compares transmission of an array of chars and a STL vector of chars in TPO++, Boost.MPI and native OpenMPI in the homogeneous configuration. We can see, that there is almost no difference in latency or throughput between TPO++ and MPI, for both data structures. The same holds for Boost.MPI, when transmitting an array. On the other hand, when sending a STL vector,

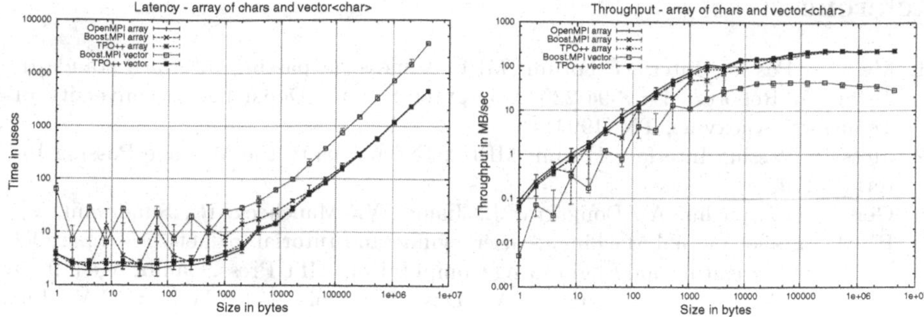

Fig. 4. Comparison of latency (left) and throughput (right) of Average latency and throughput of OpenMPI, Boost. MPI and TPO++ in the homogeneous configuration. The error bars are indicating the maximum and minimum values.

the latency of Boost.MPI is about $4\mu s$ higher, and the maximum throughput achieved is only 40 instead of 200 MB/s. This is due to the fact that Boost.MPI uses the MPI packing and unpacking mechanism, which results in an extra data copy. TPO++ can send multiple basic datatypes with a single call to `MPI_Send`, without creating a MPI datatype.

5 Conclusions and Future Work

We have shown that an object-oriented, type-safe interface to message-passing in C++ is possible, without sacrificing performance for commonly used datatypes. Our interface makes use of object-oriented concepts to simplify the implementation, and, specifically, remove the burden from the user of specifying explicitly the types of the data sent. Techniques from generic programming can be used to reduce the overhead of the abstraction.

On the other hand, the transmission of user-defined objects is considerably less efficient. In object-oriented communication tuned for homogeneous architectures, we can avoid this by sending POD without invoking serialization methods. To record and translate the datatypes, serialization cannot be avoided on heterogeneous architectures. We plan to investigate, how to reduce this overhead by caching frequently used typemaps.

Also, if the data sent has very irregular layout, the cost of recording and creating a typemap can be larger than copying the data. Therefore, another performance improvement could be to automatically switch between these two methods.

Acknowledgements. We thank the anonymous reviewers for their suggestions on how to improve this paper. This work has been partially funded by the Baden-Württemberg Ministry of Science, Research and the Arts as part of the research programme "Bioinformatics-Grid Tübingen".

References

1. Message Passing Interface Forum: MPI: A message passing interface standard. Technical Report UT-CS-94-230, Computer Science Department, University of Tennessee, Knoxville, TN (1994)
2. Message Passing Interface Forum: MPI-2: Extensions to the Message Passing Interface (1997)
3. Geist, A., Beguelin, A., Dongarra, J., Jiang, W., Mancheck, R., Sunderam, V.: PVM: Parallel Virtual Machine. A user's guide and tutorial for networked parallel computing. Scientific and Engineering Computation. MIT Press, Cambridge (1994)
4. Skjellum, A., Lu, Z., Bangalore, P.V., Doss, N.: Mpi++. In: Wilson, G.V., Lu, P. (eds.) Parallel programming using C++, pp. 465–506. MIT Press, Cambridge (1996)
5. Kafura, D.G., Huang, L.: Collective communication and communicators in mpi++. In: Proceedings of the MPI Developer's Conference, pp. 79–86 (1996)
6. Coulaud, O., Dillon, E.: Para++: A high level C++ interface for message passing. Journal of Parallel and Distributed Computing, 46–62 (1998)
7. Squyres, J.M., McCandless, B.C., Lumsdaine, A.: Object Oriented MPI: A class library for the Message Passing Interface. In: Proceedings of the POOMA conference (1996)
8. Grundmann, T., Ritt, M., Rosenstiel, W.: TPO++: An object-oriented message-passing library in C++, pp. 43–50. IEEE Computer society, Los Alamitos (2000)
9. Grundmann, T., Ritt, M., Rosenstiel, W.: Object-oriented message-passing with TPO++. In: Bode, A., Ludwig, T., Karl, W., Wissmüller, R. (eds.) Lecture notes in computer science, pp. 1081–1084. Springer, Heidelberg (2000)
10. Yao, Z., long Zheng, Q., liang Chen, G.: GOOMPI: A Generic Object Oriented Message Passing Interface. In: Jin, H., Gao, G.R., Xu, Z., Chen, H. (eds.) NPC 2004. LNCS, vol. 3222, pp. 261–271. Springer, Heidelberg (2004)
11. Kambadur, P., Gregor, D., Lumsdaine, A., Dharurkar, A.: Modernizing the C++ interface to MPI. In: Proceedings of the 13th European PVM/MPI Users' Group Meeting, pp. 266–274 (2006)
12. Pinkenburg, S., Ritt, M., Rosenstiel, W.: Parallelization of an object-oriented particle-in-cell simulation. In: Workshop Parallel Object-Oriented Scientific Computing, OOPSLA (2001)
13. Pinkenburg, S., Rosenstiel, W.: Parallel I/O in an object-oriented message-passing library. In: Kranzlmüller, D., Kacsuk, P., Dongarra, J. (eds.) EuroPVM/MPI 2004. LNCS, vol. 3241, pp. 251–258. Springer, Heidelberg (2004)
14. International Standards Organization: Programming languages – C++. ISO/IEC publication 14882:2003 (2003)

Implementation and Evaluation of an MPI-IO Interface for GPFS in ROMIO

Francisco Javier García Blas[1], Florin Isailă[1], Jesús Carretero[1], and Thomas Großmann[2]

[1] University Carlos III, Spain
{fjblas,florin,jcarrete}@arcos.inf.uc3m.es
[2] High Performance Computing Center Stuttgart (HLRS)
grossmann@hlrs.de

Abstract. This paper presents an implementation of the MPI-IO interface for GPFS inside ROMIO distribution. The experimental section presents a performance comparison among three collective I/O implementations: two-phase I/O, the default file system independent method of ROMIO, view-based I/O, a file system-independent method we developed in a previous work and a GPFS specific collective I/O implementation based on data-shipping. The results show that data shipping-based collective I/O performs better for writing, while view-based I/O for reading.

Keywords: Parallel file system, parallel programming, parallel I/O.

1 Introduction

The last years have shown a substantial increase in the amount of data produced by the parallel scientific applications and stored persistently. Parallel file systems (PFS) such as PVFS [1], Lustre [2] and General Parallel File System (GPFS) [3] offer scalable solutions to this ever increasing demand.

The large majority of large scale scientific parallel applications are written in Message Passing Interface (MPI) [4], which has become the de-facto standard for programming scalable distributed memory machines. MPI parallel applications may access parallel file system, through the MPI-IO interface [5]. The most popular MPI-IO implementation is ROMIO.

The goal of this paper is to present an implementation and evaluation of an MPI-IO interface for GPFS. A previous work [6] has presented an MPI-IO implementation for GPFS inside the IBM MPI. This implementation was proprietary and, to the best of our knowledge, has never been released to the public domain. This work targets to fill this gap and, additionally, to present an evaluation of the new implementation for two well-known benchmarks.

The remainder of the paper is organized as follows. Section 2 discusses some related work. Section 3 summarizes basic concepts of MPI-IO and the GPFS parallel file system, necessary for understanding the design and implementation our solution. Implementation details are discussed in Section 4. The experimental results are presented in Section 5. Section 6 contains our conclusions and our future plans.

A. Lastovetsky et al. (Eds.): EuroPVM/MPI 2008, LNCS 5205, pp. 159–166, 2008.

2 Related Work

A large number of researchers have contributed through the years with parallel I/O optimizations. The collective I/O techniques merge small requests into larger ones before issuing then to the file system. If the requests are merged at the I/O nodes the method is called *disk-directed I/O* [7]. If the merging occurs at intermediary nodes or at compute nodes the method is called *two-phase I/O* [8]. MPI-IO/GPFS collective I/O [6] leverages GPFS data shipping mode by defining at file open a map of file blocks onto I/O agents. A heuristic allowsMPI-IO/GPFS to optimize GPFS block prefetching through the use of GPFS multiple access range (MAR) hints. In order to mitigate striping overhead and benefit from the collective I/O accesses on Lustre [9], the authors propose two techniques: split writing and hierarchical striping.

Several researchers have contributed with optimizations of MPI-IO data operations: data sieving, non-contiguous access [10], collective caching [11], cooperative write-behind buffering [12], integrated collective I/O and cooperative caching [13]. Packing and sending MPI data types has been presented in [14].

3 Background

This section gives a short overview of GPFS and ROMIO.

3.1 Overview of GPFS File System

GPFS is a PFS for supercomputers or clusters. Its architecture is based on the Virtual Shared Disks (VSD), which are mounted at all client nodes. The data and metadata reside on VSDs and might be cached in clients cache. In order to guarantee data coherency, GPFS relies on a distributed locking manager. Locks are acquired and kept by clients while caching data. The granularity of locking in GPFS is at the byte-range level, consequently, writes to non-overlapping data blocks of the same file can proceed concurrently.

GPFS is highly optimized for large-chunk I/O operations with regular access patterns (contiguous or regularly strided). However, its performance for small-chunk, non-contiguous I/O operations with irregular access patterns (non-constant strided) is not sufficiently addressed. GPFS provides as an access alternative data-shipping. This technique binds each GPFS file block to a single I/O agent, which will be responsible for all accesses to this block. For write operations, each task sends the data to be written to the responsible I/O agents. I/O agents in turn issue the write calls to GPFS. For reads, the I/O agents read the file first, and ship the data read to the appropriate tasks.

This approach is similar to the two-phase I/O, described in Section 3.2). It is more effective than the default locking approach, when fine-grained sharing is present.

3.2 ROMIO

ROMIO is a freely available, high-performance, portable implementation of MPI-IO. The architecture of ROMIO allows the virtualization of MPI-IO routines on

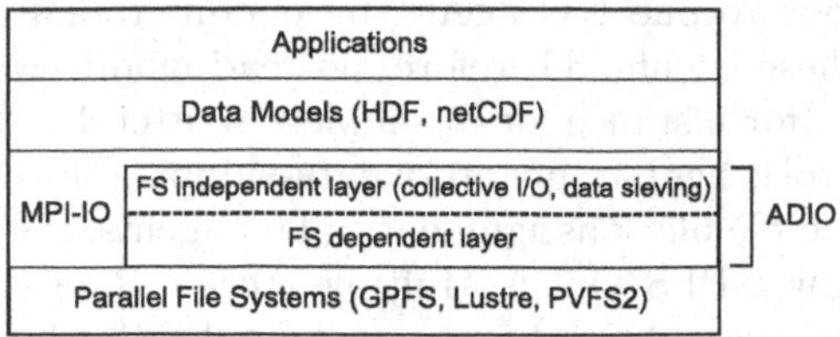

Fig. 1. Parallel I/O Software Architecture

top of concrete file systems through the abstract I/O interface ADIO [15], as shown in the lower part of Figure 1. ADIO consists of a file-system independent layer, where optimizations such as collective I/O or efficient non-contiguous I/O (data sieving) are implemented, and a file-system specific part, whose interface is a small set of basic functions for performing file I/O, such as open, close, write, read, fcntl, etc. It is this interface that a developer has to implement in order to add support for a new file system.

The collective I/O operations are typically implemented in the file-system independent layer. In this paper we will compare the collective I/O for GPFS with two collective I/O implemented at this layer: two-phase I/O, the original optimization of ROMIO and view-based I/O [16], an optimization we have described in the previous work.

Two-phase I/O [8] consists of an I/O phase and a shuffle phase. Data is gathered or scattered at a subset of compute nodes, called aggregators in the shuffle phase. The file system access is performed in the I/O phase in contiguous chunks by aggregators. View-based I/O is a collective I/O optimization that leverages the MPI-IO file view mechanism, for transferring view description information to aggregators at view declaration. In this way, view-based I/O avoids the necessity of transferring large lists of offset-length pairs at file access time as the present implementation of two-phase I/O. Additionally, this approach reduces the cost of scatter/gather operations at application compute nodes.

4 ADIO for GPFS

In this Section we describe details of the implementation of the ADIO interface of GPFS, data-shipping I/O. The whole implementation was done in the file system dependent layer. This includes the collective I/O optimization, due to the fact that it is based on data-shipping (described in Subsection 3.1), a GPFS-specific hint. Some of the file operations map on the GPFS POSIX interface in the same way as for a local POSIX-based UNIX file system. Therefore, we describe here only the file access operations, which differ, namely the collective operations.

The data-shipping mode is activated in GPFS through a blocking collective operation offered by the GPFS library. This call is issued inside ADIO if the corresponding user hint is passed when the file is opened. Subsequently, GPFS assigns each file block to one I/O agent (by default round-robin), and all file I/O goes through these agents. Therefore, no read-modify-write operations are needed at the client (for instance for incomplete written blocks).

The user can control other parameters of data-shipping through hints: the number of I/O agents, the file block assignment to I/O agents, the sizes of file blocks.

Another interesting GPFS hint is Multiple Access Range (MAR). This hint is used for enforcing a user-defined client cache policy for both prefetching and write-behind. The user can define the file ranges that are to be used for prefetching. Unlike data-shipping, which is helpful for collective access operations, MAR is more suitable to the independent read or write operations. Inside GPFS-specific ADIO calls for read and writes, the access MPI data type is converted into a list of offset-lengths, which is passed as a hint to GPFS. The evaluation of this functionality is subject of future work.

5 Evaluation

The evaluation of our implementation was performed on NEC Cacau Xeon EM64T cluster at HLRS Stuttgart. This cluster has the following characteristics: 200 biprocessor Intel Xeon EM64T CPU's (3.2GHz) compute nodes with 2 GBytes of RAM memory interconnected by Infiniband network. The file system uses the fast Infiniband network infrastructure. IBM GPFS parallel file system version 3.1.0 was configured with 8 I/O servers and 512 KBytes file block size. The MPICH2 distribution was MPICH2 1.0.5. The communication protocol of MPICH2 was TCP/IP on top of the native Infiniband communication library. We ran all tests with one process per compute node.

5.1 GpfsPerf Benchmark

First, we have used an IBM benchmark called gpfsPerf in order to evaluate the GPFS library outside ROMIO. GpfsPerf writes and reads a collection of fixed-size records to/from a file with three different types of access patterns: sequential, strided, and random. Here we show the results for strided access pattern. The benchmark uses the POSIX standard I/O interfaces in order to accomplish this. In all cases the size of the produced file was 250 MBytes per execution.

Figure 2 shows a comparison between aggregate throughputs for data-shipping and POSIX write and read operations. We show results for two representative cases: stride sizes of 256K and 512K and record sizes of 8K and 128K. Data-shipping is expected to achieve the highest throughput rate because of the absence of locking overhead. For write, we observe that data-shipping version outperforms POSIX in most of the cases. Additionally, the improvement increases with the number of compute nodes. Data-shipping shows the largest performance benefit for strided writes and small record sizes. For small records (8K), the reads show better results with data-shipping than for large records (128K). We expect that in these cases the MAR approach works better. However, for the scope of this paper we have chosen data-shipping, due to the fact that the two benchmarks issue small granularity accesses.

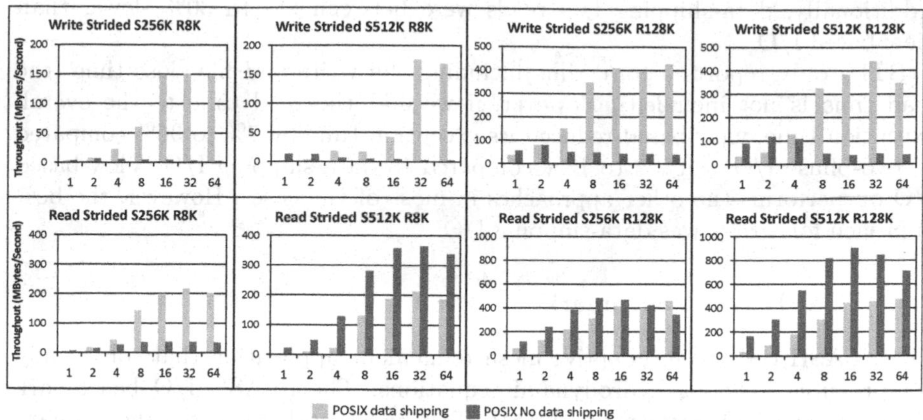

Fig. 2. Gpfsperf performance for strided pattern access

5.2 BTIO Benchmark

NASA's BTIO benchmark [17] solves the Block-Tridiagonal (BT) problem, which employs a complex domain decomposition across a square number of compute nodes. Each compute node is responsible for multiple Cartesian subsets of the entire data set. BTIO class B issues 40 collective MPI collective writes followed by 40 collective reads. We use 16 to 64 processes and a class of data set sizes B (1697.93 MBytes). For class B, the access pattern of BTIO is nested-strided with a nesting depth of 2 with a granularity of 2040 bytes.

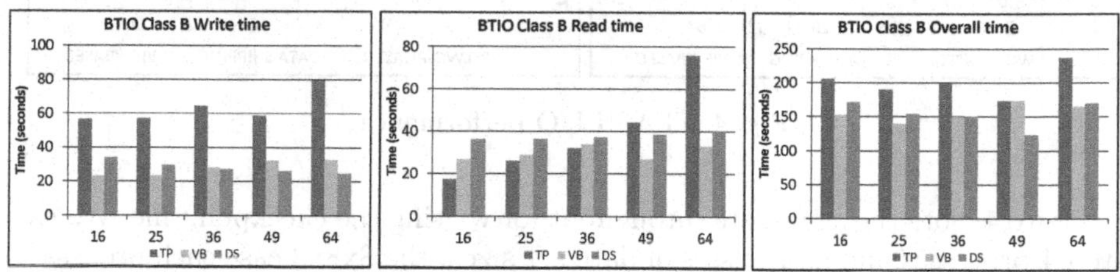

Fig. 3. BTIO performance for class B

Figure 3 compares the time results of two-phase I/O, view-based I/O and data-shipping I/O. We can see that data-shipping I/O was more effective for write operations. View-based I/O writes were between 3% to 33% slower and 40% to 69% slower for two-phase I/O. The main reason for this behavior is the fact that both view-based I/O and two-phase I/O used the POSIX interface for access to the final storage, and, therefore, they paid the overhead of locking. However, when we use a small set of clients, view-based I/O writes were between 21% to 32% faster than data-shipping I/O. Data-shipping I/O reads were between 13% to 47% faster than two-phase I/O for a large number of compute nodes.

Additionally, data-shipping I/O reads were between 9% to 30% slower than view-based I/O.

BTIO only reports a total time including the write and file close time (the read time is not included). However, if we add the read time to the overall execution time, view-based reduced its execution time by 7% to 36% compared to two-phase I/O and 2% to 13% compared to data-shipping I/O. View-based I/O outperforms the other approaches in most of the cases. However, the best approach for writes was data-shipping I/O.

5.3 FLASH I/O Benchmark

The FLASH code is an adaptive mesh refinement application that solves fully compressible, reactive hydrodynamic equations. The FLASH I/O benchmark simulates the I/O pattern of FLASH. The benchmark recreates the primary data structures in the FLASH code and produces a checkpoint file, a plot-file for centered data, and a plot-file for corner data, using parallel HDF5. The access pattern is non-contiguous both in memory and in file, making it a challenging application for parallel I/O systems.

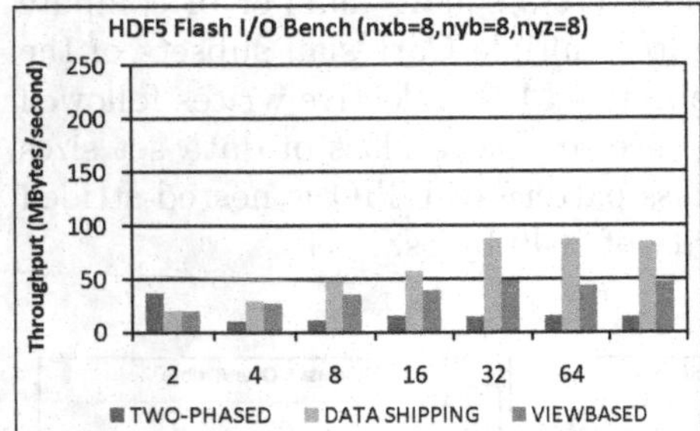

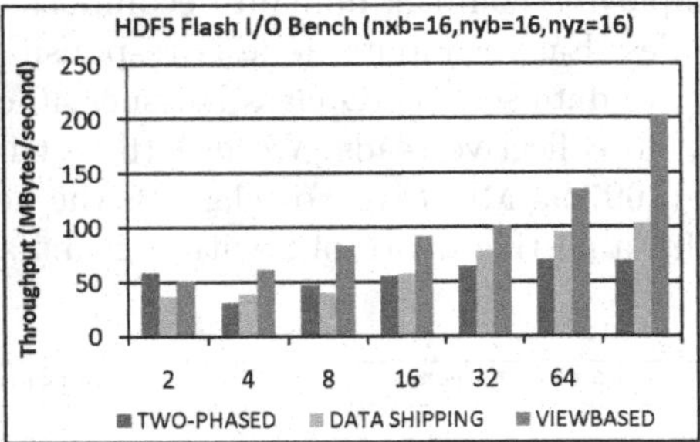

Fig. 4. FLASH I/O performance

Figure 4 plots the aggregate throughput for writing the checkpoint file. We use 1 to 64 processes and two classes of data set sizes: the 8x8x8 case each processor outputs approximately 8 MB and the 16x16x16 case approximately 60 MB per compute node.

As shown in the graph, we compared two-phase I/O, data-shipping I/O and view-based I/O. The graph on the left shows that, for a small problem size, the benchmark reaches file write throughputs of up to 83 MBytes/sec for data shipping I/O. Additionally, it is important to note that from 16 processes, the difference between data-shipping I/O and view-based I/O grows significantly. When compared to view-based I/O, data-shipping I/O improves FLASH I/O write time between 3% and 44%. Also, when compared to two-phase I/O, data-shipping I/O improves write time between 63% and 82%.

For the 16x16x16 case, we note that the view-based I/O performs better for large files. The 16x16x16 write performance results for 64 compute nodes show that view-based I/O attains a 92% improvement compared to two-phase I/O and a 94% for data-shipping I/O. This is due to the fact that all the data fit in the cache and are flushed to the file system at close time.

6 Conclusions

In this paper we presented an implementation and evaluation of a MPI-IO interface for the GPFS parallel file system. Our experimental results show that data-shipping I/O can significantly reduce the total run time of a data intensive parallel application by reducing I/O cost. For example, data shipping I/O reduced the BTIO overall execution time by 25%. The FLASH IO performance results prove that data-shipping I/O outperforms two-phase I/O significantly. In addition, we show that our previous work outperforms two-phase I/O using GPFS. Finally, the benchmarks show that our approaches bring satisfactory results for large files. Unlike indicated in [6], we showed that certain performance enhancement can be obtained for both write and read operations when data-shipping is enabled.

In the future, we plan to further evaluate the file access performance of the MPI/IO interface based on additional factors: prefetching and write-back policy, number of data-shipping agents, file block size and file block mappings.

Acknowledgment

This work has been partially funded by project TIN2007-63092 of Spanish Ministry of Education and project CCG07-UC3M/TIC-3277 of Madrid State Government. Also, we are grateful to the High Performance Computing Center Stuttgart (HLRS) for letting us perform the required benchmarks on their Cacau Xeon EM64T system.

References

1. Ligon, W., Ross, R.: An Overview of the Parallel Virtual File System. In: Proceedings of the Extreme Linux Workshop (June 1999)
2. Inc., C.F.S.: Lustre: A scalable, high-performance file system. Cluster File Systems Inc. white paper, version 1.0 (November 2002), `http://www.lustre.org/docs/whitepaper.pdf`
3. Schmuck, F., Haskin, R.: GPFS: A Shared-Disk File System for Large Computing Clusters. In: Proceedings of FAST (2002)
4. Message Passing Interface Forum: MPI: A Message-Passing Interface Standard (1995)
5. Corbett, P., Feitelson, D., Hsu, Y., Prost, J.P., Snir, M., Fineberg, S., Nitzberg, B., Traversat, B., Wong, P.: MPI-IO: A parallel file I/O interface for MPI. Technical Report NAS-95-002, NASA Ames Research Center, Moffet Field, CA (January 1995)

6. Prost, J.P., Treumann, R., Hedges, R., Jia, B., Koniges, A.: MPI-IO/GPFS, an optimized implementation of MPI-IO on top of GPFS. In: Supercomputing 2001: Proceedings of the 2001 ACM/IEEE conference on Supercomputing (CDROM), pp. 17–17. ACM Press, New York (2001)
7. Kotz, D.: Disk-directed I/O for MIMD Multiprocessors. In: Proc. of the First USENIX Symp. on Operating Systems Design and Implementation (1994)
8. del Rosario, J., Bordawekar, R., Choudhary, A.: Improved parallel I/O via a two-phase run-time access strategy. In: Proc. of IPPS Workshop on Input/Output in Parallel Computer Systems (1993)
9. Yu, W., Vetter, J., Canon, R.S., Jiang, S.: Exploiting Lustre File Joining for Effective Collective IO. In: CCGRID 2007: Proceedings of the Seventh IEEE International Symposium on Cluster Computing and the Grid, pp. 267–274. IEEE Computer Society Press, Washington (2007)
10. Thakur, R., Gropp, W., Lusk, E.: Optimizing Noncontiguous Accesses in MPI-IO. Parallel Computing 28(1), 83–105 (2002)
11. keng Liao, W., Coloma, K., Choudhary, A., Ward, L., Russel, E., Tideman, S.: Collective Caching: Application-Aware Client-Side File Caching. In: Proceedings of the 14th International Symposium on High Performance Distributed Computing (HPDC) (July 2005)
12. keng Liao, W., Coloma, K., Choudhary, A.N., Ward, L.: Cooperative Write-Behind Data Buffering for MPI I/O. In: PVM/MPI, pp. 102–109 (2005)
13. Isaila, F., Malpohl, G., Olaru, V., Szeder, G., Tichy, W.: Integrating Collective I/O and Cooperative Caching into the Clusterfile Parallel File System. In: Proceedings of ACM International Conference on Supercomputing (ICS), pp. 315–324. ACM Press, New York (2004)
14. Ching, A., Choudhary, A., Liao, W.K., Ross, R., Gropp, W.: Efficient Structured Data Access in Parallel File Systems. In: Proceedings of the IEEE International Conference on Cluster Computing (December 2003)
15. Thakur, R., Gropp, W., Lusk, E.: An abstract device interface for implementing portable paralllel-I/O interfaces
16. Blas, J.G., Florin Isaila, D.E.S., Carretero, J.: View-based collective i/o for mpi-io. In: CCGRID 2008: Proceedings of the 8th IEEE International Symposium on Cluster Computing and the Grid (2008)
17. Wong, P., der Wijngaart, R.: NAS Parallel Benchmarks I/O Version 2.4. Technical Report NAS-03-002, NASA Ames Research Center, Moffet Field, CA (January 2003)

Self-consistent MPI-IO Performance Requirements and Expectations

William D. Gropp[1], Dries Kimpe[2], Robert Ross[3], Rajeev Thakur[3], and Jesper Larsson Träff[4]

[1] Computer Science Department, University of Illinois at Urbana-Champaign Urbana, IL 61801, USA
wgropp@uiuc.edu

[2] Scientific Computing Research Group, K. U. Leuven Celestijnenlaan 200A, B-3001 Leuven, Belgium
dries.kimpe@cs.kuleuven.be

[3] Mathematics and Computer Science Division Argonne National Laboratory, Argonne, IL 60439, USA
{rross,thakur}@mcs.anl.gov

[4] NEC Laboratories Europe, NEC Europe Ltd. Rathausallee 10, D-53757 Sankt Augustin, Germany
traff@it.neclab.eu

Abstract. We recently introduced the idea of self-consistent performance requirements for MPI communication. Such requirements provide a means to ensure consistent behavior of an MPI library, thereby ensuring a degree of performance portability by making it unnecessary for a user to perform implementation-dependent optimizations by hand. For the collective operations in particular, a large number of such rules could sensibly be formulated, without making hidden assumptions about the underlying communication system or otherwise constraining the MPI implementation. In this paper, we extend this idea to the realm of parallel I/O (MPI-IO), where the issues are far more subtle. In particular, it is not always possible to specify performance *requirements* without making assumptions about the implementation or without *a priori* knowledge of the I/O access pattern. For such cases, we introduce the notion of performance *expectations*, which specify the desired behavior for good implementations of MPI-IO. I/O performance requirements as well as expectations could be automatically checked by an appropriate benchmarking tool.

1 Introduction

In [7], we introduced the notion of self-consistent performance requirements for MPI implementations. Such requirements relate various aspects of the semantically strongly interrelated MPI standard to each other. The requirements are based on meta-rules, stating for instance that no MPI function should perform worse than a combination of other MPI functions that implement the same functionality, that no specialized function should perform worse than a more general

A. Lastovetsky et al. (Eds.): EuroPVM/MPI 2008, LNCS 5205, pp. 167–176, 2008.

function that can implement the same functionality, and that no function with weak semantic guarantees should perform worse than a similar function with stronger semantics. In other words, the *library-internal* implementation of any arbitrary MPI function in a given MPI library should not perform any worse than an *external (user)* implementation of the same functionality in terms of (a set of) other MPI functions. Otherwise, the performance of the library-internal MPI implementation could trivially be improved by replacing it with an implementation based on the external user implementation. Such requirements, when fulfilled, would ensure consistent performance of interrelated parts of MPI and liberate the user from having to perform awkward and non-portable optimizations to cope with deficiencies of a particular implementation. For the MPI implementer, self-consistent performance requirements serve as a sanity check.

In this paper, we extend this idea to the MPI-IO part of MPI [1, Chapter 7]. The I/O model of MPI is considerably more complex than the communication model, with performance being dependent to a much larger extent on external factors beyond the control of both the application and the MPI library. Also, the I/O access patterns of different processes are not known beforehand. As a result, in some instances, we can only formulate performance *expectations* instead of performance *requirements.* Performance expectations are properties that are expected to hold most of the time and would be desirable for an MPI implementation to fulfill (from the perspective of the user and for performance portability).

We use the following notation in the rest of this paper. The performance relationship that MPI function $\mathsf{MPI\_}A(n)$ should perform no worse than function(s) $\mathsf{MPI\_}B(n)$ for total I/O volume n is expressed semi-formally by $\mathsf{MPI\_}A(n) \preceq \mathsf{MPI\_}B(n)$. To distinguish between requirements and expectations, we use the notation $\mathsf{MPI\_}A(n) \subseteq \mathsf{MPI\_}B(n)$ to indicate the performance relationship that MPI function $\mathsf{MPI\_}A(n)$ is expected to perform no worse than function $\mathsf{MPI\_}B(n)$ for total I/O volume n.

One value in defining expectations more formally is that they can suggest the need for additional hints to help an MPI implementation achieve the user's expectation of performance. In Section 5, we illustrate this point by describing some situations where achieving performance expectations may require additional information and suggest new standard hints that could be adopted in revisions of the MPI standard. Furthermore, this approach to defining standard hints is arguably a better approach than attempting to standardize common practice, as was done for the I/O hints in MPI 2.0. The formal definitions also help guide the development of tests to ensure that implementations meet user expectations.

As with the set of self-consistent performance requirements for MPI communication, the MPI-IO requirements and expectations can, in principle, be automatically checked with an appropriate, configurable benchmark and experiment-management and mining tool. We have not developed such a tool so far.

2 MPI-IO

MPI-IO [1] is an interface for parallel file I/O defined in the spirit of MPI and building on the same key concepts. Access patterns and data layouts in files are described by (derived) datatypes, and data is then sent (written) from memory into a region or regions in file, or received (read) from file into memory.

The MPI-IO model can be analyzed along different dimensions [1, Chapter 7, page 204]. File I/O operations can be classified as

1. independent vs. collective,
2. blocking vs. nonblocking, and
3. blocking collective vs. split collective

Positioning within a file can be done via

1. explicit offsets,
2. individual file pointers, or
3. shared file pointers.

These different classes of file I/O operations and positioning mechanisms have different semantics and performance characteristics. In addition to these I/O modes, we must also consider the side effects of other MPI-IO calls, such as those that change the bytes of a file that a process may access (by defining *file views*) or supply (implementation- and system-dependent) *hints* that may impact underlying behavior. For example, a call to MPI_File_set_info that changes the cb_nodes hint for a file could have a dramatic impact on subsequent collective I/O on that same file by limiting the number of processes that actually perform I/O. This example also helps explain the difficulty in defining performance requirements for MPI-IO.

Further details about MPI-IO can be found in [1,2].

3 Requirements Versus Expectations

We define both requirements and expectations for MPI-IO performance. Performance *requirements* are conditions that a good MPI-IO implementation should be able to fulfill. In some cases, however, it is not possible to specify requirements for a variety of reasons discussed below. For such cases, we define performance *expectations*, which would be desirable for an implementation to fulfill.

For example, it is tempting to specify that collective I/O should perform no worse than independent I/O. However, the I/O access pattern specified by the collective I/O function is not known unless the implementation analyzes the request, which may require communication among processes. If, after analysis, the implementation determines that collective I/O optimization is not beneficial for this request, and uses independent I/O instead, the cost of the analysis is still incurred, and the collective I/O function is slower than if the user had directly called the equivalent independent I/O function. Also, I/O performance is influenced by characteristics of the (parallel) file system, only some of which can be controlled by (or are even visible to) the MPI implementation. This fact also makes it difficult to specify requirements in some cases.

The situation is further complicated by hints. In the MPI-IO model, the interaction with the file system can be influenced through hints that are supplied to MPI_File_open, MPI_File_set_info, or MPI_File_set_view calls. The MPI standard defines a number of such hints that can deeply affect performance. It is the user's responsibility to use these hints sensibly, which often requires knowledge of the underlying file system. It is easy to supply hints that have a negative effect on performance. For example, Figure 1 shows the effect of an unfortunate choice for the standard-defined cb_buffer_size hint (buffer size for collective buffering) on the performance of the noncontig benchmark [8]. This benchmark generates a regular, strided, nonoverlapping access pattern perfectly suited for collective buffering. However, the optimal value for cb_buffer_size is hard to determine, and depends on many factors such as the number of processes involved and the characteristics of the file system and communication network. In this graph we see that there is a small range of values for which optimal read performance is achieved. In fact, the default value of 4 MB used by the MPI-IO implementation, ROMIO [3,5], does not happen to fall into that range on this system. Selection of an appropriate cb_buffer_size can be further complicated by the potential for interaction between alignment of these buffers during collective I/O and the granularity of locks in the (parallel) file system, if the file system uses locks.

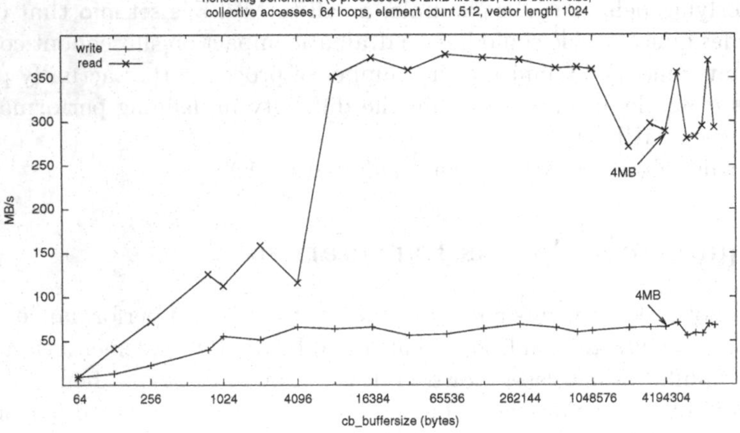

Fig. 1. Effect of the hint cb_buffer_size on the noncontig benchmark

Two features further complicate the matter. First, MPI implementations are allowed to support their own hints in addition to the ones described in the standard. We, of course, cannot say anything about the impact of such hints here. Second, a conforming MPI implementation is allowed to ignore all hints (including standard-defined hints), and their effect may therefore be void. In all, hints are a nonportable feature (performance wise). We therefore cannot formulate strict performance requirements, but by making assumptions about how an implementation could (or should) sensibly use hint information, we can nevertheless formulate reasonable performance *expectations*. Similar to the performance requirements, such expectations can be stated as rules that can be checked by a suitable tool.

4 Performance Requirements

Where an MPI-IO operation can, in an obvious way, be implemented by other, possibly more general MPI-IO operations, a self-consistent performance requirement states that this alternative implementation should not be faster than the original operation. A comprehensive, but not exhaustive set of such requirements is presented in the following.

An I/O operation with an explicit offset should be no slower than implementing it with a call to MPI_File_seek followed by the corresponding individual file pointer operation.

$$\text{MPI\_File\_\{read|write\}\_at} \preceq \text{MPI\_File\_seek} + \text{MPI\_File\_\{read|write\}} \quad (1)$$

A collective I/O operation should be no slower than the equivalent split collective operation.

$$\text{MPI\_File\_\{read|write\}\_all} \preceq \quad (2)$$

$$\text{MPI\_File\_\{read|write\}\_all\_begin} + \text{MPI\_File\_\{read|write\}\_all\_end} \quad (3)$$

A blocking I/O operation should be no slower than the corresponding nonblocking operation followed by a wait.

$$\text{MPI\_File\_\{read|write\}} \preceq \text{MPI\_File\_\{iread|iwrite\}} + \text{MPI\_Wait} \quad (4)$$

By the assumption that an operation with weaker semantic guarantees should be no slower (presumably faster) than a similar operation with stronger guarantees, we can formulate the following requirement

$$\text{Write with default consistency semantics} \preceq \text{write with atomic mode} \quad (5)$$

Preallocating disk space for a file by using MPI_File_preallocate should be no slower than explicitly allocating space with MPI_File_write.

$$\text{MPI\_File\_preallocate} \preceq \text{MPI\_File\_write\_*} \quad (6)$$

I/O access using an individual file pointer should be no slower than I/O access using the shared file pointer (because accessing the shared file pointer may require synchronization).

$$\text{MPI\_File\_write} \preceq \text{MPI\_File\_write\_shared} \quad (7)$$

Similarly, collective I/O using an individual file pointer should be no slower than collective I/O using the shared file pointer.

$$\text{MPI\_File\_write\_all} \preceq \text{MPI\_File\_write\_ordered} \quad (8)$$

I/O operations in native file format should be no slower than I/O operations in `external32` format, since `external32` may require conversion, and it provides stronger semantics guarantees (on portability). An MPI implementation where `external32` is faster than native format could be "fixed" by using the same approach to native I/O as in the `external32` implementation, only without performing any data conversion.

$$\text{I/O in native format} \preceq \text{I/O in } \texttt{external32} \text{ format} \quad (9)$$

5 Performance Expectations

We describe some examples of performance expectations, which for various reasons cannot be mandated as performance requirements.

5.1 Noncontiguous Accesses

MPI-IO allows the user to access noncontiguous data (in both memory and file) with a single I/O function call. The user can specify the noncontiguity by using derived datatypes. A reasonable performance expectation would be that a read or write with a noncontiguous datatype $T(n)$ that specifies t contiguous regions of size n in memory or file is no slower than if the user achieved the same effect by t individual, contiguous reads or writes. Ideally, an implementation should do better.

$$\text{MPI\_File\_\{read|write\}\_at}(T(n)) \subseteq t \times \text{MPI\_File\_\{read|write\}\_at}(n) \quad (10)$$

We cannot state this as a requirement because the performance in such cases can be highly dependent on the access pattern, which must be determined from the memory and file layouts specified. This analysis can itself take some time. If after the analysis it is determined that no optimization is beneficial, and instead multiple contiguous reads/writes should be performed, the above relation will not hold. A good implementation should perform the analysis as efficiently as possible to minimize the overhead.

Figure 2 shows an example where this expectation is not met. Here, for small, sparse accesses (described in Figure 3[a]), multiple contiguous writes perform better than a single noncontiguous write. In this case, the MPI library mistakenly decided to apply the data-sieving optimization [6] to the file access, when in fact multiple contiguous accesses perform better because of the sparsity of the access pattern.

5.2 Implications for Hints

These considerations suggest several possible hints that would aid an MPI implementation in achieving the performance expectations. General hints that could be used with any MPI implementation (because they describe general features of the program and data) include:

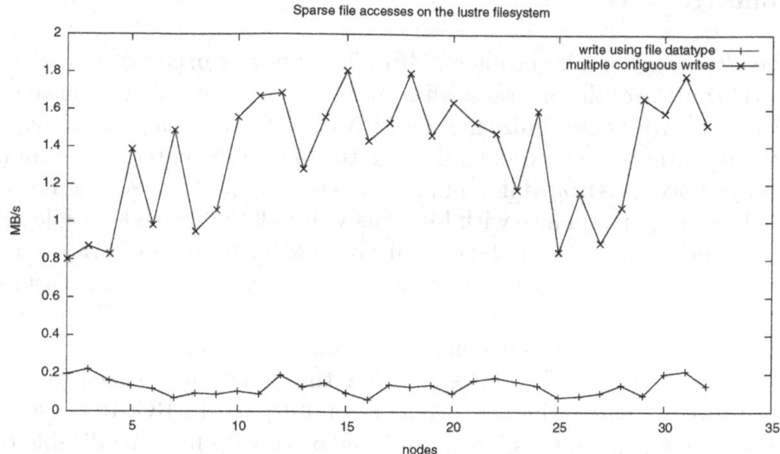

Fig. 2. Using file datatypes versus multiple contiguous accesses

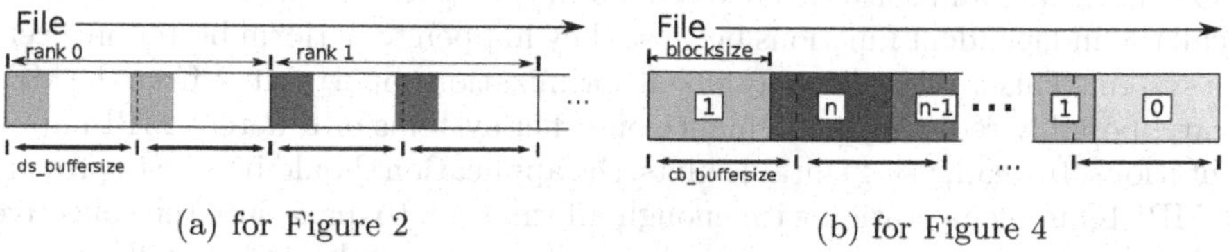

(a) for Figure 2 (b) for Figure 4

Fig. 3. Access patterns (color indicates rank)

`write_density`. A measure of the ratio between the extent and size of datatypes used for writing to a file. Implementations may want to use independent I/O if the density is low and collective I/O if the density is high.

`read_density`. Similar to `write_density`, but for reading.

Specific hints to control a particular MPI implementation can be used when the implementation is unable to meet a performance expectation without additional knowledge. These might include:

`use_only_contig_io`. Only use contiguous I/O.

The advantage of such a hint is that the MPI program is more *performance portable*: Rather than replacing one set of MPI calls with a different set everywhere they occur, the hint can allow the MPI implementation to avoid the analysis and simply choose the appropriate method. This places most of the performance tuning at the point where the hint is provided, rather than at each location where file I/O is performed. Currently, ROMIO implements the hints romio_ds_read and romio_ds_write to allow users to selectively disable the use of noncontiguous I/O optimizations, overriding the default behavior. Use of these hints could obtain desired performance in the example shown in Figure 2 without reverting to the use of multiple I/O calls.

5.3 Collective I/O

Collective I/O faces a similar problem. MPI_File_write_all can choose to make use of the collective nature of the function call to merge the file accesses among all participating ranks in order to optimize file access. Most I/O systems perform relatively well on large contiguous requests (compared to small fragmented ones), and, as a result, merging accesses from different processes usually pays off. A user may therefore expect better performance with MPI_File_write_all than with MPI_File_write.

However, merging access patterns will not always be possible (for example, when accesses are not interleaved). When this happens, the implementation of MPI_File_write_all will usually fall back to calling MPI_File_write. Because of the additional synchronization and communication performed in determining the global access pattern, MPI_File_write_all will actually have performed worse compared with directly calling its independent counterpart. In ROMIO, users have the option of using the romio_cb_read and romio_cb_write hints to disable collective I/O optimizations when they know these optimizations aren't beneficial.

Of course, merging access patterns is just one possible optimization. Given the diverse hardware and software encountered in I/O systems, we do not want users to call the independent functions because they happen to perform better on a certain system. This would take away global optimization opportunities from the MPI library, possibly reducing performance on other systems or different MPI implementations. Instead, using suitable hints, the application should be able to provide the MPI-IO implementation with enough information to make sure the collective version does not perform worse than the corresponding independent call.

Therefore, we can only state the following as an expectation.

$$\mathsf{MPI\_File\_\{read\_all|write\_all\}} \subseteq \mathsf{MPI\_File\_\{read|write\}} \tag{11}$$

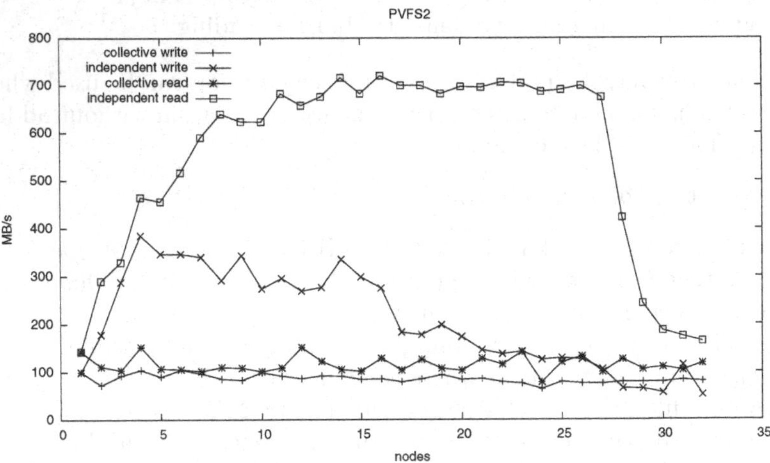

Fig. 4. Collective versus independent file access for a case that is not suitable for collective I/O

Figure 4 shows an example where this expectation does not hold. The corresponding access pattern is shown in Figure 3[b]. Here, independent reads and writes are faster than collective reads and writes. The access pattern was specially crafted to trigger bad performance with the two-phase algorithm [4] implemented in ROMIO. This algorithm, which is only enabled if interleaved access is detected, first analyzes the collective access pattern to determine the first and last offset of the access region. Next, this region is divided among a configurable subset of processes, and each process is responsible for all I/O for its portion of the file. In this example, most of the data that each process accesses is destined for another process, resulting in most data passing over the network twice.

Another performance expectation is that an MPI-IO operation for which a "sensible" hint has previously been supplied should perform no worse than the operation without the hint.

6 Conclusions

Codifying performance requirements and expectations provides users and implementers with a common set of assumptions (and goals) for working with MPI-IO. From the user's perspective, these requirements and expectations encourage the use of collective I/O and file views to combine I/O operations whenever possible, to use default consistency semantics when appropriate, to avoid shared file pointers when independent file pointers are adequate, and to use hints for tuning rather than breaking from "best practice." Implementers must strive to meet these expectations wherever possible without the use of additional hints, and also support hints that enable users to correct the implementation's behavior when it goes astray. Implementers and users should strive to improve and standardize the hints used for this purpose across multiple MPI-IO implementations.

At present, some MPI-IO hints are inherently nonportable. The lack of portability is not because conforming implementations are allowed to ignore them, but because setting them meaningfully requires intimate knowledge of the I/O layers below the MPI implementation. As demonstrated in Section 3, setting such hints carelessly can damage performance. Portable hints only give additional information about the application itself. As an example, consider a hint describing how much data is interleaved in collective accesses. Without this information, an implementation is forced to do additional calculation and communication, possibly making the collective access functions perform worse than their independent counterparts. At the same time, the availability of this information (assuming it is correct) should not degrade performance.

We have shown that some of the performance problems with the MPI-IO functions can be attributed to lack of information—the MPI implementation does not possess enough information to determine the optimal algorithm. However, many problems should be solvable by implementing smarter MPI-IO optimization algorithms. Much work still remains to be done in optimizing MPI-IO functionality!

Acknowledgments

This work was supported in part by the Mathematical, Information, and Computational Sciences Division subprogram of the Office of Advanced Scientific Computing Research, Office of Science, U.S. Department of Energy, under Contract DE-AC02-06CH11357.

References

1. Gropp, W., Huss-Lederman, S., Lumsdaine, A., Lusk, E., Nitzberg, B., Saphir, W., Snir, M.: MPI – The Complete Reference, The MPI Extensions, vol. 2. MIT Press, Cambridge (1998)
2. Gropp, W., Lusk, E., Thakur, R.: Using MPI-2: Advanced Features of the Message-Passing Interface. MIT Press, Cambridge (1999)
3. ROMIO: A high-performance, portable MPI-IO implementation, `http://www.mcs.anl.gov/romio`
4. Thakur, R., Choudhary, A.: An Extended Two-Phase Method for Accessing Sections of Out-of-Core Arrays. Scientific Programming 5(4), 301–317 (1996)
5. Thakur, R., Gropp, W., Lusk, E.: On implementing MPI-IO portably and with high performance. In: Proceedings of the 6th Workshop on I/O in Parallel and Distributed Systems, pp. 23–32. ACM Press, New York (1999)
6. Thakur, R., Gropp, W., Lusk, E.: Optimizing noncontiguous accesses in MPI-IO. Parallel Computing 28(1), 83–105 (2002)
7. Träff, J.L., Gropp, W., Thakur, R.: Self-consistent MPI performance requirements. In: Cappello, F., Herault, T., Dongarra, J. (eds.) PVM/MPI 2007. LNCS, vol. 4757, pp. 36–45. Springer, Heidelberg (2007)
8. Worringen, J., Träff, J.L., Ritzdorf, H.: Fast parallel non-contiguous file access. In: SC 2003: Proceedings of the 2003 ACM/IEEE conference on Supercomputing, p. 60. IEEE Computer Society, Washington (2003)

Performance Issues of Synchronisation in the MPI-2 One-Sided Communication API

Lars Schneidenbach, David Böhme, and Bettina Schnor

Institute for Computer Science
University of Potsdam
Germany
{lschneid,dboehme,schnor}@cs.uni-potsdam.de

Abstract. The efficient implementation of one-sided communication for cluster environments is a challenging task. Here, we present and discuss performance issues which are inherently created by the design of the MPI-2 API specification.

Based on our investigations, we propose an one-sided communication API called NEON. The presented measurements with a real application demonstrate the benefits of the NEON approach.

1 Introduction and Related Work

The message passing interface (MPI) standard is a state of the art programming interface for parallel applications in cluster computing. One-sided communication (OSC) is an extension specified in the MPI-2 standard [5]. It enables parallel applications to use remote memory access (RMA).

MPI-2-OSC defines so-called *access epochs*. An arbitrary number of communication calls can be synchronized within a single epoch. This bundling reduces the need to make synchronization calls for each transfer. MPI-2 provides several API calls to open and close such an epoch. Figure 1 shows some pseudo code of the typical use of one sided communication. The depicted post-start-complete-wait synchronization scheme is used in our experiments, later.

Recently, different implementation options were analysed in the literature [1,3,4,9]. While one would expect inefficiencies of OSC over transports without support for remote direct memory access (RDMA), interconnects like InfiniBand are expected to offer similar or better performance for OSC than for two-sided communication. However,

```
Node A:                       Node B:
- - - - - - - - - - - - - - - - - - - - - - - - -
MPI_Win_Create                MPI_Win_Create

MPI_Win_Start                 MPI_Win_Post
 <computation>                 <computation>
MPI_Put
 <computation>

MPI_Win_Complete              MPI_Win_Wait
```

Fig. 1. Pseudo code example of unidirectional asynchronous data exchange via one-sided communication from node A to node B

A. Lastovetsky et al. (Eds.): EuroPVM/MPI 2008, LNCS 5205, pp. 177–184, 2008.

measurements in the literature and our own measurements show a different picture. Asynchronous two-sided communication still offers the best performance if only a few number of communications are synchronised with one epoch [1,3,7].

One reason is the inefficient support of the pipeline model [7,8] which festlegt? to fill the bottleneck step as early as possible. The API forces the implementation to either violate the pipeline model or to send an extra synchronisation message. This is inefficient at least on networks like Gigabit Ethernet.

Also the so called *early sender problem* can occur: At the time of the put call, the sender has not received a buffer announcement (`MPI_Win_post`) from the target. This is a major issue for asynchronous operations, since the capability to overlap communication and computation is reduced. If there is no processor available to progress on these blocked operations, the communication has to be deferred and the communication pipeline is again inefficiently used.

In our earlier work, we focussed on the pipeline model and the comparison of two- and one-sided communication over Gigabit Ethernet [7]. In this paper we investigate the efficient synchronization in cluster environments over InfiniBand.

2 Synchronisation of Communication

This section discusses two major issues of synchronization of one-sided communication.

2.1 Separation of Notification and Completion

The predominant interaction between parts of parallel applications is assumed to be similar to the classical producer-consumer scheme known from synchronisation in operating systems. This is because each process has to know about two situations:

1. The origin of the communication has to know about the availability of the destination buffer.
2. The target has to know when the access epoch is completed.

To inform a remote process of these situations, two notifications have to be done by the application:

1. The target has to signal or *announce* an available buffer (*announcement*).
2. The origin has to *notify* the target when an access epoch is completed (*notification*).

Thus 4 API calls are required for both one-sided and two-sided communication. The only difference is that one-sided communication is able to synchronise more than one communication operation with a single notification signal. The announcement is always an explicit operation. Therefore, an API can not be improved here. Thus, we focus on the notification of the target process.

For the remainder of this paper, *notification* means the notification of the target process described above. By *completion* we mean the operation to wait for the initiated communication operations to finish. After the completion of the operation, the application can reuse the local buffers that were used for communication.

In the OSC-API of MPI-2, notification and completion of one-sided communication is included in a single API-call (`MPI_Win_complete` or `MPI_Win_fence`). This introduces a performance penalty because of two requirements:

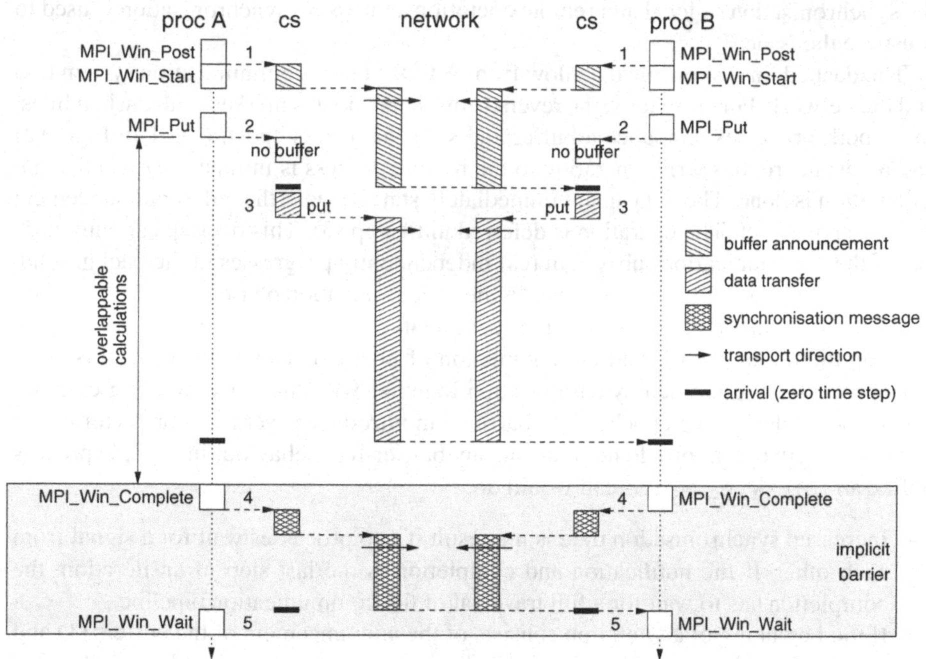

Fig. 2. Implicit barrier with bi-directional synchronisation

1. To allow a maximum overlapping of computation and communication, the completion of local operations should happen as late as possible. This is a well known technique for asynchronous two-sided communication. It also increases the ability of the application to tolerate process skew.
2. The notification message has to be sent as early as possible in order to let the target continue as early as possible to work with the new data in the destination buffer (e. g. in case of process skew).

Both requirements can not be met with a single API call.

This is not a problem for two-sided communication since the notification is implicitly included in the `MPI_Isend`-communication call and the completion is done by an `MPI_Wait` or `MPI_Test`. Notification and completion are separated, and both of the above requirements are met.

2.2 Implicit Barriers

Another problem of combined notification and completion is that programmers create an *implicit barrier* between processes with bidirectional communication. We call this barrier *implicit*, because it is not obvious to the programmer.

The problem is depicted in Figure 2. It represents one iteration of a bulk synchronous application with asynchronous one-sided communication. It shows the bi-directional interaction of two processes (proc A and B) based on the post-start-complete-wait synchronisation (pscw) in MPI-2 (see Figure 1). Since an `MPI_Win_fence`-call includes the synchronisation of local and remote operations, the pscw-synchronisation is used to illustrate the issue.

The dashed lines show the data flow from A to B via the communication system (cs) and the network. For simplicity, the reverse flow is not always marked with dashed lines. First, both processes announce a buffer and start the access epoch (see 1 in Figure 2) and an asynchronous write operation to the remote process is initiated (2). Then some calculation is done. The put can not immediately start because the buffer announcement has not arrived yet. The operation is deferred until step (3). This overlap can only happen if the communication subsystem (cs) independently progresses in the background. Otherwise, the data is transferred only in the synchronization phase.

At the end, the synchronisation is a combination of notifying the remote process and completion of put (4) and the completion of post (5). This is the critical issue if both processes use explicit synchronisation together with the corresponding completion of post (closing the epoch). This barrier can introduce several major performance penalties. Both operations do not indicate any barrier-like behaviour in a single process unlike an `MPI_Win_fence` call would do.

- Increased synchronisation time is the result if two processes wait for a signal from each other. If the notification and completion is the last step of an iteration, the completion has to wait for a full traversal of the communication pipeline.
- If the beginning of an iteration consists of the announcement of the buffers (1) and a following initiation of a data transfer (2), more *early sender problems* will occur. If the traversal of the announcement takes more time than any calculation between the post and the put, the data transfer has to be deferred. This results in inefficient usage of the communication pipeline.
- The barrier itself reduces the overall process skew tolerance of the asynchronous one-sided communication. Any delay in a process results in a delayed notification message and, thus, in increased synchronisation time. This results in increased iteration time of both processes.

3 Proposed Solution

To solve the above issues, we propose to separate notification from completion. Completion has to be a separate call since it has to be called as late as possible.

Since the separation will result in skewed processes, a leading process can announce the buffer and avoid an early sender problem at slower processes. Then, the slower process can make full use of communication overlap to potentially catch up the other process. Further, a process delay at the leading process will have less impact on the overall runtime. The implicit barrier is removed or better stretched out to be asynchronous and can be overlapped with computation.

The notification can be done in two ways: either the parameters of communication operations can be extended by a flag to mark this operation to be the last in this epoch (*implicit notification*), or by extending the API by a new routine to explicitly notify the remote process (*explicit notification*).

Some communication systems can handle implicit notification more efficiently. E. g. Gigabit Ethernet is less efficient if two messages have to be transmitted. With other transports like InfiniBand it is faster to send two messages via RDMA instead of using one message with notification of completion by events. Since implicit notification can be implemented on top of InfiniBand using two explicit messages (see Section 4), implicit notification is more efficient for different network capabilities.

Our proposal is implicit notification using a flag for one-sided communication operations. If the flag is not used by the programmer, the synchronisation can still work as long as notification is done at the completion call at the latest.

4 NEON

We designed an OSC-API called *NEw ONe sided communication interface* (NEON) which is intended to show that the performance of OSC can be improved by the design of the synchronisation calls. Therefore, only the basic synchronisation and completion calls are implemented together with a `NEON_Put` operation to transfer data. The main concepts are based on the producer-consumer synchronisation. These concepts are presented in more detail in [7]. All available routines are listed and shortly described in Table 1.

Table 1. Routines of the NEON-API

`NEON_Init`	Initialisation of the NEON environment
`NEON_Exit`	Cleanup
`NEON_Post`	Assign tags to a buffer and announce it
`NEON_Repost`	Reannounce a buffer if it is reused
`NEON_Unpost`	Tell the library that this buffer will not be reused
`NEON_Put`	Write data to a remote buffer, depending on parameters it includes synchronisation
`NEON_Wait`	Local completion of announcements and communication
`NEON_Register`	Registering memory in advance if the transport requires it
`NEON_Unregister`	De-registering memory.

A so called *non-final* flag is the important parameter of `NEON_Put` concerning the implicit notification. Usually, each put is taken as a final operation that allows the communication library to notify the remote process. If the programmer wants to put further data into the remote buffer, this flag has to be set.

5 Performance Measurements

In this section we compare our NEON prototype on top of InfiniBand with MVAPICH2 [6]. We use an application called *cellular automaton* [2] to evaluate the performance. It performs a 2D-stencil calculation similar to the popular game of life by John Horton Conway. We use a 1D-partitioning where each process calculates the same number of lines (10, 100, or 1000) and communicates the topmost and undermost line to its neighbors. Each line contains 1024 bytes of data.

We use MVAPICH2 (version 0.98) and NEON on 8 nodes of a cluster with 2.33 GHz Intel CPUs and Mellanox InfiniHost III. The latency and the bandwidth of NEON are

Table 2. Latency and Bandwidth

API	Latency (µs)	Bandwidth at 1KB (MiB/s)	Bandwidth at 1MB (MiB/s)
MPI	4.04	127.3	942.2
NEON	12.55	69.97	954.9

Table 3. Cellular automaton (iteration time) with NEON and MPI-2 over InfiniBand

API	CA10	CA100 (ms)	CA1000
MPI	0.0484	0.177	1.46
NEON	0.0309	0.147	1.43

worse than the results of MVAPICH2 (shown in Table 2). Especially, at the message size of the used cellular automaton ($\approx$ 1024 B), NEON achieves only 55 % of the bandwidth of MVAPICH2. We figured out that one major reason is the use of the *immediate-data*-feature of InfiniBand to notify the remote process in our implementation. The required event handling at the destination significantly increases the latency. This can be improved in future versions.

But the cellular automaton on top of NEON outperforms the MPI-2 version (see Table 3). The table shows the comparison with 3 different communication-computation ratios. Table 3 (CA10 means 10 lines per process). This shows the benefit of the NEON API.

We conducted two further experiments with the cellular automaton to show the weaknesses of the MPI-2 API. First, we move the `MPI_Put` towards the middle of the iteration by calculating an increasing number of lines before putting data. An iteration consists of 1000 lines to calculate. The results are shown in Figure 3 (delay put). It shows that the further the put is moved away from the announcement of the remote buffer at the beginning of the iteration, the better performance can be achieved.

The reason is the non-available buffer announcement of the target that occurs too short in advance of the communication. This is a problem that all RDMA-based implementations will have because they require the remote buffer to be announced[1]. In those cases, any implementation has to defer the transfer to the completion call. This is an appropriate solution if there's no network processor to progress on pending communication in the background. But, this increases the time for the implicit barrier and no communication overlap can happen at all. Due to the two-way interdependency between neighbouring processes, the same situation will occur in the next iteration.

The impact of the implicit barrier and the possibility to tolerate process skew is shown by a second experiment. The `MPI_Win_complete` call is gradually shifted from the beginning[2] to the end of the iteration. The results show that the best results are achieved if the completion is done in the middle of the iteration.

[1] Using intermediate buffers can solve this too, but introduces further issues that are out of focus of this paper.

[2] Right after the `MPI_Put`-call.

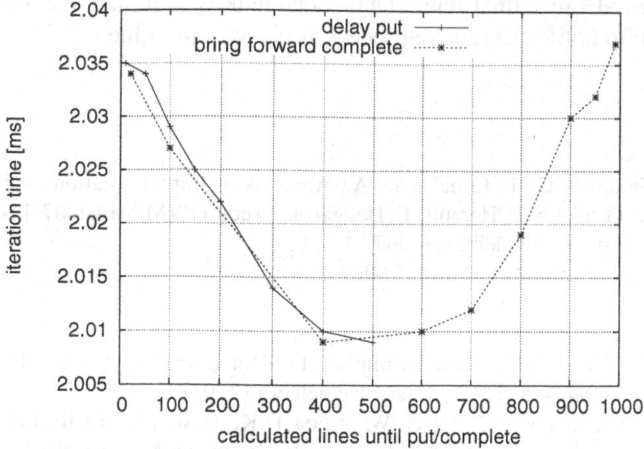

Fig. 3. Effects of shifting synchronisation or completion

Moving the completion too close to the communication call makes the communication more and more synchronous. The reason is that communication is enforced at the completion and the calculation afterwards can not overlap communication. Moving the completion too close to the end of the iteration will increase the negative impact of the barrier.

This experiment shows the benefits of the proposed separation of notification and completion in Section 3.

6 Conclusion

There are two important factors that hamper the performance of the MPI-2-OSC. First, the API forces the implementer to either violate the pipeline model or to send a late and explicit synchronisation message. Second, gluing together completion and remote notification of the target reduces the capability to overlap communication and computation.

Further, if the application uses bi-directional communication, it creates an implicit barrier. This is not obvious to the programmer as well as it reduces the tolerance to process skew.

The discovered issues can be solved by using an API that separates local completion and remote notification. We propose the use of a flag inside communication operations to increase portability. If an underlying transport performs better with implicit synchronisation, this way of notification is more efficiently implementable.

Based on this knowledge we designed NEON, a one-sided communication API that uses implicit notification. The presented measurements with a prototype implementation of NEON on top of InfiniBand prove our ideas. The benefits of NEON over Gigabit Ethenet are presented in our earlier work [7].

With a slight modification, the separation of notification and completion can also be built into the MPI-2 API. This can be done by extending the existing one-sided communication calls by a final-flag. The flag can make a call final to allow the communication system to notify the remote process as early as possible.

References

1. Barrett, B., Shipman, G.M., Lumsdaine, A.: Analysis of Implementation Options for MPI-2 One-Sided. In: Cappello, F., Herault, T., Dongarra, J. (eds.) PVM/MPI 2007. LNCS, vol. 4757, pp. 242–250. Springer, Heidelberg (2007)
2. The Cellular Automaton Application (2006), `http://www.cs.uni-potsdam.de/bs/cellularautomat/`
3. Gropp, W., Thakur, R.: An Evaluation of Implementation Options for MPI One-Sided Communication. In: Di Martino, B., Kranzlmüller, D., Dongarra, J. (eds.) EuroPVM/MPI 2005. LNCS, vol. 3666, pp. 415–424. Springer, Heidelberg (2005)
4. Huang, W., Santhanaraman, G., Jin, H.-W., Panda, D.K.: Design Alternatives and Performance Trade-Offs for Implementing MPI-2 over InfiniBand. In: Di Martino, B., Kranzlmüller, D., Dongarra, J. (eds.) EuroPVM/MPI 2005. LNCS, vol. 3666, pp. 191–199. Springer, Heidelberg (2005)
5. MPI-2: Extensions to the Message Passing Interface, Message Passing Interface Forum (July 1997)
6. MVAPICH: MPI over InfiniBand and iWARP (2008), `http://mvapich.cse.ohio-state.edu/`
7. Schneidenbach, L., Schnor, B.: Design issues in the implementation of MPI2 one sided communication in Ethernet based networks. In: PDCN 2007: Proceedings of the 25th conference on Parallel and Distributed Computing and Networks (PDCN), pp. 277–284. ACTA Press (Febuary 2007)
8. Wang, R.Y., Krishnamurthy, A., Martin, R.P., Anderson, T.E., Culler, D.E.: Modeling and Optimizing Communication Pipelines. In: Proceedings of the 1998 Conference on Measurement and Modeling of Computer Systems (SIGMETRICS). Madison (1998)
9. Woodall, T.S., Shipman, G.M., Bosilca, G., Graham, R.L., Maccabe, A.B.: High Performance RDMA Protocols in HPC. In: Mohr, B., Träff, J.L., Worringen, J., Dongarra, J. (eds.) PVM/MPI 2006. LNCS, vol. 4192, pp. 76–85. Springer, Heidelberg (2006)

Lock-Free Asynchronous Rendezvous Design for MPI Point-to-Point Communication*

Rahul Kumar, Amith R. Mamidala, Matthew J. Koop, Gopal Santhanaraman, and Dhabaleswar K. Panda

Network-Based Computing Laboratory
Department of Computer Science and Engineering
The Ohio State University
{kumarra,mamidala,koop,santhana,panda}@cse.ohio-state.edu

Abstract. Message Passing Interface (MPI) is the most commonly used method for programming distributed-memory systems. Most MPI implementations use a rendezvous protocol for transmitting large messages. One of the features desired in a MPI implementation is the ability to asynchronously progress the rendezvous protocol. This is important to provide potential for good computation and communication overlap to applications. There are several designs that have been proposed in previous work to provide asynchronous progress. These designs typically use progress helper threads with support from the network hardware to make progress on the communication. However, most of these designs use locking to protect the shared data structures in the critical communication path. Secondly, multiple interrupts may be necessary to make progress. Further, there is no mechanism to selectively ignore the events generated during communication. In this paper, we propose an enhanced asynchronous rendezvous protocol which overcomes these limitations. Specifically, our design does not require locks in the communication path. In our approach, the main application thread makes progress on the rendezvous transfer with the help of an additional thread. The communication between the two threads occurs via system signals. The new design can achieve near total overlap of communication with computation. Further, our design does not degrade the performance of non-overlapped communication. We have also experimented with different thread scheduling policies of Linux kernel and found out that the round robin policy provides the best performance. With the new design we have been able to achieve 20% reduction in time for a matrix multiplication kernel with MPI+OpenMP paradigm on 256 cores.

1 Introduction

Cluster based computing is becoming quite popular for scientific applications due to its cost effectiveness. The Message Passing Interface (MPI) is the most commonly used method for programming distributed memory systems. Many applications use MPI

* This research is supported in part by DOE grants DE-FC02-06ER25755 and DE-FC02-06ER25749, NSF Grants CNS-0403342 and CCF-0702675; grants from Intel, Sun MicroSystems, Cisco Systems, and Linux Networks; and equipment donations from Intel, AMD, Apple, IBM, Microway, PathScale, SilverStorm and Sun MircroSystems.

A. Lastovetsky et al. (Eds.): EuroPVM/MPI 2008, LNCS 5205, pp. 185–193, 2008.

point-to-point primitives to send large messages. Typical MPI implementations use a rendezvous protocol for transmitting large messages. The rendezvous protocol involves a handshake to negotiate buffer availability and then the message transfer takes place. The message transfer usually occurs in a zero-copy fashion.

One of the features desired in a quality MPI implementation is the ability to asynchronously progress the rendezvous protocol. This is important to provide potential for good computation and communication overlap to the applications. Many modern network interfaces offload network processing to the NIC and thus are capable of handling communication without the intervention of CPU. MPI provides non-blocking semantics so that the application can benefit from computation and communication overlap. The benefits of non-blocking semantics depend on the ability to achieve asynchronous progress. Thus, it is important to address this issue in the MPI implementation.

There are several designs that have been proposed previously to provide asynchronous progress. These designs typically use an additional thread to handle incoming rendezvous requests. For example, in [1], a RDMA Read based threaded design is proposed to provide asynchronous progress. Though the basic approach has been proven to achieve good computation and communication overlap, there are several overheads associated with the implementation of the design. First, the existing design uses locking to protect the shared data structures in the critical communication path. Second, it uses multiple interrupts to make progress. Third, there is no mechanism to selectively ignore the events generated. In this paper, we propose an enhanced asynchronous rendezvous protocol which overcomes these limitations. Specifically, our design does not require locks in the communication path. In our approach, the main application thread makes progress on the rendezvous transfer with the help of an additional thread. The communication between the two threads occurs via system signals.

We have incorporated our design in MVAPICH [2], a popular MPI implementation over InfiniBand. The new design can achieve almost total overlap of communication with computation. Further, our design does not reduce the performance of non-overlapped communication. We have also experimented with different thread scheduling policies of Linux kernel and found out that round robin policy provides the best performance. With the new design we have been able to achieve 20% reduction in time for a matrix multiplication kernel with MPI+OpenMP paradigm on 256 cores.

2 Background

2.1 InfiniBand Overview

The InfiniBand Architecture [3] (IBA) defines a switched network fabric for interconnecting compute and I/O nodes. InfiniBand supports two types of communication semantics. They are called *Channel* and *Memory* semantics. In channel semantics, the sender and the receiver both explicitly place work requests to their Queue Pair (QP). After the sender places the send work request, the hardware transfers the data in the corresponding memory area to the receiver end. In memory semantics, Remote Direct Memory Access (RDMA) operations are used instead of send/receive operations.

InfiniBand supports event handling mechanisms in addition to polling. In InfiniBand, the Completion Queue (CQ) provides an efficient and scalable mechanism to report completion events to the application.The CQ can provide completion notifications for both send and receive events as well as many asynchronous events. In the polling mode, the application uses an InfiniBand verb to poll the memory locations associated with the completion queue. In the asynchronous mode, the application does not need to continuously poll the CQ to look for completions. The CQ will generate an interrupt when a completion event is generated. Further, IBA provides a mechanism by which only "solicited events" may cause interrupts. In this mode, the application can poll the CQ, however on selected types of completions, an interrupt is generated. This mechanism allows interrupt suppression and thus avoid unnecessary costs (like context-switch) associated with interrupts.

2.2 Overview of MVAPICH Communication Protocols

MPI communication is often implemented using two general protocols:

Eager protocol: In this protocol, the sender process sends the message eagerly to the receiver. The receiver needs to provide buffers in advance for the incoming messages. This protocol has low startup overhead and is used for small messages.

Rendezvous protocol: The rendezvous protocol involves a handshake during which the buffer availability is negotiated. The message transfer occurs after the handshake. This protocol is used for transferring large messages. In the rendezvous protocol, the actual data can be transferred using RDMA write or RDMA read over InfiniBand. Both these approaches can achieve zero copy message transfer. MVAPICH [2] currently has both these modes for transferring data in the rendezvous protocol.

3 Existing Asynchronous Rendezvous Protocol

In this Section, we first explain the existing implementation for achieving asynchronous progress in the rendezvous protocol. The basic design was proposed in [1] and used InfiniBand's RDMA read capability together with IBA's event notification mechanism. Figure 1 (left) provides an overview of the approach. As shown in the figure, the main idea in achieving asynchronous progress is to trigger an event once a control message arrives at a process. This interrupt invokes a callback handler which processes the message and makes progress on the rendezvous. The required control messages which triggers the events in the existing scheme are: a) RNDV_START and b) RNDV_FINISH. In addition, the RDMA read completion also triggers a local completion event. This design provides good ability to overlap computation and communication via asynchronous progress. For example, if an application is busy doing computation, the callback handler can make progress via the interrupt mechanism. However, there are a couple of important details that arise in implementing the approach.

One main issue in the existing approach is the overhead of interrupt generation. As explained above, a total of three interrupts are generated for every rendezvous transfer of data. This can potentially degrade the performance for medium messages using this protocol. Further, it is not easy to provide for a mechanism to selectively ignore the events generated by the control messages. This feature can be used whenever the

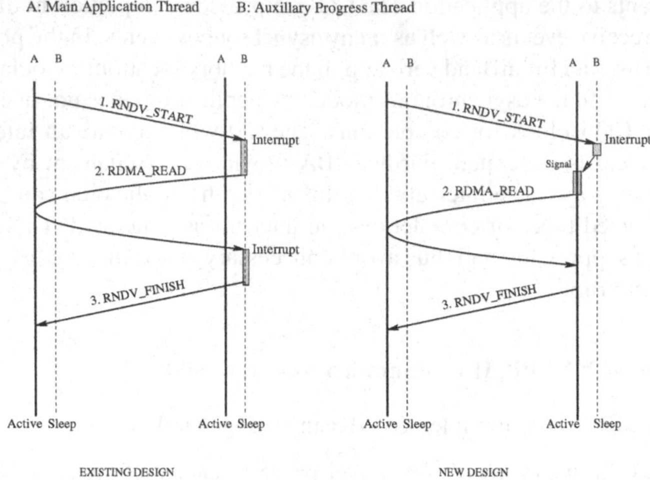

Fig. 1. Asynchronous Rendezvous Protocol Implementations

main application thread is already making progress and is expecting the control messages. Another important issue which cannot be overlooked is the overhead of locking/unlocking shared data structures. In this paper, we take into account all these issues and propose a new implementation alternative. Specifically, we aim to:

- Avoid using locks for shared data structures
- Reduce the number of events triggered by the control messages
- Provide for an ability for the process to selectively ignore the events generated

4 The Proposed Design

As explained above, the existing design has several limitations. In this Section, we explain our new approach of achieving asynchronous progress. Figure 1 (right) explains the basic idea in the new implementation. In our approach, each process creates an auxiliary thread at the beginning. The auxiliary thread waits for RNDV_START control message. As seen from the figure, the RNDV_START control message issued by the sender interrupts the auxiliary thread. This thread in turn sends a signal to the main thread to take the necessary action. This is different from the earlier approach where the auxiliary thread made progress on the rendezvous communication. Since, only one thread is involved with communication data structures, no locking mechanism is required for the data structures. In the second step, the main thread issues the RDMA_READ for the data transfer. After issuing RDMA_READ, the main thread resumes to perform the computation. Unlike the existing approach, the RDMA_READ completion does not trigger any interrupt in our design. We believe this interrupt does not help in overlap in Single Program Multiple Data (SPMD) programming model where each process performs the same task and the load is equally balanced. This was also observed in our experiments [4]. However, triggering of the interrupt on

RDMA_READ completion can be easily added to the protocol if required. In our design, the main thread sends the RNDV_FINISH message soon after it discovers the completion of RDMA_READ.

There are several benefits of this new design. First, locks are avoided thus reducing contention for shared resources. Also, in our design the signal from the auxiliary thread is disabled by the main thread when it is not expecting a message from any process. By doing so, the main thread is not unnecessarily interrupted by an unexpected message since it does not have the receive buffer address to make progress on the communication. The main thread also disables signal if it is already inside the MPI library and making communication progress. Since the main thread can disable the interruption from the auxiliary thread, the execution time of the application is unaffected if rendezvous protocol is not used by the application. Also, the signal is enabled only if a non-blocking receive has been posted and not for blocking receives. Also, at most of the time the auxiliary thread is waiting for interrupts from the NIC and does not perform any communication processing. Therefore, as the auxiliary thread is I/O bound the dynamic priority of the thread is very high which helps in scheduling it quickly. Finally, the new design also cuts down the number of interrupts to one thus improving the communication performance.

5 Experimental Evaluation

The experiments were conducted on 64 node InfiniBand Linux cluster. Each machine has a dual 2.33 GHz Intel Xeon "Clovertown" quad-core processors for a total of eight cores per node. Each node is connected by DDR network interface card MT25208 dual-port Memfree HCA by Mellanox [5] through a switch. InfiniBand software support is provided through OpenFabrics/Gen2 stack [6], OpenFabrics Enterprise Edition 1.2.

5.1 Comparison with Existing Design

Figure 2 shows the performance of basic bandwidth micro-benchmark. We used OSU Benchmarks [7] for the experiment. The legend 'no-async' refers to the basic RDMA read based rendezvous protocol without any enhancements for asynchronous progress, 'existing-async' refers to the existing asynchronous progress design proposed in [1] and 'new-async' refers to the proposed design described in Section 4. Figure 2 shows that the bandwidth of the proposed design closely matches with the base bandwidth numbers, which matches our expectations. However, with the old design the bandwidth is very low. In the bandwidth test, the receiver posts several requests and waits for the completion of all the pending messages. As several rendezvous start messages are received by the process, the auxiliary thread is continuously interrupted. Also, since the main thread is not involved in computation, both the threads concurrently poll the MPI library. The main thread cannot make any progress, however, it hinders the auxiliary thread from being scheduled on the processor. Therefore, due to exhaustion of CPU resources by the main thread the bandwidth performance is affected. The bandwidth performance is also nondeterministic as it depends on the scheduler to schedule the auxiliary process quickly. The effects of schedule is discussed in Section 5.3.

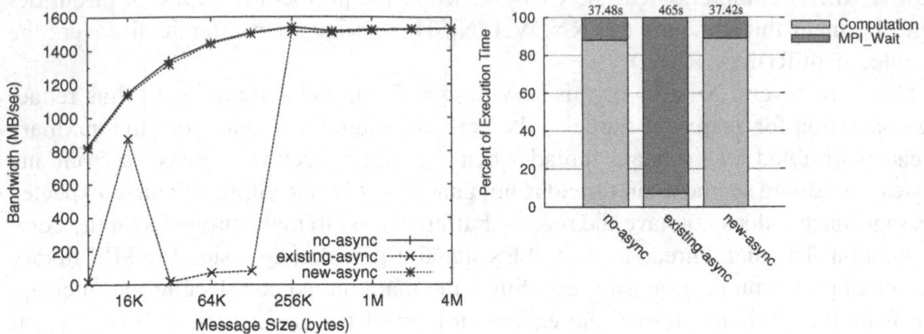

Fig. 2. Bandwidth for large messages **Fig. 3.** NAS-SP Normalized Execution Time

The performance of the new design is very similar to the base bandwidth performance since the main thread disables interrupts from the auxiliary thread when it is already inside the MPI library.

The poor performance of the existing design can be seen not only on micro-benchmarks but also in the performance of SP NAS Parallel Benchmark [8] application as can be seen in Figure 3. It can be seen from the figure that with the old design most of the execution time is wasted in MPI_Wait. In the remaining evaluations we do not show the performance of the old design. We found that the old design performs well when using an extra-core, however, it performs poorly when a single processor is assigned per process.

5.2 Overlap Performance

Figures 4 and 5 show the overlap performance of the proposed design. Sandia Benchmark [9] (SMB) has been used to evaluate the overlap capability of the implementation. Overlap potential at the receiver and at the sender have been shown in Figures 4 and 5 respectively. Since the base design and the proposed design employ RDMA read, almost total overlap is achieved at sender for both protocols. However, at the receiver the base

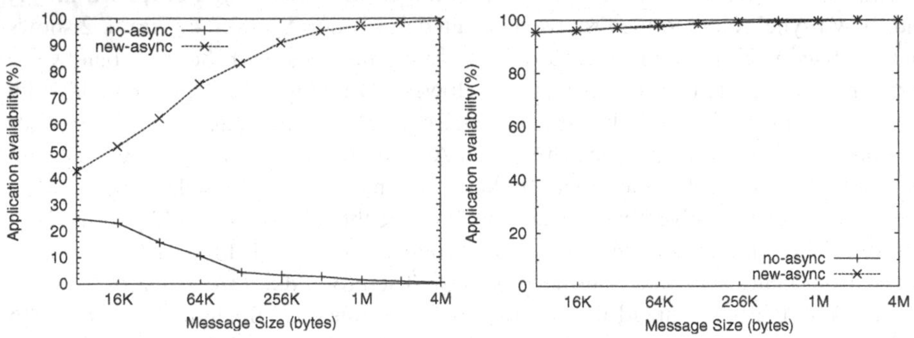

Fig. 4. Application availability at Receiver **Fig. 5.** Application availability at Sender

RDMA read based protocol offers no overlap, as expected. The proposed design is able to achieve increasing overlap with increasing message size and reaches almost 100% overlap for messages greater than 1MB.

5.3 Effect of Scheduling Algorithm

Figures 6 and 7 show the effect of scheduling algorithm on the overlap performance of the new design. Results for the default Linux scheduler, FIFO and Round robin have been compared. For each of the executions with different scheduling algorithm, the auxiliary thread is assigned the highest possible priority so that it is scheduled as soon as it is interrupted. Figure 6 shows the results for different message sizes. We observe that

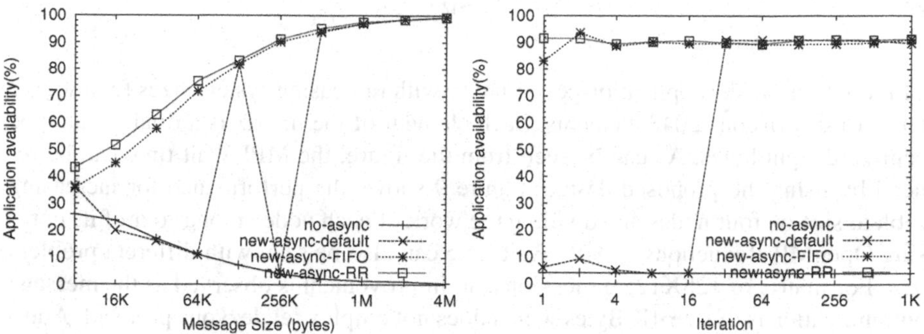

Fig. 6. Effect of scheduling algorithm

Fig. 7. Effect of increased time of execution

with the default scheduling algorithm, the performance is not consistent for all message sizes. At some message sizes the auxiliary thread is not scheduled on the processor on being interrupted. However, with FIFO scheduling algorithm the performance improves and is best for round-robin algorithm.

Figure 7 shows the overlap performance for 256KBytes message with increased number of iterations in each execution. From the figure, it is observed that with the default scheduling algorithm the performance of the design improves after a certain time interval. We feel that the improved performance is due to the dynamic priority scheme of Linux scheduling algorithm. Since the auxiliary thread hardly uses the CPU and is mostly waiting for completion events it is assigned a high dynamic priority which helps increase its performance. However, for FIFO and round robin the performance is optimal even for low number of iterations.

5.4 Application Performance

In this Section we use a matrix multiplication kernel to evaluate the application performance of the proposed design. The kernel uses Cannon's algorithm [10] and employs both MPI and OpenMP [11] programming models. In our experiment, we use only four cores per node. OpenMP programming model is used within the node and MPI is used for inter-node communication.

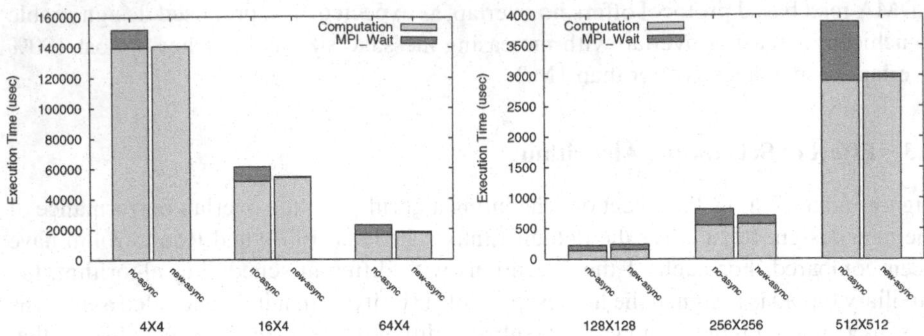

Fig. 8. Matrix Multiplication: Varying system size

Fig. 9. Matrix Multiplication: Varying problem size

Figure 8 shows the application performance with increasing system sizes for a square matrix of dimensions 2048 elements. Each element of the matrix is a double datatype occupying eight bytes. As can be seen from the figure, the MPI_Wait time can be reduced by using the proposed design. Figure 9 shows the performance for increasing problem size on four nodes and dividing the work of each node among four of its cores using OpenMP. Reductions in MPI_Wait time can also be seen with different problem sizes. For matrix of 128X128 dimensions, no improvement is observed as the message communication is of size 4K Bytes which does not employ rendezvous protocol. Additional results on application performance are included in [4].

6 Related Work

Several studies have been done to show the importance of overlap capability in MPI library. Brightwell et al. [12] show the ability of applications to benefit from such features. Eicken et al. [13] propose for hardware support for active messages to provide communication and computation overlap. In our design we provide a mechanism to achieve overlap with the current hardware capability. Schemes to achieve overlap in one-sided communication have been proposed in [14]. Surs et al. [1] propose thread based rendezvous protocol which employs locks for protection. However, in our design we propose a lock free mechanism to achieve overlap.

7 Conclusions and Future Work

There are several designs that have been proposed in the past to provide asynchronous progress. These designs typically use progress helper threads with support from the network hardware to make progress on the communication. However, most of these designs use locking to protect the shared data structures in the critical communication path. Secondly, multiple interrupts may be necessary to make progress. Further, there is no mechanism to selectively ignore the events generated during communication.

In this paper, we proposed an enhanced asynchronous rendezvous protocol which overcomes these limitations. Specifically, our design does not require locks in the communication path. In our approach, the main application thread makes progress on the rendezvous transfer with the help of an additional thread. The communication between the two threads occurs via system signals. The new design achieves almost total overlap of communication with computation. Further, our design does not reduce the performance of non-overlapped communication. We have also experimented with different thread scheduling policies of Linux kernel and found out that round robin policy provides the best performance. With the new design we have been able to achieve 20% reduction in time for a matrix multiplication kernel with MPI+OpenMP paradigm on 256 cores. In future, we plan to carry out scalability studies of this new design for a range of applications and system sizes.

References

1. Sur, S., Jin, H.W., Chai, L., Panda, D.K.: RDMA Read Based Rendezvous Protocol for MPI over InfiniBand: Design Alternatives and Benefits. In: Symposium on Principles and Practice of Parallel Programming (PPOPP 2006) (March 2006)
2. Network-Based Computing Laboratory: MVAPICH: MPI over InfiniBand and iWARP, http://mvapich.cse.ohio-state.edu
3. InfiniBand Trade Association: InfiniBand Architecture Specification, http://www.infinibandta.com
4. Kumar, R., Mamidala, A.R., Koop, M.J., Santhanaraman, G., Panda, D.K.: Lock-free Asynchronous Rendezvous Design for MPI Point-to-point Communication. Technical report, Dept. of Computer Science and Engineering, The Ohio State University (2008)
5. Mellanox: Mellanox Technologies, http://www.mellanox.com
6. OpenFabrics Alliance: OpenFabrics, http://www.openfabrics.org/
7. OSU Micro-Benchmarks, http://mvapich.cse.ohio-state.edu/benchmarks/
8. NAS Parallel Benchmarks (NPB),www.nas.nasa.gov/Software/NPB/
9. Sandia National Laboratories: Sandia MPI Micro-Benchmark Suite, http://www.cs.sandia.gov/smb/
10. Kumar, V., Grama, A., Gupta, A., Karypis, G.: Introduction to parallel computing: design and analysis of algorithms. Benjamin-Cummings Publishing Co., Inc. (1994)
11. OpenMP, http://openmp.org/wp/
12. Brightwell, R., Underwood, K.D.: An Analysis of the Impact of MPI Overlap and Independent Progress. In: ICS 2004: Proceedings of the 18th annual international conference on Supercomputing (March 2004)
13. Eicken, T.V., Culler, D.E., Goldstein, S.C., Schauser, K.E.: Active messages: a mechanism for integrated communication and computation. In: ISCA 1992: Proceedings of the 19th annual international symposium on Computer architecture. ACM, New York (1992)
14. Nieplocha, J., Tipparaju, V., Krishnan, M., Panda, D.K.: High performance remote memory access communication: The armci approach. Int. J. High Perform. Comput. Appl. (2006)

On the Performance of Transparent MPI Piggyback Messages*

Martin Schulz, Greg Bronevetsky, and Bronis R. de Supinski

Center for Applied Scientific Computing
Lawrence Livermore National Laboratory
PO Box 808, L-560, Livermore, CA 94551, USA
{schulzm,bronevetsky1,bronis}@llnl.gov

Abstract. Many tools, including performance analysis tools, tracing libraries and application level checkpointers, add piggyback data to messages. However, transparently implementing this functionality on top of MPI is not trivial and can severely reduce application performance. We study three transparent piggyback implementations on multiple production platforms and demonstrate that all are inefficient for some application scenarios. Overall, our results show that efficient piggyback support requires mechanisms within the MPI implementation and, thus, the interface should be extended to support them.

1 Motivation

Tools and support layers often must send additional information along with every message initiated by the main application. In most cases, tools must uniquely associate this additional information, often called piggyback data, with a specific message in order to capture the correct context and to avoid additional communication paths. A wide range of software systems piggyback data onto messages for diverse purposes; we detail a few here. Tracing libraries correlate send and receive events by sending vector clock information [4]. Performance analysis tools attach timing information to detect and analyze critical paths [2] or to compensate for instrumentation perturbation [7]. Application level checkpoint layers transmit epoch identifiers to synchronize global checkpoints [5].

Unfortunately, the MPI standard does not include a transparent piggyback mechanism. Instead, each system must provide its own, often ad-hoc, implementation. While a generic piggyback service could be added to an infrastructure like P^NMPI [6], the optimal solution depends on the specific usage scenario.

In this paper, we study the overhead and tradeoffs of three methods to support piggyback data:

– manual packing and unpacking the piggyback data and application payload into the same buffer;

* This work was performed under the auspices of the U.S. Department of Energy by Lawrence Livermore National Laboratory under contract DE-AC52-07NA27344 (LLNL-CONF-402937).

A. Lastovetsky et al. (Eds.): EuroPVM/MPI 2008, LNCS 5205, pp. 194–201, 2008.

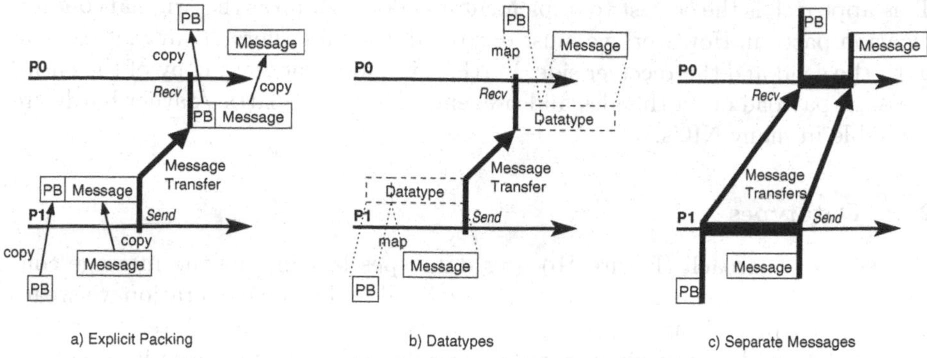

Fig. 1. Three different methods for transparent MPI piggyback messaging

– using datatypes with absolute addresses to attach piggyback data to the application payload; and
– using separate messages for the piggyback data and application payload.

All three methods are implemented as transparent PMPI modules.

Using these three mechanisms we discuss the impact on message latency and bandwidth, as well as application performance across three production platforms. Our results show that piggyback mechanisms implemented on top of MPI significantly reduce performance. In particular, communication intensive applications incur overhead that makes layered piggyback techniques inappropriate for performance critical scenarios, which significantly impedes the implementation of transparent tools. Thus, we need other mechanisms, potentially within the MPI implementation, that support this critical functionality.

2 MPI Piggyback Techniques

In the following we present three approaches to associate piggyback data with MPI messages transparently: explicitly packing piggyback data; using datatypes with absolute addresses; and sending separate piggyback messages. We focus on all combinations of MPI point-to-point messages including asynchronous and combined send/receive calls. We do not explore associating piggyback data with collective operations, which requires separate collectives tailored to the specific piggyback semantics (e.g., whether or not the data must be aggregated).

2.1 Explicit Pack Operations

Our first approach, shown in Figure 1a, uses *MPI_Pack* to pack the message payload and the piggyback data into a newly allocated buffer and transfers this buffer using the original send primitive. We receive the data into a local buffer and unpack the piggyback data and the message payload into their destinations.

This approach is the easiest to implement and does not alter the original communication pattern. However, we must create additional buffers of varying sizes on both the send and the receiver side. Further, it adds a memory copy of the entire message payload on both sides and prevents the use of scatter/gather hardware available in many NICs.

2.2 Datatypes

Our second approach (Figure 1b) uses datatypes to combine the message contents with the piggyback data. During each send and receive operation we create a new datatype with *MPI_Type_struct* that combines a global pointer to the piggyback data with the original, possibly user-defined, datatype description of the message payload. We then communicate the payload that includes the piggyback data through the new datatype with the original send/receive primitives. This approach avoids additional explicit memory copies but must construct a special datatype for each communication operation. While it is possible to reduce the number of individual datatypes created, a new datatype must be defined for each message buffer location since it must use absolute addresses. Further, many MPI implementations do not provide efficient implementations of complex datatypes and may lose the memory copy savings.

2.3 Separate Messages

Our third approach does not change the actual message. Instead, it transmits the piggyback data in a separate message, as shown in Figure 1c. This approach duplicates all send and receive calls, transmitting the original, unchanged message in one message and the piggyback data in the other. "Wildcard" receives, however, require special treatment. If the original receive operation does not specify an explicit source node or tag but instead uses `MPI_ANY_SOURCE` and/or `MPI_ANY_TAG`, we must first complete the receive with the wildcards, determine all message parameters, including the sender ID and tag, and then post the second receive without wildcards. In the case of asynchronous receives, this means that we can not post the second receive until after the test or wait operation has completed for the first receive.

3 Experimental Setup

For the following experiments we implement our three piggyback methods as separate, application transparent modules, using the PMPI profiling layer and we explore both sending or packing piggyback data before or after the message payload. To reduce complexity, to enable code reuse, and to simplify experimentation we use $P^N MPI$ [6] for common services (e.g., request tracking) and to load the different modules dynamically. We use an additional driver module that emulates a tool or library using the piggyback functionality. This driver requests predefined piggyback sizes and produces and consumes the piggyback data.

We use three different machines for our experiments: *Thunder*, a 1024 node Linux cluster with 1.4 GHz 4-way Itianium-2 nodes and a Quadrics QSNetII (elan4) network running Quadrics MPI; *Atlas*, a 1152 node Linux cluster with 2.4 GHz 4-way dual-core Opteron nodes and an Infiniband network running MVAPICH; and *uP*, a 108 node AIX cluster with 1.9 GHz 8-way Power5 nodes and a Federation Switch network running IBM's MPI implementation.

4 Results

We first evaluate the impact of our piggyback methods on point-to-point bandwidth and latency using a simple ping-pong test between two tasks on different nodes sending arrays of integers. Figure 2 shows the results for varying message sizes and a constant piggyback size of one four byte integer. This represents a typical scenario since most tools use small piggyback messages.

The results show that explicit packing always leads to the worst performance while sending two separate messages has the best. In particular, using separate piggyback messages usually incurs only a negligible bandwidth reduction. The other methods, packing or using datatypes, incur a similar, larger bandwidth reduction on both commodity platforms (*Thunder* and *Atlas*), suggesting that their respective datatype implementations perform internal packing similar to our explicit packing approach. In contrast, it appears that IBM's MPI implementation on *uP* optimizes the transfer of custom datatypes, resulting in higher bandwidth compared to the packing approach. All approaches incur significant latency penalties of up to 200%. On *Thunder*, all approaches show similar latency overhead, while on *Atlas* and *uP* the datatype method has higher latency overhead compared to the other two approaches, which suggests a higher relative cost for the repeated creation and destruction of the custom datatypes. Sending or packing the piggyback data before or after the actual payload has little or no impact on raw latency and bandwidth in general.

The remainder of our analysis focuses on *Thunder* and *uP* since the *Atlas* and *Thunder* results are similar. Figure 3 presents latency (using 4 byte messages) and bandwidth (using 512 Kbyte messages) results with varying piggyback sizes. We observe an almost constant bandwidth reduction independent of the size of the piggyback data since performance is dominated by the significantly larger message payload. Latency, on the other hand, directly depends on the piggyback data size and is further influenced by changes in the underlying message protocol triggered by the additional payload (as indicated by the changing slopes).

However, impact on bandwidth and latency does not necessarily translate into application overhead. Thus, we study the performance of two scientific applications: Sweep3D, a computation-bound 3D neutron transport code from the ASCI Blue benchmark suite [1], and SMG2000, a communication-intensive semi-coarsening multigrid solver from the ASC Purple benchmark suite [3]. We execute both codes on 16 processors using a global working set size of 120x120x120 for Sweep3D and a local working set size of 70x70x70 for SMG2000. We run each configuration five times and report the lowest execution time. We report on two versions of each code: one that uses wildcard receives; and one that precisely

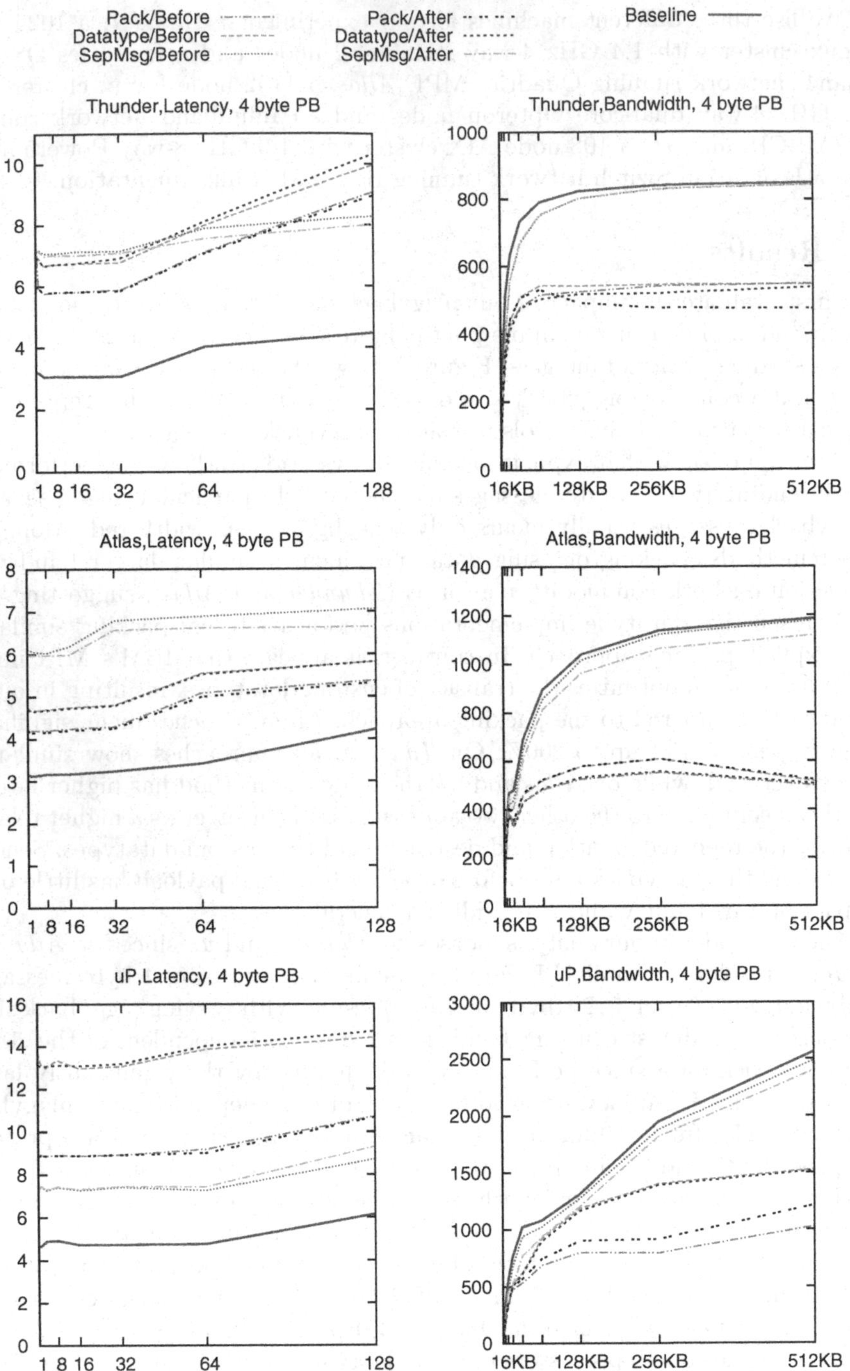

Fig. 2. Point to point transfer time in μs (left) and bandwidth in MB/s (right) with 4 byte piggyback data for varying message sizes shown on the x-axis

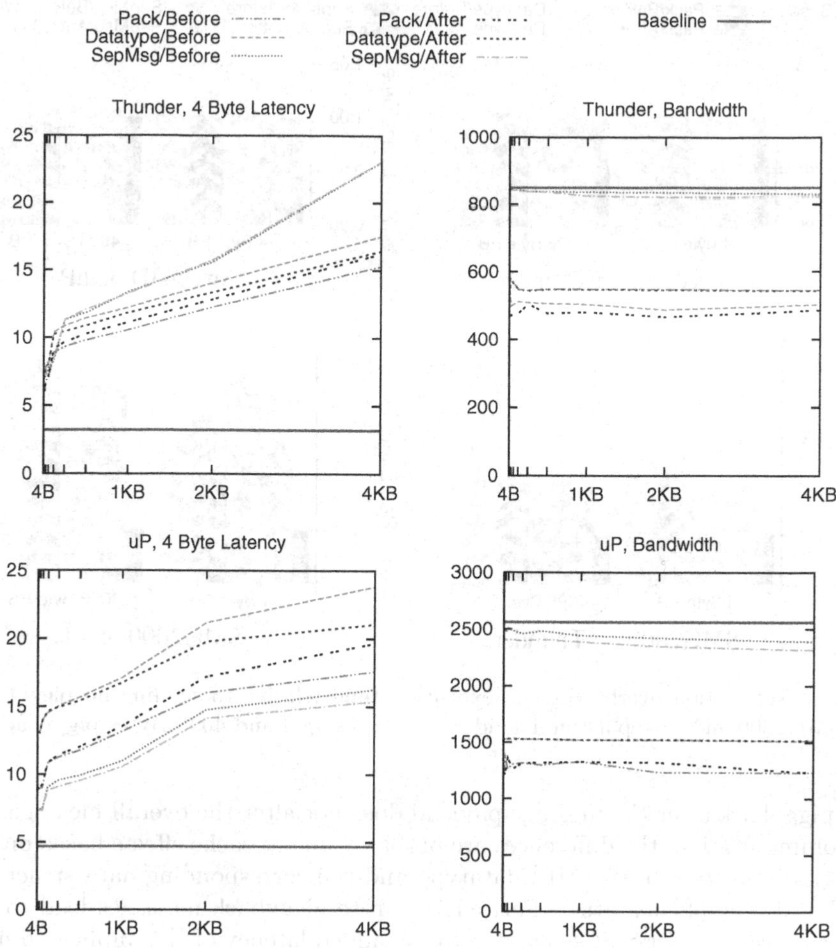

Fig. 3. Point to point latency in μs (left) and bandwidth in MB/s (right), varying piggyback size from 4 bytes to 4 Kbyte, as shown on the x-axis

specifies all receive parameters. These versions allow us to investigate the impact of the separate message method on wildcard receives, where the second receive must be postponed, as described in Section 2.3. This can have a notable effect on the performance of asynchronous wildcard communication.

The results, shown in Figure 4, reveal a negligible overhead for any piggyback implementation for the computation-bound Sweep3D. On *Thunder* the overhead is practically within the limits of measurement accuracy, while we see small overheads on *uP* (about 1% to 2%). The communication-bound SMG2000, on the other hand, presents a different picture: the average overhead on *Thunder* is around 10%, with the best performance being achieved by either packing the piggyback data after the message or using datatypes that prepend the piggyback data. Since packing the piggyback data first or using a datatype to place

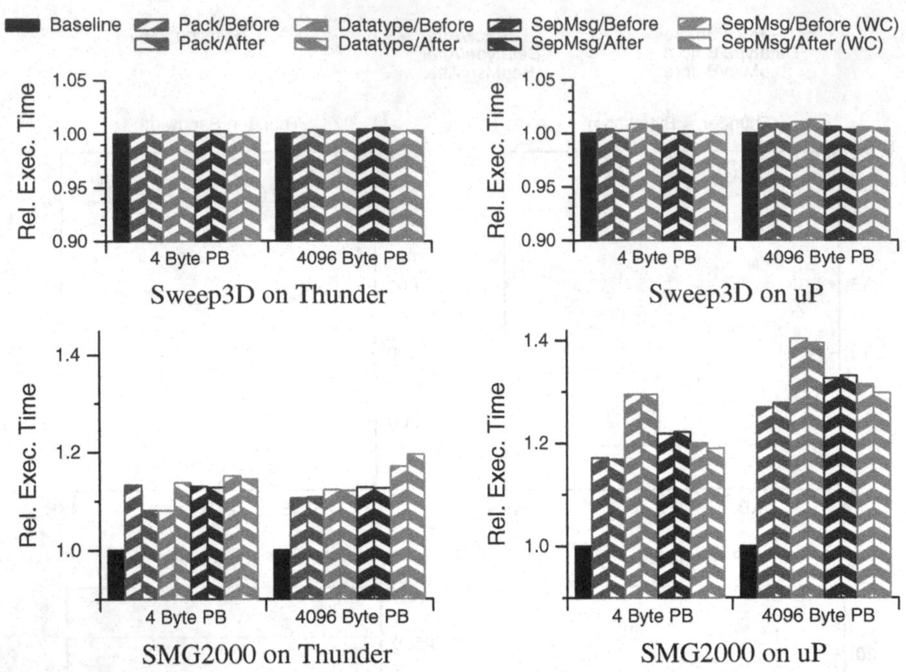

Fig. 4. Application overheads, i.e., execution time relative to sending no piggyback, for SMG2000 and Sweep3D on Thunder and uP using 4 and 4096 Bytes piggybacks

the piggyback after the message payload does not alter the overall message size or communication, the differences are probably due to cache effects based on the traversal direction of the MPI datatype and the corresponding data structures in the MPI implementation. Further, separate piggyback messages lead to the highest overhead for SMG2000, due to the added latency of the duplicated message traffic and the complex completion semantics during *MPI_Waitall*, which SMG2000 uses extensively. Using wildcard receives (marked as WC in the figure) further increases overhead to almost 20% for larger piggybacks.

We see even larger overheads on *uP* (20% for small piggybacks and 30% for large) with explicit packing achieving the best performance. In contrast to *Thunder*, using separate messages does not have the highest overhead compared to the other methods and using wildcard receives actually reduces overhead slightly. The latter most likely stems from the smaller number of outstanding requests the MPI implementation has to deal with since piggyback receives are postponed. Using datatypes exhibits the worst performance, most likely due to SMG2000's extensive use of user defined datatypes, as opposed to the simple integer arrays in the ping-pong test. Wrapping SMG200's already complex datatypes into the new piggyback datatypes prevents some of the previously observed optimizations for datatypes under IBM's MPI implementation.

5 Conclusions

Associating extra data with individual messages is an essential parallel tool technique, with uses in performance analysis, debugging and checkpointing. In this paper we have studied the performance of three different piggyback approaches — packing piggyback data and the original message into a unified buffer, using datatypes to transmit piggyback and message data in a single message, and using separate messages for piggyback and message data — and contrasted them to uninstrumented baselines.

We show that the choice of piggyback method strongly depends on the target application: simply examining the impact on bandwidth and latency can mispredict the real impact. In addition, performance depends on the MPI implementation and its optimization of advanced mechanisms such as custom datatypes or message coalescing. However, in communication intensive codes, like SMG2000, the use of piggyback data, independent of the implementation choice, leads to unacceptably high overhead for most tools.

In summary, our results show that we cannot layer a fully general piggyback solution on top of MPI without significantly harming performance for some application scenarios. However, efficient, low perturbation tools require exactly such an implementation. We therefore advocate extending the MPI standard to include a piggyback mechanism and are working with the MPI Forum towards this goal. Such extensions would allow the MPI implementation to optimize piggyback transfers, e.g., by including the data into a configurable header, and provide a truly portable and generally efficient piggyback mechanism.

References

1. Accelerated Strategic Computing Initiative. The ASCI sweep3d benchmark code (December 1995), `http://www.llnl.gov/asci_benchmarks/asci/limited/sweep3d/`
2. Barnes, B., Rountree, B., Lowenthal, D., Reeves, J., de Supinski, B.R., Schulz, M.: A Regression-Based Approach to Scalability Prediction. In: Proceedings of the International Conference on Supercomputing (June 2008)
3. Falgout, R., Yang, U.: Hypre: a Library of High Performance Preconditioners. In: Sloot, P.M.A., Tan, C.J.K., Dongarra, J., Hoekstra, A.G. (eds.) ICCS-ComputSci 2002. LNCS, vol. 2331, pp. 632–641. Springer, Heidelberg (2002)
4. Kranzlmüller, D., Volkert, J.: NOPE: A Nondeterministic Program Evaluator. In: Zinterhof, P., Vajtersic, M., Uhl, A. (eds.) ACPC 1999 and ParNum 1999. LNCS, vol. 1557, pp. 490–499. Springer, Heidelberg (1999)
5. Schulz, M., Bronevetsky, G., Fernandes, R., Marques, D., Pingali, K., Stodghill, P.: Implementation and Evaluation of a Scalable Application-level Checkpoint-Recovery Scheme for MPI Programs. In: Supercomputing 2004 (SC 2004) (November 2004)
6. Schulz, M., de Supinski, B.R.: P^NMPI tools a whole lot greater than the sum of their parts. In: Supercomputing 2007 (SC 2007) (2007)
7. Shende, S., Malony, A.D., Morris, A., Wolf, F.: Performance Profiling Overhead Compensation for MPI Programs. In: Proceedings of the 12th European PVM/MPI Users' Group Meeting, September 2005, pp. 359–367 (2005)

Internal Timer Synchronization for Parallel Event Tracing

Jens Doleschal, Andreas Knüpfer, Matthias S. Müller, and Wolfgang E. Nagel

ZIH, TU Dresden, Germany

Abstract. Performance analysis and optimization is an important part of the development cycle of HPC applications. Among other prerequisites, it relies on highly precise timers, that are commonly not sufficiently synchronized, especially on distributed systems. Therefore, this paper presents a novel timer synchronization scheme especially adapted to parallel event tracing. It consists of two parts. Firstly, recording synchronization information during run-time and, secondly, subsequent correction, i.e. transformation of asynchronous local time stamps to synchronous global time stamps.

Keywords: Event tracing, clock synchronization, performance analysis.

1 Introduction

Performance analysis and optimization is an important part of the development cycle of HPC applications. With today's complex and massively parallel hardware platforms it is most essential in order to achieve even a tolerable share of the peak performance.

Profiling and event tracing are well known approaches for performance measurement of parallel programs. For both there are many academic and commercial tools available that allow application developers to focus on the important task of optimization instead of the necessary task of performance measurement.

Both techniques rely on highly precise timers, usually CPU cycle counters, to evaluate the speed of individual activities during execution. While profiling is accumulating average run-time statistics for repeated activities, tracing records all repeated instances together with precise timing information.

Distributed profiling and tracing always rely on multiple local timers. This is sufficient to evaluate local run-time behavior, e.g. execution time of functions or waiting time due to communication or I/O. Distributed activities are measured with respect to multiple local timers. If they are not sufficiently synchronized, this will indicate incorrect performance values as measurement of speed is very sensitive to timing errors.

Even though some HPC platforms provide global high precision timers, on the majority of platforms additional synchronization is necessary, in particular on commodity cluster platforms. Widely used synchronization of *system timers*, e.g. with NTP [9], is no sufficient solution to this problem, because system timers are far to imprecise.

This paper presents a novel timer synchronization scheme for parallel event tracing. It consists of two parts. Firstly, recording synchronization information during run-time and, secondly, subsequent correction, i.e. transformation of local (asynchronous) time stamps to global time stamps.

A. Lastovetsky et al. (Eds.): EuroPVM/MPI 2008, LNCS 5205, pp. 202–209, 2008.

The rest of this paper is organized like following: After an examination of timer fluctuations and their effects on tracing a new solution will be presented, including an re-designed communication scheme and post-mortem timer correction. Next, the practical relevance is demonstrated with a real-world example. Finally there is an overview on related work and conclusions together with an outlook to future development.

2 Effects of Insufficiently Synchronized Timers

Using inaccurately synchronized timers for performance analysis of message-passing applications will result in an erroneous representation of the program trace data. The errors can be classified into two groups:

Q1 Qualitative error: Violation of the logical order of distributed events, e.g. messages that seem to be sent backwards in time because the receive event is situated before the send event.

Q2 Quantitative error: Distorted time measurement of distributed activities. This leads to skewed performance values, e.g. for message speed, when dividing by incorrect duration values.

To avoid Q1, the synchronization error of all local timers must not exceed one half of the minimum message latency. The presence of Q1 errors can always be detected by checking logic constraints of message passing behavior. Unlike the previous, Q2 errors are always present due to inevitable measurement errors. More accurate synchronization will decrease the impact of Q2. However, Q2 errors are in general undetectable unless Q1 is present at the same time or physical pre-conditions are violated.

A suitable timer synchronization algorithm needs to consider actual behavior of timers, e.g. CPU clock cycle counters. Figure 1 shows the non-linear behavior of seven separate timers C_i relative to one master timer C_0. All timers have been synchronized at begin and end, in between they are linearly interpolated. Even though the fluctuations are as small as 10^{-7} (deviation per time) they are large enough to cause errors of type Q1. This particular example was measured on a cluster of multi-core AMD Opteron nodes with 2.6 GHz CPU speed (the timer resolution). It uses Infiniband interconnect with minimum message latency of $3\ \mu s$ for small messages.

3 Parallel Synchronization Scheme

To provide a low-overhead and well scalable tracing environment the underlying synchronization scheme has to have the same properties. Timer synchronization with low disturbance to concurrent and following events while tracing a message-passing application is a complex issue. It requires several properties:

- balanced, low synchronization overhead,
- portable, scalable and robust synchronization algorithm,
- restore the relationship of concurrent events,
- accurate mapping of the event flow for an enhanced performance analysis.

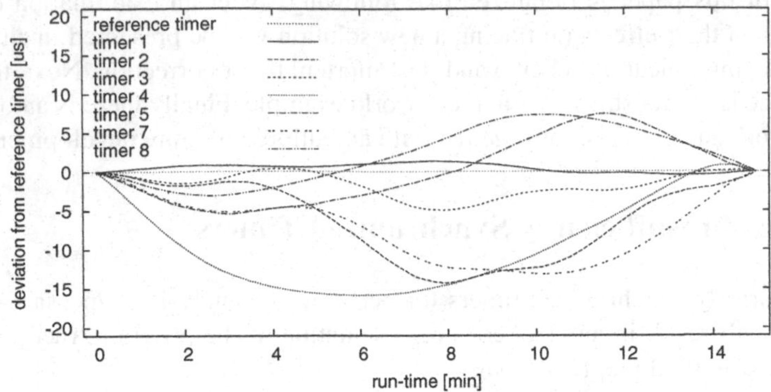

Fig. 1. Typical timer fluctuation for distributed CPU timers on a cluster platform with minimum message latency of 3 μs. The deviation from the linear behavior (y-axis in μs) is 10^{-7}, yet it is sufficient to cause Q1 errors in the course of few minutes run-time.

The overhead of the synchronization scheme while tracing an application depends on the number of exchanged messages and synchronization phases and can be reduced to a minimum if only the synchronization data will be recorded at run-time and the time stamps modified in a post-processing step. In the latter step, the actual timer parameters are determined by a linear interpolation between synchronization points. Then the local time stamps in the traces are corrected accordingly. Even though the timers are almost linear and monoton, there are small fluctuations in the drift. Therefore synchronization errors will accumulate over long intervals. This makes a linear begin-to-end correction insufficient for long trace runs, compare Figure 1.

Therefore, it is necessary to synchronize the timers frequently and to piecewise interpolate the timer parameters between the synchronization phases. The length of the interval between two successive synchronization phases has to be chosen well suited. Depending on the length, the overhead will be increased or the accuracy affected. The determination of the synchronization frequency while using post-processing can only be done by using the expectation of the clock drifts.

At appropriate points in the program flow the global synchronization with an explicit communication scheme will be inserted. This provides precise information about concurrency of local timers at a specific moment. Collective functions with an explicit or implicit barrier (e.g. most of the MPI collective functions associated with `MPI_COMM_WORLD`) are especially well suited for inserting global synchronization points, because this reduces the disturbance of concurrent events.

3.1 Communication Patterns for Parallel Synchronization

Within each synchronization phase, a specially designed message pattern monitors pairwise clock alignment. In order to establish relationships between the local timers the explicit communication uses pairwise ping-pong messages to define concurrency with low uncertainty. For a n-way parallel synchronization timer information have to distributed to and from all participants.

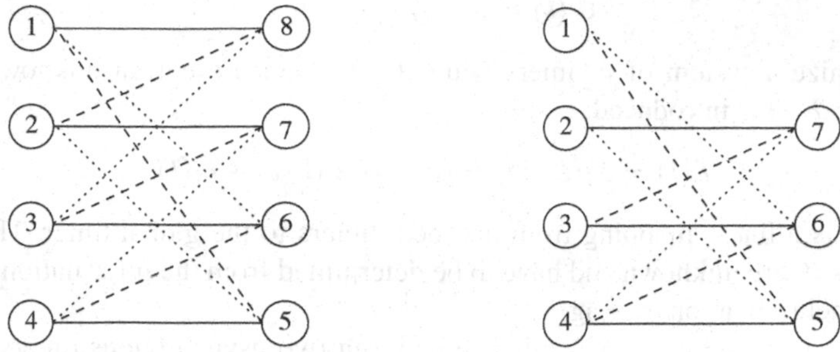

Fig. 2. Specially designed communication pattern with edge-coloring for odd and even number of clocks

An appropriate communication scheme for a parallel synchronization is formulated as a graph coloring problem with $G = (V, E)$. The synchronization graph G has to be a connected, valid edge-colored graph, i.e. all edges incident to the same node are differently colored, $G = (V, E)$ with $n \geq 3$ and

$$\Delta(G) \leq k = \lceil log_2 n \rceil, \ \delta(G) \geq 2,$$
$$|d_G(v) - d_G(w)| \leq 1 \ \forall v, w \in V, \ v \neq w.$$

With these properties the communication scheme guarantees that all processes with local timers are involved in the process of synchronization. The restriction to the vertex degrees is done to provide a load-balanced and robust communication scheme, i.e. a timer can be estimated along different paths of the graph. The edge-coloring is used to create a controlled communication scheme, such that the synchronization phase is divided into $k = \lceil log_2 n \rceil$ time slots. Within each time slots every participant has at most one communication partner. This avoids disturbances by other processes during the critical process of exchanging synchronization messages, e.g. a single process cannot be swamped by a number of synchronization messages.

Let the number of timers be even. A solution of the synchronization graph problem can be given by using a bipartite graph. With the first two colors an alternately edge-colored Hamilton cycle is constructed. With the remaining colors the graph is filled up with edges, such that the graph remains valid edge-colored. The resulting graph is a bipartite, k-regular graph colored with k colors and containing a Hamilton cycle. For odd numbers of timers a solution can be derived from the solution with $n+1$ vertices by deleting one vertex and all adjacent edges. The effort of this communication scheme uses $O(n * logn)$ synchronization messages and requires $O(\log n)$ time for n participants.

The left graph in Figure 2 shows a solution of the synchronization graph problem. By deleting vertex eight and all adjacent edges the left graph can be transformed into the right graph, that is a valid solution for odd numbers of timers.

3.2 Time Stamp Correction

For a short synchronization phase the drift of a physical timer can be assumed to be constant if the physical parameters (e.g. temperature) of the environment are constant. Therefore a timer model has a linear dependency from offset a_i and drift b_i [8]:

$$C_i(t) = a_i + b_i * t + \varepsilon_i. \tag{1}$$

To synchronize a system of n timers without a reference timer an unknown perfect global timer $\mathcal{T}(t)$ is introduced:

$$\mathcal{T}(t) = F_i(C_i(t)) = \alpha_i + \beta_i * C_i(t) + \delta_i(t) \tag{2}$$

Thereby, F is a linear mapping from the local timers to the global timer. The offsets α_i and drifts β_i are unknown and have to be determined from the information achieved from the synchronization messages.

Figure 3 shows the message exchange between two asynchronous timers within a time slot of a synchronization phase. If there is no adequate correction of the time stamps the red message leads to violate the clock condition (Q1) and the other messages will offer skewed performance values (Q2). With the global timer (2) a comparison of the logic constraints of message-passing events can be achieved (clock condition):

$$F_i(t_{ij}^{[k]}) + d_{ij}^{[k]} = F_j(r_{ij}^{[k]}) \tag{3}$$

$$F_i(r_{ji}^{[k]}) - d_{ji}^{[k]} = F_j(t_{ji}^{[k]}) \tag{4}$$

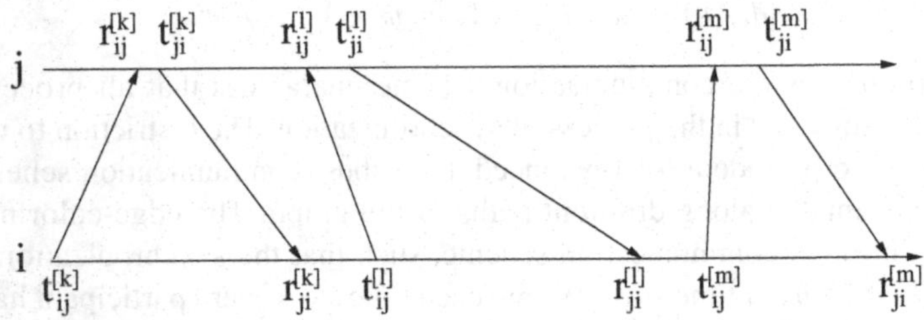

Fig. 3. Synchronization message exchange between two asynchronous clocks within a time slot. If the time stamps will not be corrected the red message leads to violate the clock condition.

While the timers are not synchronized an exact determination of the message delays $d_{ij}^{[k]}$ and $d_{ji}^{[k]}$ cannot be given. Therefore, the synchronization model uses a technique introduced in [3, 4] to estimate the message delays. The solution corresponds to the maximum likelihood estimator presented in [6].

Using the clock condition of message passing events (Equations (3), (4)) and the estimated synchronization message delays $\hat{d}_{ij}$ a global linear system for internal timer synchronization can be achieved. The estimator for this linear system uses a least-square regression analysis to estimate the offsets α_i and drifts β_i [2]. If the δ_i in (2) are independent and identical distributed then the expectation of the error of the global linear system $\delta_{ij} = \delta_j - \delta_i + d_{ij} - \hat{d}_{ij}$ is equal zero.

4 Practical Experiments

The practical relevance of the synchronization scheme is demonstrated by tracing the *130.soccoro* benchmark from the SPEC MPI suite. It was executed on an AMD Opteron

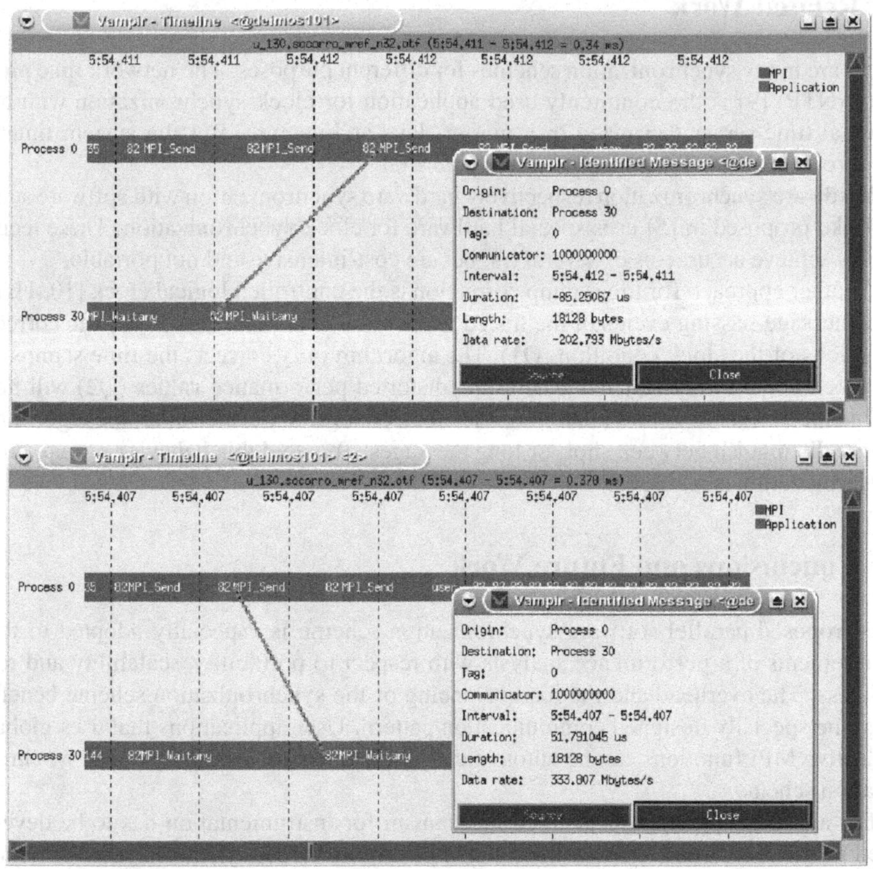

Fig. 4. Apparently backward running message from process P0 to P30 after approx. 6 minutes of run-time (top) and the same situation with frequent time re-synchronization (bottom).

cluster with 2.6 GHz clock rate connected via an Infiniband network with 3 μs latency for small messages. The application runs approximately 16.5 minutes with 32 processes distributed over 32 different cluster nodes.

Figure 4(top) shows a zoom of the timelines for the processes P0 and P30. The underlying timers were synchronized only at the very begin and the very end of the trace run [7]. A message from P0 to P30 is obviously running backwards in time, after about 6 minutes of run-time. Figure 4(bottom) shows the situation synchronized with the presented global synchronization scheme using re-synchronization after every 120 s. The overhead of the ten inserted synchronization phases was approximately 6.5 s. The violation of the clock condition (Q1) was eliminated, and the performance values of the message seem plausible (Q2).

5 Related Work

There are many synchronization schemes for different purposes. The network time protocol (NTP) [9] is the commonly used application for clock synchronization with an external time-server organized in a master-slave architecture. But the system timers synchronized by NTP have only limited precision.

Hardware synchronization respectively hardware synchronization with software support like proposed in [5] uses special hardware for clock synchronization. These techniques achieve accuracies of several ns, but are cost-intensive and not portable.

Another approach for time stamp correction is the controlled logical clock [10, 1]. It uses message-passing events of the traced application and their relationships to correct violations of the clock condition (Q1). The algorithm only corrects the time stamps if the clock condition is violated. Therefore, distorted performance values (Q2) will not be enhanced if there is no violation of the clock condition. Furthermore, the algorithm cannot distinguish between short or long messages. To avoid this behavior a weak pre-synchronization is advisable.

6 Conclusions and Future Work

The proposed parallel software synchronization scheme is especially adopted to the requirements of a performance analysis with respect to portability, scalability and robustness. The overhead and the load-balancing of the synchronization scheme benefit from the specially designed communication pattern. User applications that uses global collective MPI functions can be automatically instrumented with the parallel synchronization scheme.

For all other applications a suited mechanism for instrumentation has to be developed in the near future. Furthermore, the applicability of the synchronization scheme, in terms of accuracy, error distribution in the linear system, overhead studies and synchronization frequency, has to be investigated on various platforms in the near future.

References

[1] Becker, D., Rabenseifner, R., Wolf, F.: Timestamp Synchronization for Event Traces of Large-Scale Message-Passing Applications. In: Proceedings of the 14th European Parallel Virtual Machine and Message Passing Interface Conference (EuroPVM/MPI). Springer, Paris (September 2007)

[2] Doleschal, J.: Uhrensynchronisierung in parallelen Systemen mit expliziter Kommunikation. Master's thesis, Dresden University of Technology (2008)

[3] Gurewitz, O., Cidon, I., Sidi, M.: Network Classless Time Protocol Based on Clock Offset Optimization. IEEE/ACM Transactions on Networking 14(4) (2006)

[4] Gurewitz, O., Cidon, I., Sidi, M.: One-Way Delay Estimation Using Network-Wide Measurements. IEEE Transactions on Information Theory 52, 2710–2722 (2006)

[5] Horauer, M.: Clock Synchronization in Distributed Systems. PhD thesis, Technische Universität Wien, Fakultät für Elektrotechnik (February 2004)

[6] Jeske, D.R.: On Maximum-Likelihood Estimation of Clock Offset. IEEE Transactions on Communications 53(1), 53–54 (2005)

[7] Knüpfer, A., Brunst, H., Doleschal, J., Jurenz, M., Lieber, M., Mickler, H., Müller, M.S., Nagel, W.E.: The Vampir Performance Analysis Tool-Set. In: 2nd HLRS Parallel Tools Workshop (2008)(Submitted for publication)
[8] Maillet, E., Tron, C.: On efficiently implementing global time for performance evaluation on multiprocessor systems. Journal of Parallel and Distributed Computing 28(1), 84–93 (1995)
[9] Mills, D.L.: Improved Algorithms for Synchronizing Computer Networks Clocks. IEEE/ACM Transactions on Networking 3(3), 245–254 (1995)
[10] Rabenseifner, R.: Die geregelte logische Uhr, eine globale Uhr für die tracebasierte Überwachung paralleler Anwendungen. PhD thesis, Universität Stuttgart, Fakultät 14, Informatik (2000),
`http://elib.uni-stuttgart.de/opus/volltexte/2000/600`

A Tool for Optimizing Runtime Parameters of Open MPI

Mohamad Chaarawi[1,2], Jeffrey M. Squyres[2], Edgar Gabriel[1], and Saber Feki[1]

[1] Parallel Software Technologies Laboratory,
Department of Computer Science, University of Houston
{mschaara,gabriel,sfeki}@cs.uh.edu
[2] Cisco Systems, San Jose, CA USA
jsquyres@cisco.com

Abstract. Clustered computing environments, although becoming the predominant high-performance computing platform of choice, continue to grow in complexity. It is relatively easy to achieve *good* performance with real-world MPI applications on such platforms, but obtaining the best possible MPI performance is still an extremely difficult task, requiring painstaking tuning of all levels of the hardware and software in the system. The Open Tool for Parameter Optimization (OTPO) is a new framework designed to aid in the optimization of one of the key software layers in high performance computing: Open MPI. OTPO systematically tests large numbers of combinations of Open MPI's run-time tunable parameters for common communication patterns and performance metrics to determine the "best" set for a given platform. This paper presents the concept, some implementation details and the current status of the tool, as well as an example optimizing InfiniBand message passing latency by Open MPI.

1 Introduction

In the current top 500 list [10], clustered high performance computing systems clearly dominate from the architectural perspective. Off-the-shelf components make clusters attractive for low-end budgets as well as for large scale installations, since they offer the opportunity to customize the equipment according to their needs and financial constraints. However, the flexibility comes at the price: the performance that end-users experience with real-world applications deviates significantly from the theoretical peak performance of the cluster. This is mainly due to the fact, that each system represents a nearly unique execution environment. Typically, neither software nor hardware components have been hand-tuned to that particular combination of processors, network interconnects and software environment.

In order to optimize the performance of a particular system, research groups have turned to extensive pre-execution tuning. As an example, the ATLAS project [12] evaluates in a configure step a large number of implementation possibilities for the core loops of the BLAS routines. Similarly, the Automatically

A. Lastovetsky et al. (Eds.): EuroPVM/MPI 2008, LNCS 5205, pp. 210–217, 2008.

Tuned Collective Communication project [8] incorporates an exhaustive search in order to determine the best performing algorithms for a wide range of message lengths for MPI's collective operations. The FFTW library [1] tunes fast fourier transform operations (FFT) in a so-called planner step before executing the FFT operations of the actual application.

One critical piece of software has however not been systematically approached in any of these projects. MPI libraries represent the interface between most parallel applications and the hardware today. Libraries such as MPICH [4] and Open MPI [2] provide flexible and tunable implementations, which can be adapted either at compile or at runtime to a particular environment. In this paper, we introduce OTPO (Open Tool for Parameter Optimization), a new tool developed in partnership between Cisco Systems and the University of Houston. OTPO is an Open MPI specific tool aiming at optimizing parameters of the runtime environment exposed through the MPI library to the end-user application. These parameters might expose explicit or implicit dependencies among each other, some of them possibly even unknown to the module developers. A long term goal of OTPO is therefore to systematically discover those hidden dependencies and the effect they have on overall performance, such as the point-to-point latency or bandwidth. We present the current status of OTPO and the ongoing work.

The rest of the paper is organized as follows: Sec. 2 presents the concept and the architecture of OTPO. Sec. 3 discusses some implementation details of OTPO. In Sec. 4, we show how OTPO is used to explore the parameter space of Open MPI's short message protocol in order to minimize latency on InfiniBand networks. Finally, Sec. 5 summarizes the paper and discusses ongoing work.

2 Concept

Open MPI [2] is an open source implementation of the MPI-1 [6] and MPI-2 [7] specifications. The code is developed and maintained by a consortium consisting of 14 institutions[1] from academia and industry. The Open MPI design is centered around the Modular Component Architecture (MCA), which is the software layer providing management services for Open MPI frameworks. A framework is dedicated to a single task, such as providing collective operations (i.e., the COLL framework) or providing data transfer primitives for a particular network interconnect (i.e., the Byte Transfer Layer framework – BTL). Each framework will typically have multiple implementations available in the form of modules ("plugins") that can be loaded on-demand at run time. For example, BTL modules include support for TCP, InfiniBand, Myrinet, shared memory, and others.

Among the management services provided by the MCA is the ability to accept run-time parameters from higher level abstractions (e.g., `mpirun`) and pass them down to the according frameworks. MCA runtime parameters give system administrators, end-users and developers the possibility to tune the performance of their applications and systems without having to recompile the MPI library. Examples for MCA runtime parameters include the values of cross-over points

[1] As of January, 2008.

between different algorithms in a collective module, or modifying some network parameters such as internal buffer sizes in a BTL module. Due to its great flexibility, Open MPI developers made extensively use of MCA runtime parameters. The current development version of Open MPI has multiple hundred MCA runtime parameters, depending on the set of modules compiled for a given platform. While average end-users clearly depend on developers setting reasonable default values for each parameter, some end-users and system administrators might explore the parameter space in order to find values leading to higher performance for a given application or machine.

OTPO is a tool aiming at optimizing parameters of the runtime environment exposed through the MPI library to the end-user application. OTPO is based on a highly modular concept, giving end-user the possibility to provide or implement their own benchmark for exploring the parameter space. Depending on the purpose of the tuning procedure, most often only a subset of the runtime parameters of Open MPI will be tuned at a given time. As an example, users might choose to tune the networking parameters for a cluster, optimizing the collective operations in a subsequent run etc. Therefore, one of the goals of OTPO is to provide a flexible and user friendly possibility to input the set of parameters to be tuned. OTPO supports testing two general types of MCA parameters:

1. Single-value parameters: these parameters represent an individual value, such as an integer.
2. Multiple-value parameters: these parameters are composed of one or more sub-parameters, each of which can vary.

From a higher level perspective, the process of tuning runtime parameters is an empirical search in a given parameter space. Depending on the number of parameters, the number of possible values for each parameter, and dependencies among the parameters themselves, the tuning process can in fact be very time consuming and complex. Thus, OTPO is based on a library incorporating various search algorithms, namely the Abstract Data and Communication Library (ADCL) [3]. ADCL is a runtime optimization library giving applications the possibility to register alternative implementations for a given functionality. Using various search algorithms, the library will evaluate the performance of some/each implementation(s) provided, and choose the version leading to the lowest execution time. Furthermore, the application has the possibility to characterize each implementation using a set of attributes. These attributes are the basis of some advanced search operations within ADCL in order to speed up the tuning process. This mechanism has been used by OTPO for registering and maintaining the status of different MCA parameters.

3 Implementation

Upon start of an optimization run, OTPO parses an input file and creates a global structure that holds all the parameters and their options. OTPO then registers a function for each possible combination of MCA parameters which satisfies the

Reverse Polish Notations (RPNs) conditions specified in the parameter file with ADCL. The function registered by OTPO first forks a child process. The child process sets the parameters in the environment that need to be provided to the `mpirun` command, such as the number of processes, the MCA parameters and values, and the application/benchmark to run. Finally, the child launches `mpirun` with the argument set. The parent process waits for the child to complete, and checks if the test was successful. If the child succeeded, the parent will update the current request with the value (e.g., latency) that the child measured for the current parameter values. If the child does not complete within a user specified timeout, the parent process kills it, and updates the request with an invalid value.

When measurements for all combinations of parameter values have been updated by ADCL, OTPO gathers the results and saves them to a file. Each run of OTPO produces a file with a time stamp that contains the best attribute combinations. The result file contains the set of best measured values, the number of combinations that produced these values, and the parameter value combinations themselves. The result file might be large, having thousands of different parameter combinations.

These results files produced by the first version of OTPO are intented to be intermediate results. Currently ongoing work focuses on presenting the results in an intuitive and visual manner.

3.1 OTPO Parameter File

The OTPO parameter file describes the MCA parameters and potential values to be tested. In order to provide a maximum flexibility to the end-user, the parameters can be described by various options, e.g. depending on whether a parameter can have continues values, certain discrete values, or whether the value consists of different strings. Each line in the parameter file specifies a single parameter by giving the name of the parameter and some options, the options being one of the following:

- `d default_value`: specifies a default value for this parameter.
- `p <list of possible values>`: explicitly specify the list of possible values for the parameter.
- `r start_value end_value`: specify the start and end value for the parameter.
- `t traversal_method <arguments>`: specifies the method to traverse the range of variables for the parameter. The first version of OTPO only includes one method: "increment," which takes the operator and the operand as arguments.
- `i rpn_expression`: RPN condition that the parameter combinations must satisfy.
- `v`: specifies if the parameter is virtual, which means that it will not be set as an environment variable, but will be part of another parameter.
- `a format_string`: specifies that the parameter is an aggregate of other parameters in a certain format. Each sub parameter is surrounded by dollar signs ($) in the format string.

4 Performance Evaluation

This section presents an example using OTPO to optimize some of the InfiniBand parameters of Open MPI on a given platform. We therefore first describe Open MPI's InfiniBand support and some of its run-time tunable parameters, then present the results of the optimization using OTPO.

4.1 InfiniBand Parameters in Open MPI

Open MPI supports InfiniBand networks through a Byte Transfer Layer (BTL) plugin module named `openib`. BTL plugins are the lowest layer in the Open MPI point-to-point communication stack and are responsible for actually moving bytes from one MPI process to another. The `openib` BTL has both single- and multiple-value parameters that can be adjusted at run-time.

There are more than 50 MCA parameters that are related to the `openib` BTL module, all of which can be modified at runtime. Open MPI attempts to provide reasonable default values for these parameters, but every application and every platform is different: maximum performance can only be achieved through tuning for a specific platform and application behavior.

MPI processes communicate on InfiniBand networks by setting up a pair of queues to pass messages: one queue for sending and one queue for receiving. InfiniBand queues have a large number of attributes and options that can be used to tailor the behavior of how messages are passed. Starting with version v1.3, Open MPI exposes the receive queue parameters for short messages through the multiple-value parameter `btl_openib_receive_queues` (long messages use a different protocol and are governed by a different set of MCA parameters). Specifically, this MCA parameter is used to specify one or more receive queues that will be setup in each MPI process for InfiniBand communication. There are two types of receive queues, each of which have multiple sub-parameters. It is however outside of the scope of this paper to give detailed and precise descriptions of the MCA parameters used. The parameters are:

1. "Per-peer" receive queues are dedicated to receiving messages from a single peer MPI process. Per-peer queues have two mandatory sub-parameters (*size* and *num_buffers*) and three optional sub-parameters (*low_watermark*, *window_size*, and *reserve*).
2. "Shared" receive queues are shared between all MPI sending processes. Shared receive queues have the same mandatory sub-parameters as per-peer receive queues, but have only two optional sub-parameters (*low_watermark* and *max_pending_sends*).

The `btl_openib_receive_queues` value is a colon-delimited listed of queue specifications specifying the queue type ("P" or "S") and a comma-delimited list of the mandatory and optional sub-parameters. For example:

```
P,128,256,192,128:S,65535,256,128,32
```

will instantiate one per-peer receive queue for each inbound MPI connection for messages that are ≤ 128 bytes, and will setup a single shared receive queue for all messages that are > 128 bytes and $\leq 65,535$ bytes (messages longer than 65,535 bytes will be handled by the long message protocol).

Another good example for how to explore the parameter space by OTPO are the tunable values controlling Open MPI's use of RDMA for short messages. Short message RDMA is a resource-intensive, non-scalable optimization for minimizing point-to-point short message latency. Finding a good balance between the desired level of optimization and the resources consumed by this optimization is exactly the kind of task that OTPO was designed for. Among the most relevant parameters with regard to RDMA operations are `btl_openib_ib_max_rdma_dst_opts`, which limits the maximum number of outstanding RDMA operations to a specific destination; `btl_openib_use_eager_rdma`, a logical value specifying whether to use the RDMA protocol for eager messages; and `btl_openib_eager_rdma_threshold`, only use RDMA for short messages to a given peer after this number of messages has been received from that peer. Due to space limitations, we will not detail all RDMA parameters or present RDMA results of the according OTPO runs.

4.2 Results

Tests were run on the shark cluster at the University of Houston. Shark consists of 24 dual-core 2.2GHz AMD Opteron nodes connected by 4x InfiniBand and Gigabit Ethernet network interconnects. The InfiniBand switch is connected to a single HCAs on every node, with an `active_mtu` of 2048 and an `active_speed` of 2.5 Gbps. OFED 1.1 is installed on the nodes. A pre-release version of Open MPI v1.3 was used to generate these results, subversion trunk revision 17198. A nightly snapshot of the trunk was used, and configured with debug disabled. All the tests were run with `mpi_leave_pinned` MCA parameter set to one. The benchmark used for tuning the parameters was NetPIPE [11].

OTPO was used to explore the parameter space of `btl_openib_receive_queues` to find a set of values that yield the lowest half round trip short message latency. Since `receive_queues` is a multiple-value parameter, each sub-parameter must be described to OTPO. The individual sub-parameters become "virtual" parameters, each with a designated range to explore. OTPO was configured to test both a per-peer and a shared receive queue with the ranges listed in Table 1. Each sub-parameter spanned its range by doubling its value from the minimum to the maximum (e.g., 1, 2, 4, 8, 16, ...).

The parameters that are used are explained as follows:

- The size of the receive buffers to be posted.
- The maximum number of buffers posted for incoming message fragments.
- The number of available buffers left on the queue before Open MPI reposts buffers up to the maximum (previous parameter).
- The maximum number of outstanding sends that are allowed at a given time (SRQ only).

Table 1. InfiniBand receive queue search parameter ranges. The "max pending sends" sub-parameter is only relevant for shared receive queues.

Sub-parameter	Range	Per-peer	Shared
Buffer size (bytes)	65,536 → 1,048,576	√	√
Number of buffers	1 → 1024	√	√
Low watermark (buffers)	32 → 512	√	√
Max pending sends	1 → 32		√

Table 2. OTPO results of the best parameter combinations

Per Peer Queue		Shared Receive Queue	
Latency	Number of Combinations	Latency	Number of Combinations
3.78μs	3	3.77μs	1
3.79μs	3	3.78μs	4
3.80μs	15	3.79μs	18
3.81μs	21	3.80μs	32
3.82μs	31	3.81μs	69
3.83μs	34	3.82μs	69

The parameter space from Table 1 yields, 275 for per peer queue and 825 for shared queue valid combinations (after removing unnecessary combinations that would lead to incorrect results). These combinations stressed buffer management and flow control issues in the Open MPI short message protocol when sending 1 byte messages. It took OTPO 3 minutes to evaluate the first case by invoking NetPIPE for each parameter combination and 9 minutes for the second case. Note that NetPIPE runs several ping-pong tests and reports half the average round-trip time. OTPO sought parameter sets that minimized this value.

The results are summarized in Table 2, and reveal a small number of parameter sets that resulted in the lowest latency (3.78μs and 3.77μs). However, there were more parameter combinations leading to results within 0.05μs of the best latency. These results highlight, that typically, the optimization process using OTPO will not deliver a single set of parameters leading to the best performance, but will result in groups of parameter sets leading to similar performance.

5 Summary

In this paper we presented OTPO, a tool for optimizing Open MPI runtime parameters. The tool gives interested end-users and system administrators the possibility to "personalize" their Open MPI installation. OTPO has been successfully used to optimize the network parameters of the `openib` InfiniBand communication module of Open MPI in order to minimize the communication latency.

The currently ongoing work on OTPO includes multiple areas. As of today, OTPO only supports NetPIPE as the application benchmark. However, we plan to add more benchmarks to be used with OTPO such as the IMB [5]and

SKaMPI [9] benchmarks in order to optimize collective modules. Some of the benchmarks will also require OTPO to support additional optimization metrics, such as bandwidth or memory usage. The foremost goal however is to develop a result gathering tool that takes the results file produced by OTPO and presents it to the user in a more readable and interpretable manner.

Acknowledgments. This research was funded in part by a gift from the Silicon Valley Community Foundation, on behalf of the Cisco Collaborative Research Initiative of Cisco Systems.

References

1. Frigo, M., Johnson, S.G.: The Design and Implementation of FFTW3. Proceedings of IEEE 93(2), 216–231 (2005)
2. Gabriel, E., Fagg, G.E., Bosilca, G., Angskun, T., Dongarra, J.J., Squyres, J.M., Sahay, V., Kambadur, P., Barrett, B., Lumsdaine, A., Castain, R.H., Daniel, D.J., Graham, R.L., Woodall, T.S.: Open MPI: Goals, Concept, and Design of a Next Generation MPI Implementation. In: Proceedings, 11th European PVM/MPI Users' Group Meeting, Budapest, Hungary, September 2004, pp. 97–104 (2004)
3. Gabriel, E., Huang, S.: Runtime optimization of application level communication patterns. In: 12th International Workshop on High-Level Parallel Programming Models and Supportive Environments, Long Beach, CA, USA (March 2007)
4. Gropp, W., Lusk, E., Doss, N., Skjellum, A.: A high-performance, portable implementation of the MPI message passing interface standard. Parallel Computing 22(6), 789–828 (1996)
5. Intel MPI Benchmark, `http://www.intel.com/cd/software/products/asmo-na/eng/219848.htm`
6. Message Passing Interface Forum. MPI: A Message Passing Interface Standard (June 1995), `http://www.mpi-forum.org/`
7. Message Passing Interface Forum. MPI-2: Extensions to the Message Passing Interface (July 1997), `http://www.mpi-forum.org/`
8. Pjesivac-Grbovic, J., Angskun, T., Bosilca, G., Fagg, G.E., Gabriel, E., Dongarra, J.J.: Performance Analysis of MPI Collective Operations. Cluster Computing 10(2), 127–143 (2007)
9. Reussner, R., Sanders, P., Prechelt, L., Muller, M.: SKaMPI: A Detailed, Accurate MPI Benchmark. In: Alexandrov, V.N., Dongarra, J. (eds.) PVM/MPI 1998. LNCS, vol. 1497, pp. 52–59. Springer, Heidelberg (1998)
10. TOP 500 webpage (2007), `http://www.top500.org/`
11. Turner, D., Chen, X.: Protocol-dependent message-passing performance on linux clusters. In: Cluster Computing, 2002. Proceedings. 2002 IEEE International Conference on Linux Clusters, pp. 187–194. IEEE Computer Society Press, Los Alamitos (2002)
12. Whaley, R.C., Petite, A.: Minimizing development and maintenance costs in supporting persistently optimized blas. Software: Practice and Experience 35(2), 101–121 (2005)

MADRE: The Memory-Aware Data Redistribution Engine*

Stephen F. Siegel[1] and Andrew R. Siegel[2]

[1] Verified Software Laboratory, Department of Computer and Information Sciences, University of Delaware, Newark, DE 19716, USA
siegel@cis.udel.edu

[2] Mathematics and Computer Science Division, Argonne National Laboratory, 9700 South Cass Avenue, Argonne, IL 60439, USA
siegela@mcs.anl.gov

Abstract. A new class of algorithms is presented for efficiently carrying out many-to-many parallel data redistribution in a memory-limited environment. Key properties of these algorithms are explored, and their performance is compared using idealized benchmark problems. These algorithms form part of a newly developed MPI-based library MADRE (Memory-Aware Data Redistribution Engine), an open source toolkit designed for easy integration with application codes to improve their performance, portability, and scalability.

1 Introduction

A large class of massively parallel scientific applications are memory-bound. To achieve their scientific aims, considerable care must be taken to reduce their memory footprint via careful programming techniques, e.g. avoiding unnecessary data copies of main data structures, not storing global meta-data locally, and so forth. The need for efficient many-to-many parallel data movement arises in a wide variety of such scientific applications (e.g., [1]). Adaptive algorithms, for example, require extensive load balancing to ensure an optimal distribution of mesh elements across processors (both in terms of equal balance and spatial locality) [2]. Time-dependent particle-tracking codes are another good example, with frequent rebalancing of the main data structures as particles cross processor boundaries [3].

Typically, such load balancing operations consist of two parts: 1) computing the *map* that specifies the new destination for each block of data and 2) efficiently carrying out the movement of data blocks to their new location without exceeding the memory available to the application on any process. There has been much research into the first problem, resulting, for example, in numerous

* This research was supported by the National Science Foundation under Grants CCF-0733035 and CCF-0540948 and used resources of the Argonne Leadership Computing Facility at Argonne National Laboratory, which is supported by the Office of Science of the U.S. Department of Energy under contract DE-AC02-06CH11357.

A. Lastovetsky et al. (Eds.): EuroPVM/MPI 2008, LNCS 5205, pp. 218–226, 2008.

```
MADRE_Object MADRE_create(void* data, char *strategy, int numBlocks,
    int blockLength, MPI_Datatype datatype, MPI_Comm comm);

/* Redistributes the blocks based on the given map information.
 * destRanks and destIndices are both integer arrays of the given
 * length.  We must have 0 <= length <= madre->numBlocks.
 * destRanks[i] is the rank of the proc to which block i is to be
 * moved, or -1 if block i is dead.  destIndices[i] is the position to
 * which block i is to be moved (this integer is ignored if
 * destRanks[i] is -1).  The blocks in positions length, length+1,
 * ..., madre->numBlocks-1 are assumed to be dead. */
void MADRE_redistribute(MADRE_Object madre, int length,
    int* destRanks, int* destIndices);

void MADRE_destroy(MADRE_Object madre);

void MADRE_setDataPointer(MADRE_Object madre, void* data);

/* Current number of bytes allocated by MADRE */
long MADRE_getMem();
```

Fig. 1. Excerpt of MADRE interface, file `madre.h`

efficient algorithms for computing space-filling curves [4]. The second problem, though, has received little attention, especially in the critical case where memory resources are severely limited and memory use must be completely transparent to the application developer.

One of the few explorations into memory-efficient solutions to the data redistribution problem was undertaken by A. Pinar and B. Hendrickson [5]. They formulated a "phase"-based framework for the problem, showed that the problem of finding a minimal-phase solution is NP-hard, and studied a family of algorithms for approximating minimal solutions. Yet there are many very different solutions to the redistribution problem that do not necessarily fit into the phase-based framework and much experimentation will be necessary to compare their performance and ascertain their various qualities.

We describe here a new class of memory-aware parallel redistribution algorithms that extends the pioneering work of Pinar and Hendrickson. Properties of these algorithms are explored, and some initial performance tests are carried out. The algorithms presented form only a small part of the larger MADRE effort (Memory-Aware Data Redistribution Engine) [6], an open source, C/MPI-based toolkit that includes a wide range of strategies that allow the client to control tradeoffs between buffer space, performance, and scalability. Moreover, MADRE is architected as well to serve as a testbed for continued research in the area of parallel data redistribution, where developers can easily integrate new techniques/strategies within the framework. An excerpt of the MADRE interface is shown in Fig. 1.

2 Algorithms

In this section, we define the data redistribution problem precisely, and describe two of the algorithms implemented in MADRE for solving it.

Assume we are given a distributed-memory parallel program consisting of n processes. Each process maintains in its local memory a contiguous, fixed-sized array of data blocks, each block comprising l bytes. Let m_i be the number of blocks maintained by process i ($0 \leq i < n$). A block is specified by its *address* (i, j), where i is the process rank and j ($0 \leq j < m_i$) is the array index. Let A denote the set of all addresses. A *redistribution problem* is specified by a subset $B \subseteq A$ and an injective map $f\colon B \to A$. A redistribution algorithm solves the problem if for all $b \in B$, the contents of $f(b)$ after redistribution equal the contents of b before redistribution. The problem may be thought of as an "in-place" version of MPI_ALLTOALLV. The blocks in $A \setminus B$ are said to be *free*; there is no requirement that their data be preserved, so the algorithm is free to use them as "scratch space."

The algorithms implemented in MADRE are expected to satisfy certain requirements. The first is that each should be a complete algorithmic solution to the data redistribution problem, i.e., it should terminate in finite time with the correct result on any redistribution problem, even one with no free blocks on any process. Second, the additional memory required by the algorithm should depend only linearly on the problem size. To be precise, there must be (reasonably small) constants c_1, c_2, and c_3 such that the memory required by the algorithm on process i is at most $c_1 n + c_2 m_i + c_3 l$. Algorithms with a quadratic dependence on the problem size will not scale on current high-end machines. In particular, no single process "knows" the global map; it is only given the destination addresses for its own blocks. Finally, each algorithm should consume other resources frugally; e.g., the number of outstanding MPI communication requests on a process must stay within reasonably small bounds.

The MADRE library includes a module for the management of blocks on a single process. This module maintains the destination addresses for each block and provides a function to sort the blocks on a single process by increasing destination rank (with all free spaces at the end). All of the MADRE redistribution algorithms depend heavily on these services, in particular because it is often necessary to create contiguous send and receive buffers. For the most part, however, we will elide these details in our descriptions of the algorithms.

2.1 A Pinar-Hendrickson Parking Algorithm

Our first algorithm is an instance of the family of "parking" algorithms described in [5]. We will briefly summarize the algorithm and refer the reader to [5] for details.

The algorithm proceeds in a series of global phases. Within each phase, each process receives as much data as possible into its free space while sending out as much data as possible to other processes. When this communication completes, the space occupied by the sent data is reclaimed as free, the blocks are re-sorted, and the next phase begins. This continues until all blocks have arrived at their final destinations.

The protocol for determining how many blocks a process p will send and receive to/from other processes in a phase proceeds as follows. First, p informs each process q of the number of blocks it has to send to q. These can be thought of as "requests to send" on the part of p. At the same time, p receives similar requests from the processes that wish to send it data. If p does not have sufficient free space to receive all the incoming data, it uses some heuristic to apportion its free space among its sources. In any case, p sends each source a message stating how much of the request will be granted, i.e., the number of blocks it will receive from that source in the current phase. (Our implementation uses the *first-fit* heuristic of [5].) At the same time, p receives similar granting messages from its targets. At this point, p knows how many blocks it will send and receive to/from each process. The data is then transferred by initiating nonblocking send and receive operations for the specified quantities.

There are situations in which the basic algorithm described above does not terminate. This can happen, for example, if a cyclic dependency occurs among a set of processes with no free space. Moreover, when it does complete it may take many more phases than necessary. For these reasons, Pinar and Hendrickson introduce "parking." The idea is that if p does not have enough free space to receive all of its incoming data in the next phase it can temporarily "park" data on processes with extra free space. The parked data will be moved to its final destination when space becomes available. A root process is used in each phase to match processes with data to park with those with extra free space. A parking algorithm will always complete, as long as there is at least one free space on at least one process. (Our implementation allocates a single free space if there are no free spaces at all so that the algorithm will complete in all cases.) Parking can also significantly reduce the number of phases, approximating to within a factor of 1.5 a minimal-phase solution, though this does not always reduce the run-time.

2.2 Cyclic Scheduler

The parking algorithm described above exhibits several potential weaknesses. First, the total quantity of data moved is greater than necessary, because parking blocks are moved more than once. Second, the division into phases, with an effective barrier between each phase, may limit the possible overlap of communication with communication. For example, if some processes require very little time to complete their work in a phase, they will block, waiting for other processes to complete the phase, when they could be working on the next phase. Third, the number of phases may blow up as the total number of free spaces decreases, and there is significant overhead required to execute each phase.

Like the parking algorithm, the cyclic scheduler algorithm uses a root process to help schedule actions on other processes. However, it differs in several ways. First, all blocks are moved directly to their final destinations. Second, the cyclic algorithm separates the process of *schedule creation* from the process of *schedule execution*. The schedule is created in the first stage of the algorithm, which relies heavily on the root. Once that stage completes, each process has the complete sequence of instructions it must follow in order to complete the redistribution. In the schedule execution stage, each process executes these instructions. In this stage, the root plays no special role, there are no effective barriers, and minimal overhead is required. Since the length of a schedule is bounded above by m_i, the memory overhead required to store the schedule is within our required limits. In all of our experiments to date, the time required to complete the schedule creation stage has been insignificant compared to the time required for the execution stage.

A *schedule* is a sequence of actions for a process to follow. There are three types of actions. The first has the form "Send q blocks to process p_1 in increments of c blocks," meaning that the process should send a message to p_1 containing c blocks, then another message of c blocks, and so on, until a total of q blocks have been sent. (The final message will consist of $q\%c$ blocks if c does not divide q.) The other action types are "Receive q blocks from p_2 in increments of c" and "Send q blocks to p_1 and receive q blocks from p_2 in increments of c." The last form requires that c blocks be sent and c received, and after those operations have completed, another c are sent and received, and so on. The blocks must be sorted once before an action begins, but do not have to be sorted again until the next action is executed. This is an important point which reduces the overhead associated to executing an action. To create the schedule, the root essentially performs a depth-first search of the *transmission graph*. This is the weighted directed graph in which the nodes are the process ranks, and there is an edge from i to j of weight $k > 0$ if i has k blocks to send to j. It is not possible to store the entire transmission graph on the root in linear memory. Instead, each process p sends the root one outgoing edge, i.e., the rank r of a single process to which p has data to send and the number of blocks p has to send to r. As soon as the root has received an edge from each process, it begins scheduling actions. This involves decrementing edge weights, and sending actions back to the non-root processes. When the weight of an edge from p reaches 0, the root requests a new edge from p. Hence edges are continually flowing in to the root and actions are continually flowing out, in a pipelined manner, which is how the memory required by the root is bounded by a constant times the number of processes.

The main part of the scheduling algorithm on the root is shown in Fig. 2. The root uses a stack to perform the search of the graph. The stack holds nodes in the graph, i.e., process ranks. If p_0 and p_1 are two consecutive elements in the stack then there is an edge from p_0 to p_1. The stack grows until either (a) a cycle is reached, i.e., the destination rank of the edge departing from the last node on the stack is already on the stack, or (b) a node is reached with no remaining outgoing edges (i.e., no data to send). In case (a), an action is scheduled on each process in the cycle, while in case (b), an action is scheduled on each process in the stack. The algorithm for case (a) is shown. The functions for sending actions and retrieving edges are not shown; these involve parameters, such as the number of actions/edges to include in a single message, that provide some control over the memory/time tradeoff.

symbol	meaning
degree[i]	number of remaining edges departing from proc i in transmission graph
stack[i]	i^{th} element in DFS stack containing nodes of transmission graph
stackSize	current size of DFS stack
weight[i]	weight of current edge departing from proc i
dest[i]	destination node for current edge departing from i
free[i]	current number of free spaces on proc i

```
procedure main is
  for i ← 0 to n − 1 do
    while degree[i] > 0 do
      push(i);
      while stackSize > 0 do
        r ← peek();
        if degree[r] = 0 then
          scheduleShift();
        else
          d ← dest[r];
          p ← stackPos[d];
          if p < 0 then
            push(d)
          else
            scheduleCycle(p)

procedure scheduleCycle(s) is
  q ← min_{i≥s}{weight[stack[i]]};
  c ← min{q, max{1, min_{i≥s}{free[stack[i]]}}};
  l ← peek();
  while stackSize > s do
    p1 ← pop();
    p2 ← dest[p1];
    if stackSize > s then
      p0 ← peek()
    else
      p0 ← l
    schedule send-recv on p1 with source
    p0, dest p2, quantity q, and
    increment c;
    weight[p1] ← weight[p1] − q;
    if weight[p1] = 0 then nextEdge(p1)
```

Fig. 2. The cyclic scheduler root scheduling algorithm

In case (a), the number of blocks q to be sent and received by each process is the minimum weight of an edge in the cycle. The increment is usually the minimum number of free spaces for a process in the cycle. However, if there is a process with no free space, the increment is set to 1: in this case an extra free space reserved for this situation will be used to receive each incoming increment. When the cycle has been executed the extra space will again be free. This is the essential fact that guarantees the algorithm will always terminate, though it does require the allocation of an additional block on each process. The actions scheduled in case (b) are similar, though the first element in the stack performs only a send and the last performs only a receive.

The algorithm is most effective when it finds many long cycles or shifts with large quantities. The idea is that all of the actions comprising a single cycle/shift can execute in parallel and should take approximately the same amount of time. Moreover, as pointed out above, no local redistribution needs to take place during the execution of an action.

3 Experiments

We report here on a few experiments comparing the performance of the different algorithms. The experiments are designed to test the algorithms in situations

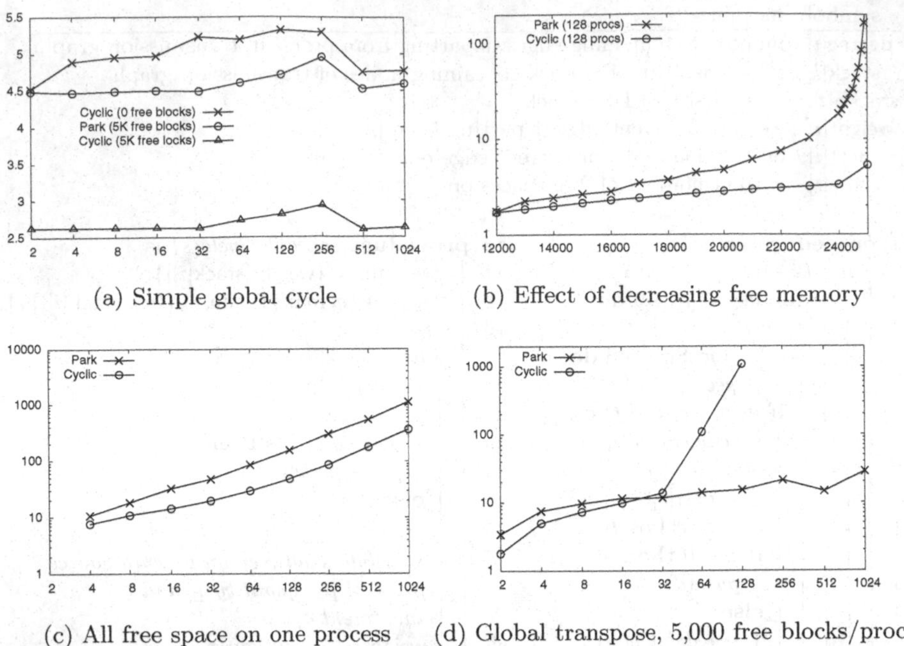

(a) Simple global cycle (b) Effect of decreasing free memory

(c) All free space on one process (d) Global transpose, 5,000 free blocks/proc

Fig. 3. Time to redistribute data. In all cases, each process maintains 25,000 memory blocks of 16,000 bytes each and the y-axis is the time in seconds to complete a data redistribution. In (a), (c), and (d), the x-axis shows the number n of processes in logarithmic scale. In (b), (c), and (d) the time is logarithmic.

of very high data movement and very limited memory. All experiments were executed on the 1024-node IBM Blue Gene/L at Argonne National Laboratory. In each experiment, each process maintains an array of 25,000 blocks, and each block consists of 16,000 bytes. Thus 400 MB are consumed by data, which is a significant portion of the 512 MB total RAM available on each node.

In the first experiment, each process has no free space, and wishes to send all 25K blocks to the next process in the cyclic ordering $i \mapsto (i+1)\%n$. The data redistribution was timed for various values of n. The second experiment is similar, except that each process has 5K free spaces and sends 20K blocks. The results of these two experiments are shown in graph (a) of Fig. 3. The time for the parking algorithm to complete the first experiment is not shown because it did not terminate after 30 minutes; the reasons for this will be explained below. Otherwise, the times range from 2.5 to 5.5 seconds and remain relatively constant as the number of processes is scaled in each case, suggesting near-perfect parallelism in the data transfer.

The third experiment is similar to the two above, except that the number of processes was fixed at 128 and the number of blocks to be transferred was scaled from 12K to 25K. (Equivalently, the number of free spaces was decreased from 13K to 0.) The results are shown in graph (b) of Fig. 3, and illustrate how the time of the parking algorithm blows up with the decreasing amount of free space, in contrast with the cyclic algorithm. Our performance analyses reveal that this behavior is not due to the communication patterns, which are essentially the same in both cases. Rather, the difference arises because the parking algorithm requires that data be re-sorted at the end of each phase. In this experiment, the distribution of the blocks on process $n-1$ at the end of each phase presents a worst case scenario for the sort: the entire array of blocks must be shifted, requiring on the order of 25K calls to `memcpy` (of 16 KB) as the free space approaches 0. Moreover, the number of phases approaches 25K as the free space approaches 0, and so the time dedicated to sorting quickly blows up. For the cyclic algorithm, each process need only execute a single send-receive action of 25K blocks with increment equal to the number of free spaces. Because a sort is only required at the end of each action, the cyclic algorithm performs only one sort for the entire execution.

In the fourth experiment, one process has 25K blocks of free space and all others have no free space. Each of the processes with data wishes to send an approximately equal portion of its data to all other processes. The times are shown in Fig. 3(c). Both algorithms complete for every process count, though the time clearly grows exponentially for both.

In the fifth experiment, each process has 5K free spaces and the map sends block j of process i to position $(mi+j)/n$ of process $(mi+j)\%n$, where $m = 20\text{K}$. (The map is essentially a global transpose of the data matrix.) The results are shown in Fig. 3(d). In this case the cyclic algorithm blows up with process count, and cannot scale beyond 128 nodes without exceeding our 30-minute limit. In contrast, the parking algorithm scales reasonably well and completes the 1024-node redistribution in only 28 seconds.

Further investigation revealed the reason for the discrepancy. For this redistribution problem, the transmission graph is the complete graph in which all edges have approximately equal weight. Moreover, in our implementation each process orders its outgoing edges by increasing destination rank. The result is that all of the actions scheduled by the root are cycles of length 2. These are scheduled in the "dictionary order"

$$\{0,1\}, \{0,2\}, \ldots, \{0,n-1\}, \{1,2\}, \{1,3\}, \ldots, \{1,n-1\}, \ldots, \{n-1,n\}.$$

The execution of one of these pairs involves the complete exchange of data between the two processes. If all exchanges take approximately the same time, execution will proceed in $2n+3$ phases: in phase m, the exchanges for all pairs $\{i,j\}$, where $i+j = m$ and $i \neq j$, will take place. This means that many processes will block when they could be working. For example, in phase 1 only processes 0 and 1 will exchange data, while all other processes wait. In contrast, the parking algorithm completes in 4 stages for any n. Hence, in this case, the parking algorithm achieves much greater overlap of communication.

4 Conclusion

For a certain class of scientific applications, the limited-memory data redistribution problem will become an increasingly important component of overall performance and scalability as we move toward the petascale. Solutions to this problem are difficult and subtle. Solution algorithms can behave in ways that are difficult or impossible to predict using purely analytical means, so experimentation is essential for ascertaining their qualities. We have implemented a practical framework with multiple solutions to the problem, two of which are explored in this paper. A series of simple experiments helped us identify some potential weaknesses in a variant of the Pinar-Hendrickson parking algorithm. We addressed these in a new algorithm, only to discover different situations in which that algorithm also performs poorly.

We continue to refine the algorithms presented here and to explore entirely new ones. In addition, we are exploring heuristics capable of predicting which algorithm will perform well on a given problem. It may also be possible to combine different algorithms, so that in certain states the algorithm will be switched dynamically, in the middle of the redistribution. Finally, we are preparing another set of experiments in which we integrate MADRE directly into scientific applications, as opposed to the synthetic experiments reported on here.

Acknowledgements. We are grateful to Ali Pinar for answering questions about the parking algorithm and making the code used in [5] available to us. We also thank Anthony Chan for assistance with the Jumpshot performance visualization tool [7].

References

1. Ricker, P.M., Fryxell, B., Olson, K., Timmes, F.X., Zingale, M., Lamb, D.Q., MacNeice, P., Rosner, R., Tufo, H.: FLASH: A multidimensional hydrodynamics code for modeling astrophysical thermonuclear flashes. Bulletin of the American Astronomical Society 31, 1431 (1999)
2. de Cougny, H.L., Devine, K.D., Flaherty, J.E., Loy, R.M., Özturan, C., Shephard, M.S.: Load balancing for the parallel adaptive solution of partial differential equations. Appl. Numer. Math. 16(1-2), 157–182 (1994)
3. Nieter, C., Cary, J.R.: Vorpal: a versatile plasma simulation code. J. Comput. Phys. 196(2), 448–473 (2004)
4. Catalyurek, U., Boman, E., Devine, K., Bozdag, D., Heaphy, R., Riesen, L.: Hypergraph-based dynamic load balancing for adaptive scientific computations. In: Proc. of 21st International Parallel and Distributed Processing Symposium (IPDPS 2007). IEEE, Los Alamitos (2007)
5. Pinar, A., Hendrickson, B.: Interprocessor communication with limited memory. IEEE Transactions on Parallel and Distributed Systems 15(7) (July 2004)
6. Siegel, S.F.: The MADRE web page (2008), `http://vsl.cis.udel.edu/madre`
7. Zaki, O., Lusk, E., Gropp, W., Swider, D.: Toward scalable performance visualization with Jumpshot. Int. J. High Perform. Comput. Appl. 13(3), 277–288 (1999)

MPIBlib: Benchmarking MPI Communications for Parallel Computing on Homogeneous and Heterogeneous Clusters

Alexey Lastovetsky, Vladimir Rychkov, and Maureen O'Flynn

School of Computer Science and Informatics, University College Dublin,
Belfield, Dublin 4, Ireland
{alexey.lastovetsky,vladimir.rychkov,maureen.oflynn}@ucd.ie
http://hcl.ucd.ie

Abstract. In this paper, we analyze existing MPI benchmarking suites, focusing on two restrictions that prevent them from a wider use in applications and programming systems. The first is a single method of measurement of the execution time of MPI communications implemented by each of the suites. The second one is the design of the suites in the form of a standalone executable program that cannot be easily integrated into applications or programming systems. We present a more flexible benchmarking package, MPIBlib, that provides multiple methods of measurement, both operation-independent and operation-specific. This package can be used not only for benchmarking but also as a library in applications and programming systems for communication performance modeling and optimization of MPI operations.

Keywords: MPI, benchmark, parallel computing, computational cluster, communication performance model.

1 Introduction

Accurate estimation of the execution time of MPI communication operations plays an important role in optimization of parallel applications. *A priori* information about the performance of each MPI operation allows a software developer to design a parallel application in such a way that it will have maximum performance. This data can also be useful for tuning collective communication operations and for the evaluation of different available implementations. The choice of collective algorithms becomes even more important in heterogeneous environments. In addition to general timing methods that are universally applicable to all communication operations, MPIBlib includes methods that can only be used for measurement of some particular operations. Where applicable, these operation-specific methods work faster than their universal counterparts and can be used as time-efficient alternatives. The efficiency of timing methods will be particularly important in self-adaptable parallel applications using runtime benchmarking of communication operations to optimize their performance on the executing platform.

A. Lastovetsky et al. (Eds.): EuroPVM/MPI 2008, LNCS 5205, pp. 227–238, 2008.

A typical MPI benchmarking suite uses only one timing method to estimate the execution time of the MPI communications. The method provides a certain accuracy and efficiency. The efficiency of the timing method is particularly important in self-adaptable parallel applications using runtime benchmarking of communication operations to optimize their performance on the executing platform. In this case, less accurate results can be acceptable in favor of a rapid response from the benchmark. In this paper, we analyze different timing methods used in the benchmarking suites and compare their accuracy and efficiency on homogeneous and heterogeneous clusters. Based on this analysis, we design a new MPI benchmarking suite called MPIBlib that provides a variety of timing methods. This suite supports both fast measurement of collective operations and exhaustive benchmarking.

In addition to general timing methods that are universally applicable to all communication operations, MPIBlib includes methods that can only be used for measurement of one or more specific operations. Where applicable, these operation-specific methods work faster than their universal counterparts and can be used as their time-efficient alternatives.

Most of the MPI benchmarking suites are designed in the form of a standalone executable program that takes the parameters of communication experiments and produce a lot of output data for further analysis. As such, they cannot be integrated easily and efficiently into application-level software. Therefore, there is a need for a benchmarking *library* that can be used in parallel applications or programming systems for communication performance modeling and tuning communication operations. MPIBlib is such a library that can be linked to other applications and used at runtime.

The rest of the paper is structured as follows. Section 2 outlines existing benchmarking suites. We analyze different methods of measuring MPI communication operations, which are implemented in the suites or described in work on MPI benchmarking. Section 3 describes main features of MPIBlib, the benchmarking library that provides a variety of operation-independent and operation-specific methods of measurement. In Section 4, we discuss application of the library. The results of experiments on homogeneous and heterogeneous clusters are presented in Section 5. We compare the results and costs of different methods of measurement and focus on measuring point-to-point, scatter and gather communication operations as their results are used in the estimation of parameters of advanced heterogeneous communication performance models.

2 Related Work

There are several commonly used MPI benchmarking suites [1]-[5]. The aim of all these suites is to estimate the execution time of MPI communication operations as accurate as possible. In order to evaluate the accuracy of the estimation given by different suites, we need a unified definition of the execution time. As not all of the suites explicitly define their understanding of the execution time, we suggest the following as a natural definition. The execution time of a communication operation is defined as the real (wall clock) time elapsed from the start of the operation, given all the participating processors have started the operation simultaneously, until the successful completion of the operation by the last participating processor. Mathematically, this time can be defined as the minimum execution time of the operation, given that the participating processors do not synchronize their start and are not participating in any other communication operation. It is important to note that the definition assumes that we estimate the execution time for a single isolated operation.

Practically, the execution time of the communication operation is estimated from the results of an experiment that, in addition to the operation, includes other communications and computations. As parallelism introduces an element of non-determinism, there is a problem of reproducibility of such experiments. The methodology of designing reproducible communication experiments is described in [1]. It includes:

- Repeating the communication operation multiple times to obtain the reliable estimation of its execution time,
- Selecting message sizes adaptively to eliminate artifacts in a graph of the output of the communication operation, and
- Testing the communication operation in different conditions: cache effects, communication and computation overlap, communication patterns, non-blocking communication etc.

In the **mpptest** suite [1], these ideas were implemented and applied to benchmarking point-to-point communications.

The execution time of communication operations depends on the MPI library, native software, and hardware configurations. NetPIPE [2] provides benchmarks for different layers in the communication stack. It is based on the ping-pong communication experiments that are implemented over **memcpy**, TCP, MPI etc. In addition to evaluation of communication performance, this suite helps us identify where inefficiencies lie.

Regarding both the reproducibility of communication experiments and the dependency on communication layers, we focus on benchmarking not only point-to-point operations but also collective ones. We analyzed several MPI benchmarking suites that include tests for collective operations. Despite the different approaches to what and how to measure, they have several common features:

- computing an average, minimum, maximum execution time of a series of the same communication experiments to get accurate results;
- measuring the communication time for different message sizes – the number of measurements can be fixed or adaptively increased for messages when time is fluctuating rapidly;
- performing simple statistical analysis by finding averages, variations, and errors.

The MPI benchmarking suites are also very similar in terms of the software design. Usually, they provide a single executable that takes a description of communication experiments to be measured and produces an output for plotting utilities to obtain graphs.

As more than two processors are involved in collective communications and connected in different ways (communication trees), there are two main issues concerned with the estimation of execution time of MPI collective operations:

- measuring the execution time, and
- scheduling the communication experiments.

2.1 Measuring the Execution Time of MPI Collective Operations

Estimation of the execution time of the communication operation includes the selection of two events marking the start and the end of the operation respectively and measuring the time between these events. First of all, the benchmarking suites differ in what they measure, which can be:

- The time between two events on a single designated processor,
- For each participating processor, the time between two events on the processor, or
- The time between two events but on different processors.

The first two approaches are natural for clusters as there is no global time in these environments where each processor has its own clock showing its own local hour. The local clocks are not synchronized and can have different clock rates, especially in heterogeneous clusters. The only way to measure the time between two events on two different processors is to synchronize their local clocks before performing the measurement. Therefore, the third approach assumes the local clocks to be regularly synchronized. Unlike the first two, this approach introduces a measurement error as it is impossible to keep the independent clocks synchronized all the time with absolute accuracy.

In order to measure time, most of the packages rely on the MPI_Wtime function. This function is used to measure the time between two events on the same processor (the local time). For example, the execution time of a roundtrip can be measured on one process and used as an indication of the point-to-point communication execution time [3], [5]. The execution time of a collective communication operation can also be measured at a designated process. For collective operations with a root, the root can be selected for the measurement. As for many collective operations the completion of the operation by the root does not mean its completion by all participating processes, short or empty messages can be sent by the processors to the root to confirm the completion. A barrier, reduce, or empty point-to-point communications can be used for this purpose. The result must be corrected by the average time of the confirmation. The drawback of this approach is that the confirmation can be overlapped with the collective operation and hence it cannot simply be subtracted. As a result, this technique may give negative values of the execution time for very small messages.

The accuracy of this approach (***root timing***) is strongly dependent on whether all processes have started the execution of the operation simultaneously. To ensure the more or less accurate synchronization of the start, a barrier, reduce, or empty point-to-point communications can be used. They can be overlapped with the collective operation to be measured and previous communications as well. To achieve even better synchronization, multiple barriers are used in the benchmarking suites [3]-[5].

The local times can be measured on all processes involved in the communication and the maximum can be taken as the communication execution time. This approach (***maximum timing***) is also dependent on synchronization of the processes before communication, e.g. with a barrier.

To measure the time between two events on different processors, the local clocks of the processors have to be synchronized. Such synchronization can be provided by the MPI global timer if the MPI_WTIME_IS_GLOBAL attribute is defined and true. Alternatively, local clocks of two processors A and B can be synchronized by the following simple algorithm implemented in MPIBench [4]. Processor A sends a message to processor B, which contains the current time plus a half of the previously observed minimum roundtrip time. Processor B receives the message and returns it to A, which calculates the total time that the roundtrip took to complete. If the roundtrip time is the fastest observed so far, then the estimated time of arrival of the initial message is the most accurate yet. If so, processor B calculates the current approximation of the time offset as the message's value received in the next iteration. The processors repeat this procedure until a new minimum roundtrip time has not been observed for a prearranged number of repetitions. Given A being a base processor, this synchronization procedure is performed sequentially for all pairs (A, B_i). A similar procedure is implemented in SKaMPI [5] to find offsets between local times of the root and the other processes.

As local clocks can run at different speeds, especially in heterogeneous environments, the synchronization has to be regularly repeated. The synchronization procedures are quite costly and introduce a significant overhead in benchmarking when used. As soon as the global time has been set up, the time between two events on different processors can be measured [4], [5]. The accuracy of this approach will depend on the accuracy of the clock synchronization and on whether processors start the communication simultaneously. The ***global timing*** usually gives a more accurate estimate because its design is closer to the natural definition of the communication execution time given in the beginning of this section. However, while being more time-efficient, the methods based on local clocks can also provide quite accurate results for many popular platforms and MPI implementations. Therefore, it makes sense to allow a choice of different methods so the user may choose the most efficient for benchmarking with a required accuracy. This is especially important if the benchmarks are to be used in the software that requires the runtime results of the benchmarking.

2.2 Scheduling the Communication Experiments

To obtain a statistically reliable estimate of the execution time, a series of the same experiments are typically performed in the benchmarking suites. If the communications are not separated from each other in this series, the successive

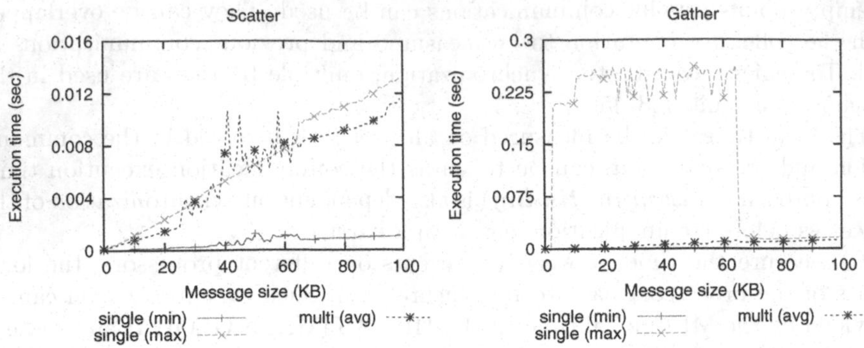

Fig. 1. IMB benchmark on a 16-node heterogeneous cluster: single/multiple scatter/gather measurements

executions may overlap, resulting in a so-called pipeline effect [6], when some processes finish the current repetition earlier and start the next repetition of the operation before the other processes have completed the previous operation. The pipeline affects the overall performance of the series of the operations, resulting in inaccurate averaged execution time. This is the case for the IMB (former PMB) benchmark [3], where the repetitions in a series are not isolated in the attempt to prevent participation of the processes in third-party communications. The IMB measures the communication execution times locally on each process, and the minimum, maximum, and average times are then returned. Fig. 1 shows the results returned by the IMB on a 16-node heterogeneous cluster for scatter and gather operations when single and multiple repetitions are used in the experiments. One can see that for the scatter experiments with a single repetition, the minimum time represents the execution time of a non-blocking send on the root and is therefore relatively small. In the gather experiments with a single repetition, the maximum time is observed on the root, reflecting the communication congestion. The difference between the minimum and maximum times decreases with an increase in the number of repetitions. In both cases, we observe a clear impact of the pipeline effect on the measured execution time of the operation:

- **Scatter:** For small and large messages, the execution time of a repetition in the series is smaller than that measured in a single experiment. For medium-sized messages, escalations of the execution time are observed that do not happen in single experiments.
- **Gather:** Escalations of the execution time for medium-sized messages, observed for single experiments, disappear with the increase of the number of repetitions due to the pipelining.

Thus, the pipeline effect can significantly distort the actual behavior of the communication operation, given that we are interested in accurate estimation of the time of its single and isolated execution.

In order to find the execution time of a communication operation that is not distorted, it should be measured in isolation from other communications. A barrier, reduce, or point-to-point communications with short or empty messages can be used between successive operations in the series. The approach with isolation gives results that are more accurate.

Some particular collective operations and their implementations allow for the use of more accurate and efficient methods that cannot be applied to other collective operations. One example is the method of measurement of linear and binomial implementations of the MPI broadcast on heterogeneous platforms proposed in [7]. It is based on measuring individual tasks rather than the entire broadcast and therefore it does not need the global time. An individual task is a part of the broadcast communication between the root and the i-th process. In each individual task, the pipelining effect is eliminated by sending an acknowledgement message from the i-th process to the root. The execution time of the task is then corrected by the value of the point-to-point execution time.

The acquisition of detailed knowledge of the implementation of collective operations can prove useful towards improving the efficiency of measurement methodologies. This becomes particularly important for benchmarking performed at runtime with on-the-fly optimization of communication operations.

3 MPIBlib Benchmarking Suite

This work is motivated by the absence of an MPI benchmarking suite that would satisfy the following requirements:

- The suite is implemented in the form of library allowing its integration into application-level software.
- The suite provides a wide range of timing methods, both universal and operation/implementation specific, allowing for the choice of the optimal (in terms of accuracy and efficiency) method for different applications and executing platforms.

We have developed such a benchmarking library, MPIBlib, the main goal of which is to support accurate and efficient benchmarking of MPI communication operations in parallel applications at runtime. The main features of MPIBlib can be summarized as follows.

MPIBlib is implemented in the form of library and includes the benchmarks for point-to-point and collective communication operations. The MPIBlib design was influenced by our work on development of the software tool for automated estimation of parameters of the heterogeneous communication performance model proposed in [8]. The software tool widely uses MPIBlib at runtime for accurate and efficient estimation of the execution time of point-to-point, scatter, and gather communications, which is required to find the parameters of the model.

To provide reliable results, the communication experiments in each benchmark are repeated either fixed or variable number of times. The latter allows the user to control the accuracy of the obtained estimation of the execution time. Namely, the definition of each benchmarking function includes the following arguments:

– Input: the minimum and maximum numbers of repetitions, *min_reps* and *max_reps* (*min_reps* $\leq$ *max_reps*), and a maximum error, *alpha* ($0 <$ *alpha* < 1);
– Output: the actual number of repetitions, *reps*, and error, *a*.

Assigning to *min_reps* and *max_reps* the same values results in the fixed number of repetitions of the communication operation, with the error arguments being ignored. As communication operations in a series are isolated from each other, we suppose that the measurement errors are distributed according to the normal law, which enables estimating within a confidence interval, ($1 -$ *alpha*). If *min_reps* $<$ *max_reps*, the experiments are repeated until the sample satisfies the Student's t-test or the number of repetitions reaches its maximum. In this case, the number of repetitions the benchmark has actually taken, reps, and the final error, a, are returned. For statistical analysis, the GNU Scientific Library [9] is used.

The point-to-point benchmarks can be run either sequentially or in parallel on the communicator consisting of more than two processors. The point-to-point benchmark estimates the execution time of the roundtrips between all pairs of processes in the MPI communicator, $i \overset{M_1}{\underset{M_2}{\longleftrightarrow}} j, i < j$. It returns three arrays, each of which contains C_n^2 values corresponding to each pair: estimations of execution time, numbers of repetitions and errors. Several point-to-point communications as well as statistical analysis can be performed in parallel, with each process being involved in no more than one communication. This allows us to significantly reduce the overall execution time of the point-to-point benchmark code and gives us quite accurate results on the clusters based on switched networks. Network switches are capable of inspecting data packets as they are received, determining the source and destination device and forwarding it appropriately. By delivering each message only to the original intended device, a network switch conserves network bandwidth and offers a generally better performance for simultaneous point-to-point communications.

The use of the results of the point-to-point benchmarks can be various. They can be used for the estimation of parameters of the analytical communication performance models, such as Hockney, LogGP. For example, the parameters of the Hockney model can be found from the execution times of two roundtrips with empty and non-empty messages. In practice, due to noises in measurements, they are found from the execution times averaged in two series of such roundtrips. MPIBlib point-to-point benchmark provides this accurate estimation.

The set of communication operations that can be benchmarked by MPIBlib is open for extensions. The definition of operation-independent benchmark functions includes a data structure argument that includes the function pointer referencing to an MPI collective operation. MPIBlib provides a choice of different implementations of MPI collective operations (for example, linear and binomial MPI_Scatter/MPI_Gather). Any of those functions as well as user-defined versions of MPI collective operations can be passed as an argument to the benchmarking subroutines.

Three timing methods that are universally applicable to all MPI communication operations are provided; these are global, maximum, and root timings. MPIBlib provides API to all timing methods described in Section 2. The user is responsible for building the MPI communicator and mapping the processes to processors. To use benchmarks on SMP/multicore processors, an accurate MPI_Wtime implementation is required, as intra-processor communications may take very short time.

MPIBlib provides both operation-specific and implementation-specific methods of measurement. One example is a method of measuring the linear and binomial scatter, which is based on the method of measuring broadcast proposed in [7].

4 Application

The results of benchmarking the collective operations can be used for evaluation of their different implementations, for building of the communication performance models, and for optimization of collective operations.

With help of MPIBlib, we managed to observe the escalations of the execution time of linear scatter/gather on the clusters based on Ethernet switch [10]. It was possible due to the isolation of collective operations and the use of the maximum timing method for scatter and the root timing method for gather.

The library was also integrated into the software tool that automates the estimation of parameters of an advanced heterogeneous communication performance model [8]. The software tool calls the MPIBlib functions for estimation of the execution time of the $i \overset{0}{\underset{0}{\longleftrightarrow}} j$ and $i \overset{M}{\underset{0}{\longleftrightarrow}} j$ roundtrips, scatter and gather communications. We used this tool on a 16-node heterogeneous cluster with a single switch, with parallel point-to-point benchmarking and the root timing of collective operations, which proved efficient and quite accurate on heterogeneous clusters with a single switch. To estimate the parameters of the heterogeneous communication performance model, we carried out additional (neither point-to-point, nor scatter/gather) communication experiments, namely, point-to-two communications $i \overset{M}{\underset{0}{\longleftrightarrow}} j, k$ [11]. The function measuring the execution time of this communication experiment was implemented on the top of the MPIBlib library. In this function, the communication experiments between non-overlapping triplets of processors were performed in parallel on the cluster.

The MPIBlib benchmarking library can also be used to tune MPI communications either upon installation of an application (or a programming system) or at runtime. For example, the results of the scatter and gather benchmarks carried out upon installation of HeteroMPI are used for optimization of collective operations [10].

The following fragment shows an example of the use of MPIBlib for finding the fastest scatter implementation. In the beginning of the program, MPIB_measure_scatter_root function is used to find estimates of the execution time of different scatter implementations for different message sizes. Then these results are used in the optimized scatter, Opt_Scatter, in order to pick the fastest implementation for each particular message size. Opt_Scatter calculates the message size in bytes, compares the estimated execution times of all implementations for this message size and invokes the fastest implementation.

```
//initialization, in the beginning of main()
for (i=0; i<N; i++)
  MPIB_measure_scatter_root(comm, algs[i], n, M,
    min_reps, max_reps, alpha, T[i], &reps, &a);

//globals
MPIB_Scatter* impls[N];//N scatter implementations
int M[n];//n message sizes
double T[N][n];//estimated times for each impl/msg

//optimized scatter, to be used instead of MPI_Scatter
int Opt_Scatter(list of MPI_Scatter arguments){
  //calculate message size m
  //find i such that M[i]<=m and M[i+1]>m
  //find j such that T[j][i]=min(T[0..N-1][i])
  return impls[j](list of MPI_Scatter arguments);
}
```

5 Experiments

In addition to the library, the MPIBlib suite provides a standalone application for benchmarking point-to-point and collective MPI communications, and a set of **gnuplot** scripts for visualization of the results of measurements. We performed experiments with point-to-point, scatter and gather benchmarks on homogeneous and heterogeneous clusters with different MPI implementations. In this paper, we present the results for a heterogeneous 16-node cluster: 11 x Xeon 2.8/3.4/3.6, 2 x P4 3.2/3.4, 1 x Celeron 2.9, 2 x AMD Opteron 1.8, Gigabit Ethernet, LAM 7.1.3. They demonstrate the effects of pipelining (Section 2) and the importance of benchmarking collective operations for different message sizes (in [10], we reported on the escalations of the execution time of gather caused by the use of TCP/IP layer in the communication stack with switched networks, see Fig. 1).

Table 1. The execution time of scatter and gather benchmarks with different timing methods on 16 node heterogeneous cluster

Timing method	Scatter, 0..100KB, 1KB stride, 1 rep (sec)	Gather, 0..100KB, 1KB stride, 1 rep (sec)
Global	28.7	44.7
Maximum	0.8	15.6
Root	0.8	15.7

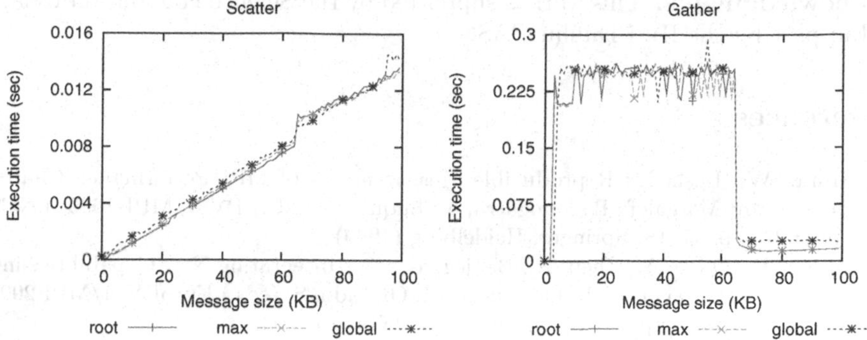

Fig. 2. Comparison of different timing methods for native (linear) LAM scatter and gather on 16 node heterogeneous cluster

We compared the results of sequential and parallel point-to-point benchmarks. In the sequential mode, while two processes are communicating, a barrier blocks all other processes. The experiment included 100 repetitions of 4KB ping-pong and took 3.5 sec. In the case of parallel roundtrips, the benchmarking procedure took significantly less time, 0.5 sec, with the same accuracy of estimation, which was possible due to the nature of the experimental network.

In the next experiment, we use MPIBlib to compare the cost and accuracy of different methods of the measurement of native MPI scatter and gather operations on the target platform. Table 1 shows the overall execution time of the benchmarks that use different timing methods and consist of one collective communication for each message size from 0 to 100 KB, with 1 KB stride. One can see that the global-time approach is very costly. The maximum and root methods are as accurate as that with global-time (see Fig. 2) but much more efficient. The difference between overall scatter and gather execution times is caused by escalations of the execution time of gather for messages of middle sizes.

In summary, the experimental results demonstrate that the use of MPIBlib can significantly speed up the estimation of the execution time of MPI communication operations without the loss of its accuracy.

6 Conclusion

In the paper, we have analyzed the commonly used MPI benchmarking suites and the methods of measurement of communication execution time. We have presented MPIBlib, the new MPI benchmarking library, which provides various operation-independent and operation-specific methods of measurement. MPIBlib is aimed at accurate and efficient runtime benchmarking of MPI communication operations in parallel applications. We have also presented an experimental demonstration showing that the use of MPIBlib can significantly speed up the benchmarking of MPI communication operations, not compromising the accuracy of the estimation. The library is freely available at `http://hcl.ucd.ie/project/MPIBlib`.

Acknowledgments. This work is supported by the Science Foundation Ireland and in part by the IBM Dublin CAS.

References

1. Gropp, W., Lusk, E.: Reproducible Measurements of MPI Performance Characteristics. In: Margalef, T., Dongarra, J., Luque, E. (eds.) PVM/MPI 1999. LNCS, vol. 1697, pp. 11–18. Springer, Heidelberg (1999)
2. Turner, D., Oline, A., Chen, X., Benjegerdes, T.: Integrating New Capabilities into NetPIPE. In: Dongarra, J., Laforenza, D., Orlando, S. (eds.) EuroPVM/MPI 2003. LNCS, vol. 2840, pp. 37–44. Springer, Heidelberg (2003)
3. Intel MPI Benchmarks. User Guide and Methodology Description (2007)
4. Grove, D., Coddington, P.: Precise MPI performance measurement using MPIBench. In: Proceedings of HPC Asia (September 2001)
5. Worsch, T., Reussner, R., Augustin, W.: On Benchmarking Collective MPI Operations. In: Kranzlmüller, D., Kacsuk, P., Dongarra, J., Volkert, J. (eds.) PVM/MPI 2002. LNCS, vol. 2474, pp. 271–279. Springer, Heidelberg (2002)
6. Bernaschi, M., Iannello, G.: Collective communication operations: experimental results vs. theory. Concurrency: Practice and Experience 10(5), 359–386 (1998)
7. Supinski, B., de Karonis, N.: Accurately measuring MPI broadcasts in a computational grid. In: The 8th International Symposium on High Performance Distributed Computing, pp. 29–37 (1999)
8. Lastovetsky, A., Mkwawa, I., O'Flynn, M.: An Accurate Communication Model of a Heterogeneous Cluster Based on a Switch-Enabled Ethernet Network. In: Proceedings of ICPADS 2006, Minneapolis, MN, pp. 15–20 (2006)
9. GNU Scientific Library (2007), `http://www.gnu.org/software/gsl/manual/`
10. Lastovetsky, A., O'Flynn, M., Rychkov, V.: Optimization of Collective Communications in HeteroMPI. In: Cappello, F., Herault, T., Dongarra, J. (eds.) PVM/MPI 2007. LNCS, vol. 4757, pp. 135–143. Springer, Heidelberg (2007)
11. Lastovetsky, A., Rychkov, V.: Building the Communication Performance Model of Heterogeneous Clusters Based on a Switched Network. In: Proceedings of the 2007 IEEE International Conference on Cluster Computing (Cluster 2007), pp. 568–575. IEEE Computer Society, Los Alamitos (2007)

Visual Debugging of MPI Applications

Basile Schaeli[1], Ali Al-Shabibi[2], and Roger D. Hersch[1]

[1] Ecole Polytechnique Fédérale de Lausanne (EPFL), School of Computer and Communication Sciences, CH-1015 Lausanne, Switzerland
[2] University of Heidelberg, Heidelberg, Germany
basile.schaeli@epfl.ch, ali.al-shabibi@cern.ch

Abstract. We present the design and implementation of a debugging tool that displays a message-passing graph of the execution of an MPI application. Parts of the graph can be hidden or highlighted based on the stack trace, calling process or communicator of MPI calls. The tool incorporates several features enabling developers to explicitly control the ordering of message-passing events during the execution, and test that reordering these events does not compromise the correctness of the computations. In particular, we describe an automated running mode that detects competing sends matching the same wildcard receive and enables the developer to choose which execution path should be followed by the application.

1 Introduction

Parallel applications are subject to errors that do not occur in single-threaded sequential applications. Such errors include deadlocks, when conflicts over resources prevent the application from moving forward, and message races, when changing the order of reception of messages changes the result of the computation. Parallel application debuggers should therefore enable explicitly testing and analyzing such errors and provide multiple abstraction levels that filter and aggregate the large amount of information displayed to the developer.

Several contributions, e.g. [4,6], focus on record and replay techniques to enable reproducing a race once it has been detected. For instance, Retrospect [4] enables the deterministic replay of MPI applications, but the lack of control on the application execution may force the developer to run its application many times until an error is revealed. To our knowledge, ISP [11] is the only tool that explicitly tests different orderings of events within MPI applications. While it could produce a suitable trace for a replay tool, being able to replay an erroneous execution deterministically is only a first step in identifying a bug. The ability to visualize and to test slightly different executions often helps understanding the origin of an error and correcting it.

Full-featured parallel debuggers such as TotalView [10] and DDT [1] support the isolation of specific processes, the inspection of message queues and are able to attach a sequential debugger to remote application instances. The debugger for the Charm++ framework [7] takes advantage of its integration within

A. Lastovetsky et al. (Eds.): EuroPVM/MPI 2008, LNCS 5205, pp. 239–247, 2008.

the Charm++ parallel runtime to provide higher-level features such as setting breakpoints on remote entry points. While these tools provide the developer with detailed information about the running processes, none of them provides an instantaneous high-level picture of the current state of the application execution.

In previous work, we described a debugger targeting applications developed using the Dynamic Parallel Schedules (DPS) parallelization framework [2]. The parallel structure of these applications is described as an acyclic directed graph that specifies the dependencies between messages and computations. The debugger may therefore display the current state of the graph very naturally and provide the application developer with much information in a compact form. Different event orderings can be explicitly tested by reordering messages in reception queues or by setting high level breakpoints.

The present contribution applies the concepts presented in [2] to MPI applications, and introduces a few MPI specific features. A graphical user interface displays the message-passing graph of the application and provides a high-level view of its communication patterns. Within the message-passing graph, we can hide or highlight MPI calls based on various criteria such as the originating process, the communicator on which the communication occurred, or the source code file or function that generated the call. We propose various types of high-level breakpoints to control the evolution of the participating processes. Execution scenarios that occur only rarely in actual executions can thereby be explicitly tested. Variants may be executed using an interactive replay functionality. Potential conflicts over *MPI_ANY_SOURCE* receives may be detected automatically. Possible matches are then drawn on the message-passing graph, enabling the developer to decide which execution path must be followed by the application. The debugger also integrates object visualization support for the *autoserial* library [3], which provides MPI function wrappers that are able to send and receive regular C++ objects.

The paper is organized as follows. Section 2 describes the general architecture of the debugger and Section 3 describes features for controlling the application execution. The debugger's impact on applications is evaluated in Section 4 and Section 5 draws the conclusions.

2 Architecture

The debugging functionality is provided via two independent components. The first, the interception layer, is a library that intercepts the MPI function calls performed by the application using the MPI Profiling Interface (PMPI [5]). When the MPI initialization function *MPI_Init* is intercepted, every process opens a TCP connection to the debugger, a standalone Java program that receives and displays information about the current state of the application.

Processes first identify themselves to the debugger by sending their rank and their process identifier. During the application execution, the interception layer then sends a notification to the debugger for every point-to-point and collective MPI function called. Notifications are also generated for the various *MPI_Wait* and *MPI_Test* functions, as well as for functions creating new communicators. With the exception of the message content, each notification contains a copy of all the parameters of the called function. These parameters may be MPI defined constants, such as *MPI_COMM_WORLD*, *MPI_INT* or *MPI_ANY_SOURCE*, whose actual value is specific to MPI implementations. The debugger therefore also receives a copy of these constants when the application starts, so as to be able to translate parameter values into human readable form when displaying information to the developer.

Notifications are sent before calling the actual MPI function. Once it has sent a notification, a process suspends its execution and waits for an acknowledgment from the debugger. By withholding specific acknowledgments, the debugger may thus delay the execution of the associated processes while letting the rest of the application execute.

Receive calls that specify *MPI_ANY_SOURCE* as the source of the expected message may potentially match send calls from multiple sources. In this paper, we refer to such calls as *wildcard receives*. Since in the general case the debugger cannot automatically determine which source is actually matched by a wildcard receive, this information is provided separately by the interception layer via a *matched* notification. If the wildcard receive is blocking, the *matched* notification is sent immediately after the reception of the message by the receive function call. For non-blocking wildcard receives, the *matched* notification is sent when an *MPI_Wait* or *MPI_Test* call successfully queries the status of the non-blocking receive. In both cases, the rank of the matched source is read from the *MPI_Status* parameter of the querying call.

The user interface of the debugger consists of a single window that provides control elements to influence the application execution, and displays the current status of the application as a message-passing graph. The vertices of the graph represent the MPI calls performed by the application. Unlike most tracing tools that display time from left to right, our representation matches the one used within the MPI standard, where time flows from top to bottom. Vertices associated to notifications from a same process are therefore displayed one below the other, similarly to successive lines of code within a source file.

The debugger draws edges between successive vertices from a same process. It also draws edges of a different color between vertices associated to matching send and receive calls. For this purpose, the debugger maintains one *unmatched sends* and one *unmatched receives* queue. Upon receiving a notification for a send (resp. receive) call, the debugger looks for a matching receive (resp. send) call within the unmatched receives (resp. unmatched sends) queue. If none is found, the incoming notification is pushed at the end of the corresponding queue. When looking for matches, the queues are explored in a FIFO manner in order to respect the FIFO property of MPI communication channels. New vertices and edges are dynamically added to the graph as the debugger receives new notifications from the application. When the debugger receives a notification for a wildcard receive from a process p, it stops matching send calls destined to p until the reception of the corresponding *matched* notification. For non-blocking

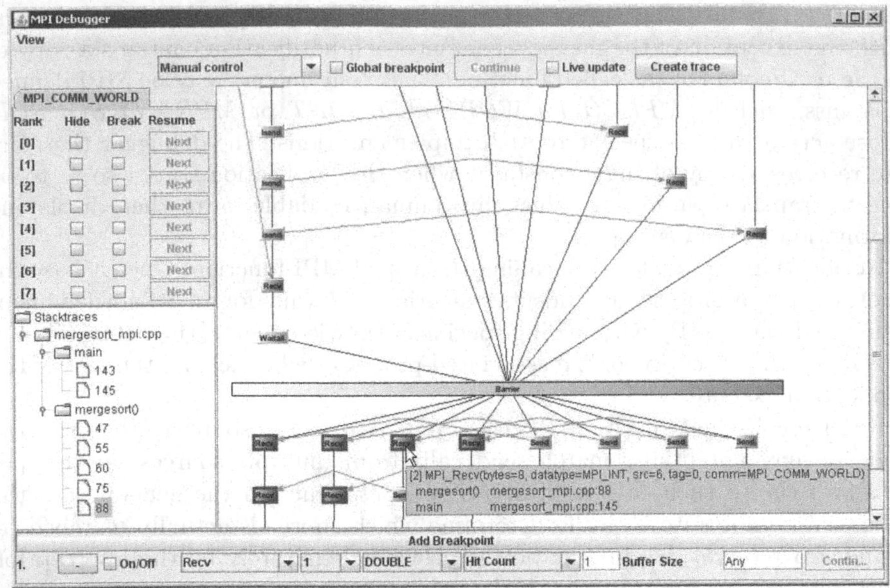

Fig. 1. Debugger window. The left panes contain the list of processes and the stack trace tree. Tooltips display detailed information about MPI calls.

wildcard receives, graph updates are therefore delayed until the application successfully queries the status of the receive call.

Single-threaded processes cannot send more than one notification at a time to the debugger. The order in which the debugger receives notifications from a given process therefore matches the order of occurrence of events within that process, and the graph displays the temporal dependencies between these calls. In case of multithreaded processes where multiple threads may simultaneously call MPI functions (e.g. using *MPI_THREAD_MULTIPLE*), the message-passing graph no longer accurately represents the temporal dependencies between events. However, the interception layer makes sure that no two threads call MPI functions simultaneously, and that the ordering of the calls matches the order in which acknowledgments are received. The ordering of messages within communication channels is therefore known to the debugger, which may accurately display send-receive matches.

On Linux, the interception layer is able to determine the stack trace of every MPI call. A panel in the debugger window displays a tree containing the files, functions and line numbers from which the MPI functions were called. Selecting a node of the tree then highlights all the associated vertices in the message-passing graph, illustrating how and when the selected file or function is used within the application. Another panel displays the list of processes involved in the computation and enables hiding the graph vertices belonging to specific processes. When the application uses multiple communicators, the list of processes belonging to each one of them appears in additional tabs. When switching to a given communicator tab, the developer may choose to display a partial message-passing graph that includes only the vertices associated to MPI calls performing communications on the selected communicator.

We provide the ability to zoom in and out of the graph in order to adapt its level of detail to the needs of the developer. The label and color of every vertex indicates the type of MPI operation executed, and tooltips display detailed information about the parameters of the call, as well as its stack trace if available. Collective operations are grouped into a single vertex and are represented as a rectangle that spans all participating processes. When the developer double-clicks the graph vertex of a suspended MPI call, the debugger attaches a user-specified sequential debugger to the calling application process, and uses the stack trace information to set a breakpoint to the source code line that immediately follows the MPI function call. The debugger then acknowledges the notification, the process is resumed and the new breakpoint is hit, enabling the developer to inspect the application code.

The *autoserial* [3] library allows sending and receiving complex C++ objects instead of simple memory buffers. It does so by providing wrappers around the *MPI_Send* and *MPI_Recv* functions. When the wrapper functions are used, the interception layer sends the full serialized object to the debugger, which may then display its content using a tree view similar to the ones found in traditional sequential debuggers. For objects to be understood by the debugger, the serialization is performed by a specialized textual serializer which includes the necessary variable name and type information within the serialized data. The interception layer also provides functions for registering serializable objects representing the user-space state of the running application. The developer may retrieve and display these objects when a process is suspended by the debugger. The request is piggybacked on the acknowledgment for the pending notification of the selected process, causing the interception layer to send a copy of the registered objects.

3 Controlling the Application Execution

A *global breakpoint* may be activated. It causes the debugger to withhold all acknowledgments, thereby suspending all processes. Clicking a button then simultaneously acknowledges all pending notifications and resumes the execution of all processes up to the next MPI call. The global breakpoint allows quickly stepping through the execution of all processes at the message-passing level rather than at the instruction level, while maintaining the opportunity to take action on every notification. *Process breakpoints* cause the debugger to systematically withhold the notifications sent by particular processes. This feature may for instance be used to arbitrarily delay specific processes in order to provoke message races. The developer may also explicitly test different execution orderings by breakpointing multiple processes and by resuming them in different orders. A finer control is provided via *conditional breakpoints.* They enable withholding acknowledgments for notifications matching one or several criteria such as the rank of the calling process, the type of MPI call, the message size or data type, or the destination rank for send calls. Moreover, the developer can specify a hit count to indicate how many times the breakpoint must be hit before it becomes active.

The use of wildcard receives leads to non-determinism within the application execution. It may be difficult to identify potential conflicts, and manually controlling the application execution may be error prone. We therefore implemented a procedure that detects potential ordering variations on wildcard receives and allows the developer to choose the send call that matches a specific wildcard receive. When this mode is active, the debugger automatically acknowledges all notifications that are not associated to send calls. Upon receiving a notification for a send call, the debugger checks whether it already received a notification for a matching receive call. If not, it holds the send notification until it receives a matching receive notification. If the matching receive explicitly specifies the source of the expected receive, the debugger acknowledges the send notification, thereby resuming the process execution. If the matching receive is a wildcard receive, the debugger draws one large arrow between the graph vertices corresponding to the potentially matching send and the wildcard receive. Since the other processes keep running, more arrows may be added as the debugger receives other potentially matching send notifications. Clicking on a send vertex then acknowledges the associated notification. The resumed process then sends its message, which matches the wildcard receive under consideration.

Since this scheme makes no assumption about whether calls should be blocking or not, it is able to reveal potential message races stemming from the buffering of messages within MPI calls. In Fig. 2a, the debugger will acknowledge the broadcast notification from process 0. If process 0 buffers the broadcasted message, the debugger eventually receives a notification for the subsequent send call, which may match the first wildcard receive of process 1 if process 2 is delayed.

On the other hand, some executions involving non-blocking or buffered sends cannot be enforced. For instance, in Fig. 2b the debugger cannot detect that a race could occur until the non-blocking send from process 0 is acknowledged.

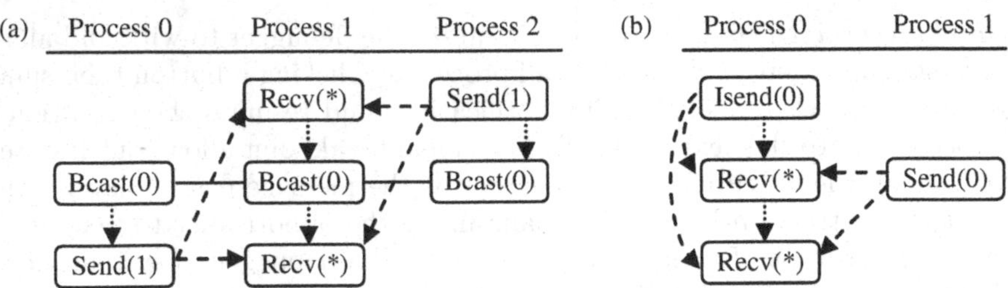

Fig. 2. (a) Send calls from both process 0 and process 2 may match any of the wildcard receives from process 1 if broadcast and send calls are buffered; (b) the debugger cannot receive the notification for the first receive from process 0 without previously acknowledging the non-blocking send from process 0. Numbers between brackets respectively indicate the rank of the destination, source or root process depending on the type of MPI call. A '*' denotes a wildcard receive.

In such cases, reliably enforcing different orderings would require the ability to reorder incoming messages within MPI reception queues. Automatically holding all send calls may also suspend the execution of the debugger when non-blocking sends are used. This is the case in Fig. 2b, where both processes are suspended by the debugger and none of them has any receive to match. Such cases must be manually resolved by clicking on one of the send vertices (in this example, on the Isend call from process 0) to acknowledge the associated notification and resume the execution.

At any moment, the developer has the possibility of generating a trace file. When the application restarts, loading the trace file causes the debugger to set internal breakpoints that will reproduce the traced (and potentially incorrect) execution. During replay, the developer may set additional breakpoints to test execution variants.

4 Impact on Applications

Running an application under the control of a scheduler may alter the ordering of events within the application. Races appearing in regular executions may thus disappear when running under control of the debugger. Nevertheless, any race stemming from nondeterministic orderings of MPI calls can be explicitly induced by appropriately setting breakpoints within the application.

A second concern is the application slow down. Since it must process all the notifications sent by the processes, the debugger becomes a bottleneck when the rate of incoming notifications increases. In order to evaluate that impact, we performed measurements on the Pleiades cluster at EPFL, which consists of 132 single-processor nodes connected with a Fast Ethernet switch. We ran the High Performance Linpack (HPL [8]) benchmark on four nodes, with one process per node and a 100×100 matrix decomposed into 10×10 blocks. This run called 664 MPI communication functions during its running time of 0.03 seconds, leading to a call rate of 22 thousand calls per second. Once connected to the debugger, the same application ran in 30 seconds, or about 1000 times slower. For this test, the debugger was therefore able to process and display about 22 notifications per second. While such a display rate is sufficient to manually step through the application execution, the developer often wants the application to execute quickly up to the point where he wants to start his analysis.

The major portion of the slow down is due to the display of the events to the application developer however. If we run the debugger without layouting and displaying the graph, the running time falls to 0.16 seconds, reducing the overhead compared to the original application running time to a factor of 5. Figure 3 presents the running time overheads as a function of the average number of MPI calls per second performed by all processes during the execution. It displays results for HPL running on 4, 8 and 16 nodes with one process per node for various matrix and block sizes (from 2000 to 8000 and from 25 to 500 respectively). For a given number of nodes, the running time overhead can be approximated fairly well using a linear function. The slope becomes less steep as the number of nodes

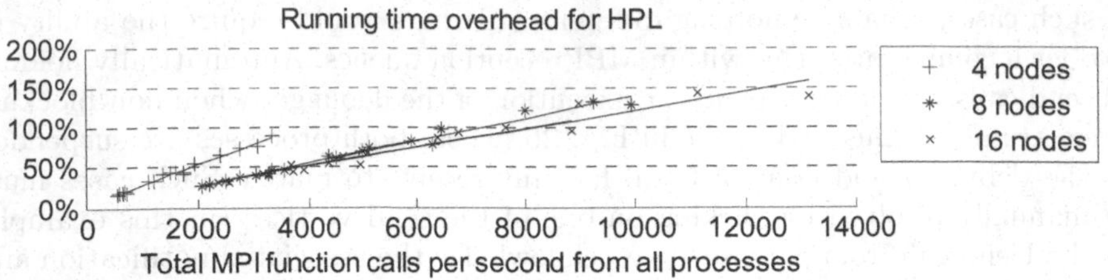

Fig. 3. Debugger overhead when the display of the graph is disabled. The call rate of MPI functions is computed when the application is not connected to the debugger. Results for 4, 8 and 16 nodes are well approximated with a linear function.

increases, due to the fact that the debugger uses one thread per MPI process to receive and acknowledge the notifications. Since the notifications are well balanced between the processes, the multithreading improves the overlapping of processing and communication within the debugger.

These results show that high notification rates may occur, and it is therefore crucial that we optimize our layout and display code to achieve better performance. The performance can currently be slightly improved by disabling the live updating of the message-passing graph, which is then refreshed at once when a breakpoint is hit and when the developer explicitly requests an update.

In addition to increasing the running time overhead, high notification rates lead to large graphs that are difficult to analyze. The interception layer implements the *MPI_Pcontrol* function to enable and disable the sending of notifications to the debugger. The developer may thus limit both the runtime overhead and the size of the considered message-passing graph by enabling notifications only during critical parts of the execution. The scheme could be extended in order to provide a finer control over the types of MPI calls that should be transferred to the debugger. The ability to collapse parts of the graph, e.g. between two barrier synchronizations, would also facilitate the visualization of large graphs.

5 Conclusion and Future Work

We have presented a debugger for MPI applications that provides the developer with a graphical view of the current status of the application execution. It dynamically draws the message-passing graph of the application, and graph vertices can be highlighted according to specific criteria in order to ease the analysis. Several types of breakpoints enable controlling the execution of the parallel application. All breakpoints operate at the level of message-passing calls rather than code instructions. They enable the developer to focus on the communication patterns of the application, and provide entry points for attaching a sequential debugger to individual processes. The debugger is also able to run the application such that the developer is able to choose how send and receive calls should be matched in the presence of wildcards.

The ability to influence the application by suspending processes and reordering message matches provides the developer with full control over its execution. This control can be used to execute cases that occur only rarely in practice, for example for testing the presence of message races or deadlocks within the parallel application.

It would be very interesting to integrate checkpoint/restart capabilities [4,9] into the message-passing graph based debugger. Combined with the provided control on the application execution, this feature would enable interactively testing multiple execution scenarios without requiring reexecuting the application from the beginning.

References

1. Allinea, the distributed debugging tool (DDT), `http://www.allinea.com`
2. Al-Shabibi, A., Gerlach, S., Hersch, R.D., Schaeli, B.: A Debugger for Flow Graph Based Parallel Applications. In: Proceedings of the ACM International Symposium on Software Testing and Analysis (ISSTA 2007), Parallel and Distributed Systems: Testing and Debugging workshop (PADTAD 2007), London, UK (2007)
3. Automatic reflexion and serialization of C++ objects, `http://home.gna.org/autoserial/`
4. Bouteiller, A., Bosilca, G., Dongarra, J.: Retrospect: Deterministic Replay of MPI Applications for Interactive Distributed Debugging. In: Cappello, F., Herault, T., Dongarra, J. (eds.) PVM/MPI 2007. LNCS, vol. 4757, pp. 297–306. Springer, Heidelberg (2007)
5. Gropp, W., Huss-Lederman, S., Lumsdaine, A., Lusk, E., Nitzberg, B., Saphir, W., Snir, M.: MPI: The Complete Reference, vol. 2. MIT Press, Cambridge (1998)
6. Hong, C.-E., Lee, B.-S., On, G.-W., Chi, D.-H.: Replay for debugging MPI parallel programs. In: Proceedings of the MPI Developer's Conference, pp. 156–160 (1996)
7. Jyothi, R., Lawlor, O.S., Kalé, L.V.: Debugging Support for Charm++. In: Procs. of the 18th International Parallel and Distributed Symposium (IPDPS 2004), Parallel and Distributed Systems: Testing and Debugging Workshop (PADTAD), p. 294 (2004)
8. Petitet, A., Whaley, R.C., Dongarra, J., Cleary, A.: HPL - A Portable Implementation of the High-Performance Linpack Benchmark for Distributed-Memory Computers, `http://www.netlib.org/benchmark/hpl/`
9. Sankaran, S., Squyres, J.M., Barrett, B., Lumsdaine, A., Duell, J., Hargrove, P., Roman, E.: The LAM/MPI Checkpoint/Restart Framework: System-Initiated Checkpointing. Int'l Journal of High Performance Computing Applications 19(4), 479–493 (2005)
10. TotalView Technologies, the TotalView Debugger, `http://www.totalviewtech.com`
11. Vakkalanka, S.S., Sharma, S., Gopalakrishnan, G., Kirby, R.M.: ISP: a tool for model checking MPI programs. In: Proc. ACM SIGPLAN Symposium on Principles and Practice of Parallel Programming (PPoPP 2008), pp. 258–256 (2008)

Implementing Efficient Dynamic Formal Verification Methods for MPI Programs*

Sarvani Vakkalanka[1], Michael DeLisi[1], Ganesh Gopalakrishnan[1], Robert M. Kirby[1], Rajeev Thakur[2], and William Gropp[3]

[1] School of Computing, Univ. of Utah, Salt Lake City, UT 84112, USA
[2] Math. and Comp. Sci. Div., Argonne Nat. Lab., Argonne, IL 60439, USA
[3] Dept. of Computer Sci., Univ. of Illinois, Urbana, Illinois, 61801, USA

Abstract. We examine the problem of formally verifying MPI programs for safety properties through an efficient dynamic (runtime) method in which the processes of a given MPI program are executed under the control of an interleaving scheduler. To ensure full coverage for given input test data, the algorithm must take into consideration MPI's out-of-order completion semantics. The algorithm must also ensure that nondeterministic constructs (e.g., MPI wildcard receive matches) are executed in all possible ways. Our new algorithm rewrites wildcard receives to specific receives, one for *each* sender that can potentially match with the receive. It then recursively explores each case of the specific receives. The list of potential senders matching a receive is determined through a runtime algorithm that exploits MPI's operation ordering semantics. Our verification tool ISP that incorporates this algorithm efficiently verifies several programs and finds bugs missed by existing informal verification tools.

1 Introduction

With the increasing use of MPI for the distributed programming of virtually all high-performance computing clusters in the world, it is important that MPI programs be verified to be free of bugs. With the need to re-verify MPI programs after each optimization step, the process of verification must involve only modest computing resources and limit manual tedium. As MPI programs can contain many types of bugs, including deadlocks, resource leaks, and numerical inaccuracies, it is practically impossible for a single tool to guarantee the coverage of bugs in all these classes. Therefore, approaches that focus on a limited bug class and guarantee full coverage for that class are preferred.

In this paper, we present our C MPI program verification tool, named Insitu Partial Order (ISP), that incorporates a novel scheduling algorithm called POE (Partial Order reduction avoiding Elusive interleavings). ISP guarantees to

* Supported in part by NSF CNS-00509379, Microsoft HPC Institutes Program, and the Mathematical, Information, and Computational Science Division subprogram of the Office of Advanced Scientific Computing Research, Office of Science, U.S. Department of Energy, under Contract DE-AC02-06CH11357.

A. Lastovetsky et al. (Eds.): EuroPVM/MPI 2008, LNCS 5205, pp. 248–256, 2008.

detect *all* deadlocks and local assertion violations in MPI programs containing 24 of the most commonly used MPI functions. For these MPI programs, the ISP tool will explore close to the minimal number of interleavings. Furthermore, ISP does not require any modeling effort on part of users, allowing it to be easily rerun during program development. ISP enjoys the same ease of use as the dynamic verification tools Umpire [2], Marmot [3], ConTest [4], and Jitterbug [5] (to name a few). However, the POE algorithm offers the formal guarantee of finding *all* deadlocks. As shown by experiments on our web site [13], of the 69 Umpire tests, 30 contain deadlocks, and ISP detects all of them, while exploring a very small number of interleavings. In contrast, Marmot fails to find deadlocks in eight of these tests, despite being run multiple times. When these tests were run with MPICH2 repeatedly, the deadlock detection success was unpredictable. Tools that rely on perturbing schedules simply cannot guarantee coverage.

The main feature of POE is that it explores only *relevant* interleavings, using a technique known as *partial order reduction* [6]. Without this idea, any exploration method for MPI will go out of hand. For instance, consider the short MPI program in Figure 1 that begins with two sends in P0 and P2, and a wildcard receive in P1. The total number of interleavings of all these MPI calls is 210.[1] However, to trigger `error1`, we need to consider the interleavings in which the send of `P2` matches the wildcard receive. Testing-oriented tools may easily miss these interleavings. Thanks to partial order reduction, ISP will: (i) pick an arbitrary order for executing `P0`'s first send and `P1`'s first receive, (ii) pick an arbitrary order to execute `P2`'s first send and `P1`'s second receive, and then (iii) consider *both* the `Send` matches with the wildcard receive (shown by `*`). ISP has built-in

```
P0: MPI_Send(to P1...);     MPI_Send(to P1, data = 22);
P1: MPI_Recv(from P0...);  MPI_Recv(from P2...);
    MPI_Recv(*, x);         IF (x==22) THEN error1 ELSE MPI_Recv(*, x);
P2: MPI_Send(to P1...);     MPI_Send(to P1, data = 33);
```

Fig. 1. A Simple MPI Example with Wildcard Receives

knowledge of the commuting properties of MPI functions. For example, consider an MPI program in which `MPI_Barrier` is invoked by N processes. ISP would, in general, explore only one of the $N!$ ways in which to have invoked the barrier calls. In our implementation of the 24 MPI functions, alternate interleavings are explored for wildcard receives, `WAIT_ANY`, and `TEST_ANY`.

Overview of ISP's use of PMPI: We use the well-known PMPI mechanism, normally used for performance studies, to support runtime model checking in ISP. We introduce an extra process called the *verification scheduler*. ISP provides its own version of "MPI_f" for each MPI function f. Within MPI_f, we arrange for handshakes with the scheduler that realizes the POE algorithm. When the scheduler finally gives permission to fire f, we invoke PMPI_f from within our version of MPI_f. The MPI runtime only sees the PMPI_f calls.

---
[1] (7!)/((2!).(3!).(2!)).

Related Work: Techniques for eliminating nondeterminacy for testing parallel programs were studied in [14]. A dynamic verification approach for reactive C programs was first proposed by Godefroid [7]. Flanagan and Godefroid [8] extend this work, incorporating a more efficient *dynamic partial order reduction* (DPOR) algorithm. In [1], we presented the first DPOR-based verification method for MPI programs that employ one-sided communication. In [9], we reported a preliminary implementation of DPOR for MPI's two-sided operations. This algorithm did not address the full range of out-of-order behaviors of MPI. It also proved incapable of controlling the MPI runtime to force the desired wildcard receive matches (see Section 2). POE overcomes both these limitations, and replaces DPOR – the former algorithm implemented within ISP. Exploiting MPI's semantics, POE employs a strategy of *lookahead computation* to discover how sends and receives in an MPI program match. A formal presentation of the POE algorithm is given in [10].

Roadmap: The remainder of this introduction presents in detail the three new ideas used in POE: *Forcing Wildcard Matches* (Section 1.1), *Handling Out-of-order Completion* (Section 1.2), and *Discovering Match-Sets* (Section 1.3). Section 2 presents the POE algorithm in detail, focusing on sends, receives, and barriers. Section 2.2 describes how many additional MPI commands are smoothly handled by the extended POE algorithm implemented in ISP. We also discuss how the user interface of a Visual Studio integration of POE works: we strive to preserve the users' view of their MPI program, despite the fact that our POE algorithm changes the internal computation through dynamic rewriting. Section 3 presents experimental results and Section 4 concludes.

1.1 Forcing Wildcard Matches

Consider the example in Figure 2, with line 2 containing a wildcard receive. A match between the `Isend` on line 6 and `Irecv` on line 2 (wildcard) will enable `Recv` on line 3 to match with the `Isend` on line 9. However, if the `Isend` on line 9 were to match the `Irecv` on line 2, a *deadlock* would result, with `Recv` (line 3) no longer able to match `Isend` (line 6). Clearly, we cannot leave out this second option (process interleaving) during testing.

The role of a dynamic verification tool for MPI is to determine, at runtime, the specific matches possible, and explore *all relevant* ones - that is, a representative of each equivalence class of equivalent interleavings. This method must be carried out at runtime: (i) the outcomes of control branches through conditional statements will be known only at runtime and (ii) the send/receive targets/sources, and other details (communicator, tag, etc.) may be values that are computed at runtime.[2]

We now explain briefly why DPOR does not work for MPI. Suppose a DPOR-based algorithm is able to determine that `Isend` (line 6) matched `Irecv` (line 2), and that `Isend` (line 9) is also a potential alternate match for this `Irecv`. According to the algorithm of [8], the dynamic verification scheduler must now somehow force this alternative match – say by firing the `Isend` (line 9) in real-time order before firing `Isend` (line 6). However, we know from MPI's semantics that the

[2] In this paper, we suppress details pertaining to communicators and tags.

MPI runtime environments *do not guarantee that this alternative matching will occur.* We call these scenarios *(potentially) elusive matches.* Tricks such as inserting 'padding' delays that can perturb schedules may make elusive matches more likely, but still provide no guarantees. Therefore, we need an algorithm different from DPOR, and POE is our answer.

POE solves the problem of elusive matches *without requiring changes to the MPI library* and *without adding padding delays.* It *dynamically rewrites* wildcard receives into specific receives, one for each actual sender that it computes to be a *certain* match. In the context of the example in Figure 2, if we can force two recursive explorations, with `MPI_Irecv(buffer, from 1, &req);` and `MPI_Irecv(buffer, from 2, &req);` used successively in lieu of the existing line 2, we would have force-matched both the sends. The crucial fact is, of course, to *never* force-match with a send that is not going to be issued – this can cause a deadlock that does not exist. POE employs a strategy to discover all potential senders precisely, as outlined in Section 1.2, and Section 2.

```
0 : // * means MPI_ANY_SOURCE
1 : if (rank == 0)
2 : { MPI_Irecv(buff1, *,
                          &req);
3 :   MPI_Recv(buff2, from 2);
4 :   MPI_Wait(&req) }
5 : else if (rank == 1)
6 : { MPI_Isend(buff1, to 0,
                          &req);
7 :   MPI_Wait(&req); }
8 : else if (rank == 2)
9 : { MPI_Isend(buff2, to 0,
                          &req);
10:   MPI_Wait(&req); }
```

Fig. 2. Relevant Interleavings and Elusive Matches during Dynamic Verification of MPI Programs

1.2 Handling Out-of-Order Completion

In MPI, (i) two `Isends` targeting two different processes may finish out of order (with respect to issue order), while two `Isends` targeting the same process must match in order. Likewise, (ii) two non-wildcard receives sourcing from the same source process must also match sends in order. Similarly, (iii) if the first receive or both receives are wildcards, even then they must match in issue order. As for waits and tests, (iv) they must not complete before their corresponding send/receive operations. Finally, (v) operations appearing *after* MPI barriers and MPI waits must not finish before the barrier or wait. *Notice that we did not say that operations before a barrier must finish before the barrier!* Section 2 will show that operations issued before a barrier can linger even after crossing the barrier.

1.3 Discovering Match-Sets

POE employs an approach to bound the scope of search for locating potential matching sends for a wildcard receive. It relies on a formal notion of *fences* to determine when two operations issued by a dynamic verification scheduler through the PMPI layer will be carried out (i) by the MPI runtime, (ii) in that order. We are not saying that MPI has "fence instructions" akin to how CPUs have assembly instructions to order intra-core execution. However, there are still conceptually equivalent ordering points defined by the MPI semantics! Based on a formulation of MPI fences, we can form *match-sets* – sets of MPI operations that can be issued out of order by a dynamic verification scheduler. This is the idea of POE's *lookahead computation* alluded to earlier.

2 Basic POE Algorithm

Consider Figure 3. Note that although the `Isend` on line 8 is issued *after* the barrier on line 7, it is a potential match for the `Irecv(*)` on line 2. This is precisely because MPI's `Isend` can linger across a `Barrier`. The only ordering that MPI guarantees is that functions *after* a barrier will not be called until all functions before (and including) the barrier have been called on any process (rank). The following steps describe how the dynamic verification scheduler implementing the POE algorithm handles this example. Our POE scheduler will intercept every MPI operation `MPI_f` issued from every MPI process. It will often not issue these operations (through `PMPI_f`) immediately – but only make a note of it, and *later* issue them. We employ a central scheduler process which helps issue MPI operations in a serialized manner, and currently replays executions by re-execution from MPI_Init.[3]

Illustration of POE on the example of Figure 3

- Collect `Irecv` (line 2), and do not issue.
- Collect `Barrier` (line 3), and do not issue.
- Since `Barrier` is a fence, do not collect anything more from rank 0; switch to rank 1.
- Collect `Barrier` (line 7), and do not issue; switch to rank 2.
- Collect `Isend` (line 11), and do not issue. Then collect `Barrier` (line 12), and do not issue.
- A fence has been reached in every rank. Now, form a *match set* in priority order, with the following priority order followed: barriers first, then non wildcard send/receives, and finally wildcard send/receives.
- In our current state, there is indeed a highest-priority match set formed by the barriers. *Now, POE sends these Barriers* into the MPI runtime through `PMPI_Barrier` calls.
- The next ordering points (fences) are attained at `Wait`.
- No match-sets of non wildcard receives exist. Skip this priority order.
- At this point, we know the full list of senders that can match the wildcard receive.
- Dynamically rewrite `Irecv(*)` into `Irecv(1)` and `Irecv(2)`, in two different executions.
- Form the first match set of `Irecv(1)` and `Isend()` of line 8. Pursue this interleaving.
- Form the second match set of `Irecv(2)` and `Isend()` of line 11. Pursue this interleaving though re-execution of the MPI program.

Note that for MPI programs with no wildcards, POE will examine the entire program under exactly one interleaving.

[3] A distributed strategy allowing concurrent issues is slated for development; also a more efficient re-execution method is reported in [9].

2.1 Semi-formal Description of POE

```
if (rank == 0)
  { MPI_Irecv (&buf0, *, &req);
    MPI_Barrier ();
    MPI_Wait (&req);
    MPI_Recv (&buf1, from 2); }
 else if (rank == 1)
  { MPI_Barrier ();
    MPI_Isend(buf1, to 0, &req);
    MPI_Wait (&req); }
 else if (rank == 2)
  { MPI_Isend(buf0, to 0, &req);
    MPI_Barrier ();
    MPI_Wait (&req); }
```

Fig. 3. Ordering Semantics and Operation Lifetimes

The POE algorithm works by finding *match-sets* of MPI operations and issuing them (possibly out-of-order) to the MPI runtime (using the PMPI versions of these operations). An MPI operation can essentially be in one of the two states: *issued* and *completed.* When an MPI operation is *issued*, it means that the MPI runtime is aware of the MPI operation. When an MPI operation is *completed*, it means that the operation has no presence in the MPI runtime. For example, when we say that an MPI receive operation is complete, we mean that a matching send has been found for that receive. For simplicity, we only deal with the following MPI operations in this section: MPI_Barrier, MPI_Isend, MPI_Wait, MPI_Irecv. We also assume that the operations have the same tag and that the communicator is MPI_COMM_WORLD for simplicity.

Since MPI semantics allow for nonblocking operations to linger across barriers, POE needs to emulate this out-of-order completion behavior of the MPI runtime. In addition, POE must also respect MPI's send and receive ordering guarantees. Therefore, rather than emulating the issue order of MPI operations, POE must emulate the *completion order* of MPI operations. Before going into more detail, we first define what we call *fence MPI operations.*

MPI Fence Operations: A *fence* is an MPI operation that must be completed before any following MPI operations from the same process can be issued. Any blocking MPI operation is a fence, as are MPI_Barrier, MPI_Wait, and MPI_Recv.

POE executes all C statements in program order; however, it issues MPI operations to the MPI runtime only when they are guaranteed to complete immediately. For example, an MPI receive (send) is issued only if a matching send (receive) is found. This is the idea of POE forming *match-sets* as introduced in Section 1.3. In order to correctly emulate the out-of-order completion inherent within the MPI semantics (Section 1.2 presents it through examples; our web page [13] has details), POE builds a graph data structure of *completes-before* edges across MPI operations within the same process. We call these edges as *intra completes-before* (IntraCB) edges.

In addition to IntraCB, POE also maintains a *conditional completes before* (CCB) edge which is added as follows. The purpose of this edge is to model how wildcard receives may *trump* non wildcard receives. For example, suppose an MPI process P0 has the code sequence Recv(from 1); Recv(from *); and MPI process P2 has code sequence Send(to 0);. Then this Send matches Recv(from *) because the first offered match "from 1" requires a send from

P1 which is not present. In this case, a CCB edge is not introduced between the receives in `P0`. However, now if we consider the same P0 process, but a P1 process which is `Send(to 0);`, then this `Send` matches `Recv(from 1);`. In this case, a CCB edge is introduced between the receives in `P0`.

If there is an IntraCB or CCB edge from i to j, then we call i as the *ancestor* of j. The POE algorithm described on Page 252 guarantees that no PMPI operation will be issued contrary to the IntraCB or CCB edges, thus guaranteeing the correctness of message matches within the MPI runtime.

It must be observed that the code snippet in Figure 1 can be verified with DPOR if the technique of dynamic rewriting of the wildcard receives is employed. However, the code snippet in Figure 3 cannot be verified with DPOR even with dynamic rewriting of wildcard receives employed. Due to the presence of the barrier, the `MPI_Isend` at line 8 can never be executed *before* the `MPI_Isend` at line 11, whereas in DPOR, we will need dependent actions to be replayable in both orders. In any interleaving of this example, however the send at line 11 is always issued before the send at line 8. The POE algorithm overcomes this problem by executing the big-step move of `MPI_Barrier` of the three processes, and then forming match-sets of the wildcard receive with `MPI_Isends` by recurively employing dynamic rewriting for both the match-sets each in a different interleaving.

2.2 Implementing `WAIT_ANY` and `TEST_ANY`

ISP implements the POE algorithm that allows for executing MPI operations in an order different from the actual program order. Hence, when ISP traps an MPI request such as `MPI_Irecv(buffer, count, datatype, source, mpi_request)`, ISP stores the arguments for later issuance. Let *op* be an MPI operation.

When *op* is one of `MPI_Wait`, `MPI_Waitall`, `MPI_Test`, or `MPI_Testall`, the out-of-order issuance does not cause any problems since the POE algorithm's IntraCB edges ensure that all ancestors, *i.e*, the `MPI_Isends` and `MPI_Irecvs` corresponding to the requests are issued before *op* itself is actually issued. When *op* is one of `MPI_Testany` or `MPI_Waitany`, all `MPI_Irecv` and `MPI_Isend` ancestors of *op* are not necessarily issued before *op* itself is issued. Hence, when ISP invokes *op*, an error is thrown by the MPI runtime that the request structure is invalid (since the MPI runtime is not aware of the as yet unissued `MPI_Isend` or `MPI_Irecv` requests). In order to circumvent this problem, ISP issues *op* with `MPI_REQUEST_NULL` for those send and receive requests that are not yet issued and hence are ignored by the MPI runtime.

This allows the `MPI_Testany` and `MPI_Waitany` to work with POE's out of order issue when the MPI runtime does not know *all* the requests it is supposed know as it would during an in-order execution.

3 Experimental Results

We have experimented with all 69 Umpire [2] test cases, and in all 30 tests that have deadlocks, ISP finds the deadlocks, generating the fewest number of interleavings. We have also run ISP on the Monte-Carlo calculation of Pi, and the Game of Life example used in the EuroPVM/MPI 2007 Tutorial [12]. In all examples that do not employ wildcard receives, `WAIT_ANY`, or `TEST_ANY`, *ISP examines exactly one interleaving.* Some of these examples were instrumented to detect resource leaks (*e.g.*, `MPI_Isend` or `MPI_Irecv` without an `MPI_Wait`, `MPI_Comm_create` without an `MPI_Comm_free`, etc.). For these examples, a successful verification run using ISP implies a complete absence of these types of issues in the program (more discussions under 'data dependent control' below).

Since ISP works by re-executing the given MPI program, the restart time of the MPI system can become a significant overhead. This price is being paid because as opposed to existing model checkers which maintain state hash-tables, we cannot easily maintain a hash-table of visited states including the state of the MPI program as well as the MPI run-time system. (Note: In resorting to re-execution, we are, in effect, banking on deterministic replay.) One very promising approach to eliminate restart overheads is the following. At `MPI_Finalize`, one can reasonably assume that the MPI run-time state is equivalent to the one just after `MPI_Init`, and therefore simply reset user state variables and transition each process to the label after `MPI_Init`. We are further looking into when it is appropriate to use this technique (see [9] for details).

Data Independent Control Flow: In most MPI programs, control flows are unaffected by 'data' variables. For such MPI programs, a successful verification using ISP on a fixed input data set is tantamount to verifying the program for all possible input data. Also, for such programs, one can eliminate data variables, and their associated update functions, since they would not contribute either to control flow decisions or to the truth of the local assertions being checked. A preliminary implementation exists to detect and eliminate such data variables from MPI programs.

A preliminary Microsoft Visual Studio integration of ISP has also been implemented. A problem faced in this implementation was due to the fact that the actual run that occurs under ISP does not ever send wildcard receives into the MPI runtime. Visual Studio issued wildcard receives would not necessarily match with the correct sends, and mask the deadlock ISP found. This problem was solved through a novel technique that (i) obtains trace information from ISP, and (ii) mimics the dynamic rewriting of wildcard receives while making the Visual Studio debugger step through error traces. With this approach, the user's view of their program is preserved (more details on our web page [13]).

4 Concluding Remarks

We described our dynamic verification approach for MPI C programs that incorporates partial order reduction and dynamic rewriting based scheduling of MPI function call interleavings. ISP guarantees to detect *all* deadlocks and local assertion violations in C MPI programs that fall within ISP's range of supported commands (the commands and our verification results are documented on our website). MPI programs with additional calls may also be checked using ISP if they do not interfere with the commands currently supported (these commands will directly issue into the MPI runtime, without going through the PMPI mechanism). We detailed how we solved special problems posed by `WAIT_ANY` and `TEST_ANY`, and also how we reconcile a user-interface view with our dynamic rewriting process. We plan to release the full sources of ISP for experimentation, parallelize ISP itself using MPI, and make ISP widely available.

References

1. Pervez, S., Palmer, R., Gopalakrishnan, G., Kirby, R.M., Thakur, R., Gropp, W.: Practical model checking method for verifying correctness of MPI programs. In: EuroPVM/MPI, pp. 344–353 (2007)
2. Vetter, J.S., de Supinski, B.R.: Dynamic Software Testing of MPI Applications with Umpire. In: Proc. of SC 2000, pp. 70–79 (2000)
3. Krammer, B., Resch, M.M.: Correctness checking of MPI one-sided communication using Marmot. In: Mohr, B., Träff, J.L., Worringen, J., Dongarra, J. (eds.) PVM/MPI 2006. LNCS, vol. 4192, pp. 105–114. Springer, Heidelberg (2006)
4. Edelstein, O., et.al.: Framework for testing multi-threaded Java programs. Concurrency and Computation 15(3-5), 485–499 (2003)
5. Vuduc, R., Schulz, M., Quinlan, D., de Supinski, B., Saebjornsen, A.: Improved distributed memory applications testing by message perturbation. In: PADTAD 2006 (2006)
6. Clarke, E.M., Grumberg, O., Peled, D.A.: Model Checking. MIT Press, Cambridge (2000)
7. Godefroid, P.: Model checking for programming languages using Verisoft. In: POPL, pp. 174–186 (1997)
8. Flanagan, C., Godefroid, P.: Dynamic partial-order reduction for model checking software. In: POPL, pp. 110–121. ACM, New York (2005)
9. Vakkalanka, S., Sharma, S.V., Gopalakrishnan, G., Kirby, R.M.: ISP: A tool for model checking MPI programs. In: PPoPP 2008, pp. 285–286 (2008)
10. Vakkalanka, S., Gopalakrishnan, G., Kirby, R.M.: Dynamic Verification of MPI programs with Reductions in Presence of Split Operations and Relaxed Orderings. In: CAV 2008 (2008)
11. Holzmann, G.J.: The Spin Model Checker. Addison-Wesley, Reading (2004)
12. Gropp, W.D., Lusk, E.: Using MPI-2: A Problem-based Approach, Tutorial (2007)
13. `http://www.cs.utah.edu/formal_verification/europvm-mpi08`
14. Oberhuber, M.: Elimination of Nondeterminacy for Testing and Debugging Parallel Programs. Automated and Algorithmic Debugging, 315–316 (1995)

ValiPVM - A Graphical Tool for Structural Testing of PVM Programs

Paulo Lopes de Souza[1], Eduardo T. Sawabe[1], Adenilso da Silva Simao[1], Silvia R. Vergilio[2], and Simone do Rocio Senger de Souza[1]

[1] Universidade de Sao Paulo, ICMC, Sao Carlos - SP, 668, Brazil
{pssouza,sawabe,adenilso,srocio,sarmanho}@icmc.usp.br
[2] Universidade Federal do Parana, UFPR, Curitiba - PR, Brazil
silvia@inf.ufpr.br

Abstract. This work presents ValiPVM, a testing tool for C/PVM parallel programs. This tool implements structural coverage criteria, using an architecture already employed for MPI programs. It supports generation and evaluation of test sets and considers the control, data and communication flows of PVM programs. ValiPVM has a graphical user interface, designed to facilitate the test execution, analysis of results and to guide the user during the execution of the testing activity.

Keywords: testing tool, PVM, coverage testing.

1 Introduction

This paper presents a new graphical software tool for testing PVM parallel programs called ValiPVM. It is a test tool able to support test criteria specific to message passing environments in the context of PVM programs. A test criterion [1] is a predicate to be satisfied by a test case set and can be used as guideline for generating test data, offering a coverage measure that indicates whether enough test cases have been executed. The tool proposed is designed to facilitate the test execution, analysis of results and to guide the user during the execution of the testing activity. It supports a family of testing criteria for control, data and communication flows [2].

ValiPVM automates the required elements generation and it provides facilities for coverage analysis. These features are the major contributions of this tool, allowing the application of the structural testing in PVM parallel programs. ValiPVM can also be used to support the test data selection and to offer coverage measures that can be used to evaluate quality of the test sets.

This work is related with a major-project called ValiPar. ValiPar introduces new criteria, models and graphs for testing parallel programs in message passing environments and shared memory [3, 4]. New parallel software testing tools have been proposed in order to support these new criteria, including ValiMPI, a tool for testing of MPI programs [5]. The goal is to increase the coverage obtained on the source code and, consequently, also to increase the probability of finding

A. Lastovetsky et al. (Eds.): EuroPVM/MPI 2008, LNCS 5205, pp. 257–264, 2008.

faults. Experimental studies accomplished with ValiMPI demonstrated that the coverage criteria are applicable and that typical source errors can be revealed [6].

ValiPVM tool, described in this paper, can be used through a graphical interface or from command line. The graphical user interface has facilities to help the test execution and the analysis of results. One important resource available is the visualization of the Parallel Control Flow Graph with information about required element covered and not covered by test execution. This information also can be visualized in source code, faciliting on-the-fly the coverage analysis and evolution of test activity.

ValiPVM also presents facilities to guide the user during the execution of the testing activity. At each step the tester can know which information must be provided, and so, he can better understand the execution sequence of this tool.

This paper is organized as follows. Section 2 presents previous work that introduces the criteria implemented by ValiPVM. Section 3 describes ValiPVM tool. Section 4 shows a use example of the tool. Section 5 contains the conclusion.

2 Previous Work

The literature describes several projects to extend testing criteria to test parallel programs [7, 8, 9]. However, only [2] addresses message passing environments, in spite of the increasing use and popularization of this kind of parallel software. In addition to this, most works do not address supporting tools, although the application of a testing criterion is only practical if a tool is available. Regarding PVM and MPI, we identify some tools that aid the simulation and debugging, but do not offer support for testing criterion and evaluation of test sets.

In [2, 6], we proposed a set of criteria specific to message passing environments and an architecture, named ValiPar [3], to implement them. These criteria were based on testing criteria for sequential programs. They are extensions that consider communication, synchronization and nondeterminism. A test model based on control, data and communication flows of the parallel program was defined to extract the relevant information for test activity. This model considers that the number of parallel processes is statically known.

A parallel program P is represented by a PCFG (Parallel Control Flow Graph), which is composed by the CFGs (Control Flow Graphs) of each process. The CFG is composed by a set of nodes and edges. Each node corresponds to a statement of the program and an edge links a node to another one. A node i in the process p is represented by n_i^p. A node can be associated to a communication primitive: a *send* or *receive*. A synchronization edge (n_i^a, m_j^b) links a *send* node in a process a to a *receive* node in a process b. These edges represent the possibility of communication and synchronization between processes. Based on PCFG, a set of coverage testing criteria were defined [2]: (AN) *All-Nodes*; (AE) *All-Edges*; (AR) *All-Nodes-R* and (AS) *All-Nodes-S* related on nodes with *receives* and *sends* respectively; and (AES) *All-Edges-S* related on possible synchronizations edges. Other proposed data-flow based criteria are: (ACU) *All-C-Uses*; (APU) *All-P-Uses*; (ASU) *All-S-Uses*; (ASCU) *All-S-C-Uses* and (ASPU) *All-S-P-Uses*.

They consider definitions and uses of variables. A variable x is defined when a value is stored in the correspondent memory position (for instance, assignments and input commands) and, when it is passed as an output parameter (reference) to a function. In the message passing context, we must also consider the communication functions, such as *receive*, since it sets a variable with the value received in the message. A use of x occurs when the value associated to x is referred. A use can be: 1) a computational use (c-use), which occurs in a computational statement related to a node in the CFG; 2) a predicative use (p-use), which occurs in a condition (predicate) associated to a control flow statement, related to an edge in the CFG; and 3) a communicational use (s-use), which occurs in a synchronization statement, related to a synchronization edge in PCFG.

3 ValiPVM Tool

The ValiPVM tool (Figure 1) supports the testing coverage criteria mentioned in last section. It implements the architecture of ValiPar [3] for PVM parallel programs, written in C language.

The **ValiInst** module extracts control and data flow information and instruments the program. These tasks are supported by IDeL (Instrumentation Description Language) [10], a meta-language for program instrumentation. The PCFG is generated in text files, one for each function, with information about definitions and uses of variables in the nodes, as well as about occurrences of send and receive commands. The instrumented program is obtained by inserting check-point statements in the program being tested. These statements do not change the program semantics; they only write necessary information in a trace

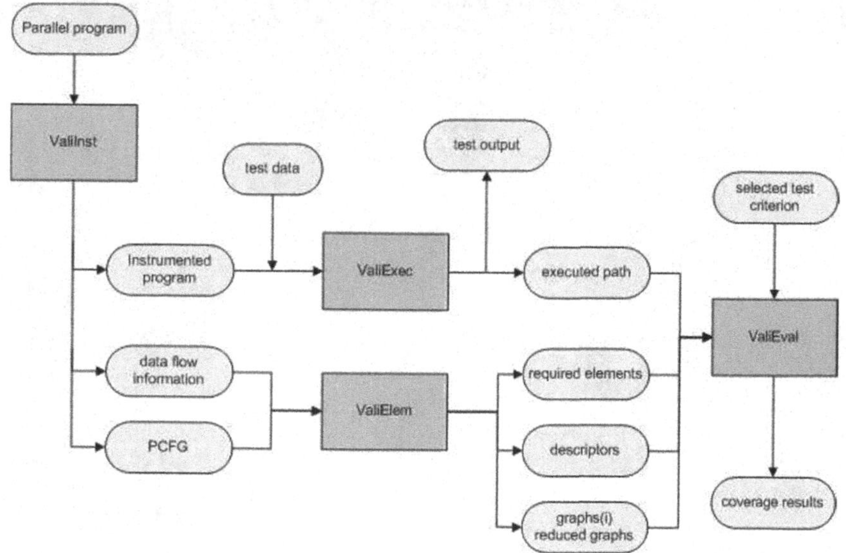

Fig. 1. ValiPVM architecture

file, by registering the node and the process identifier of the send and receive commands. The instrumented program will produce the paths executed in each process, as well as the synchronization sequence produced within a test case.

The **ValiElem** module generates the elements required by the coverage testing criteria. These elements are generated from the PCFG using the information produced by ValiInst. Besides the PCFG, two other graphs are also used: (1) a heirs reduced graph: to minimize the number of required edges; and (2) graph(i): to establish associations of definitions and uses of variables, which are the required elements of the data-flow based testing criteria.

The **ValiExec** module executes the instrumented program by using the test data provided by the tester. During the execution, the inputs and outputs of the program, command lines, execution traces and synchronization sequences are stored in separated files. The execution trace includes the path executed in each process by the test input and it is used during the evaluation of test cases to determine which required elements were covered. After the program execution, the tester can visualize the outputs and also the execution trace to determine whether the obtained output is the same as expected. If it is not, an error was identified and must be corrected before continuing the test activity.

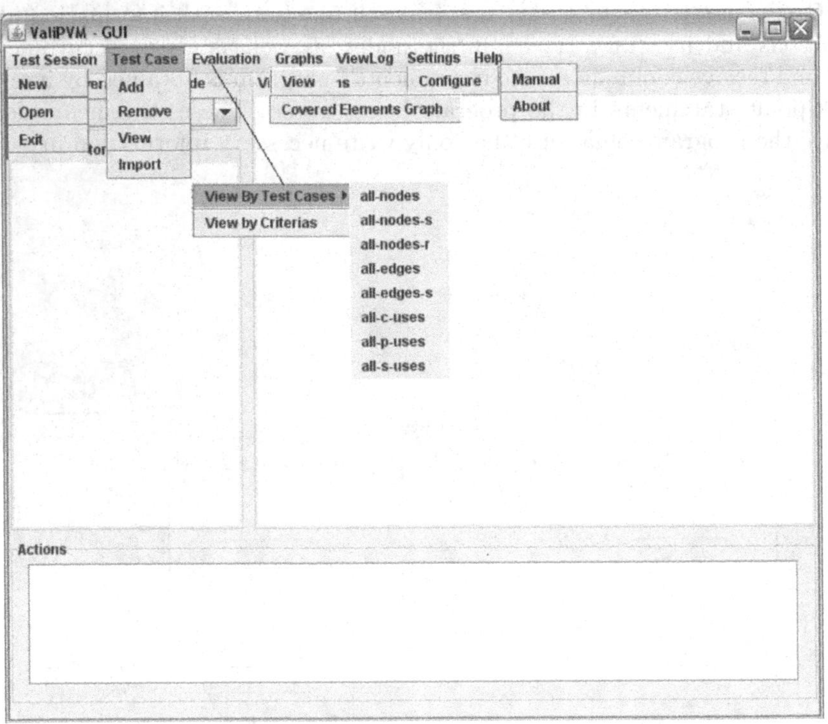

Fig. 2. ValiPVM - main options

The **ValiEval** module evaluates the coverage achieved by test sets with respect to a criterion selected by the tester. ValiEval uses the elements generated by ValiElem and the paths executed by the test cases. The coverage score and the list of covered elements for the selected test criterion are provided as output.

4 Using ValiPVM Tool

In this section we illustrate the functionalities of ValiPVM. This tool is guided by test sessions. In each one, the tester can create a test for one program, interrupt it and resume it later. The main functionalities are presented in Figure 2. As the session test is created, the tool options (e.g., *Test Session*, *Test Case*, *Evaluation*, *Graphs* and *View Log*) are enabled to the tester. The gcd (greatest common divisor) application is used in this section to illustrate some functionalities available in ValiPVM. This program calculates the gcd from three numbers, using four parallel processes: one master and three slaves. Our objective with this simple program is to explain how the tool works.

ValiPVM tool can be applied following two basic procedures: (1) to guide the selection of test cases to the program, and (2) to evaluate the test set quality, in terms of code and communication coverage.

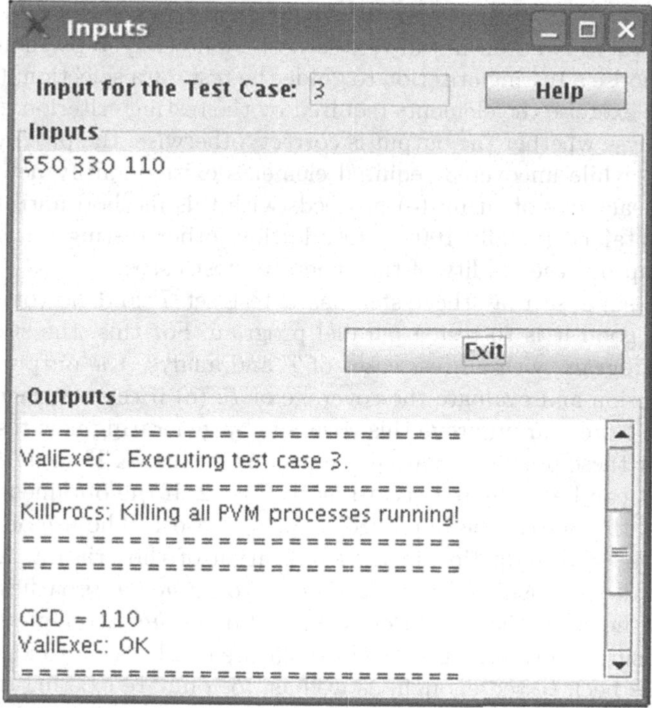

Fig. 3. ValiPVM - adding a test case

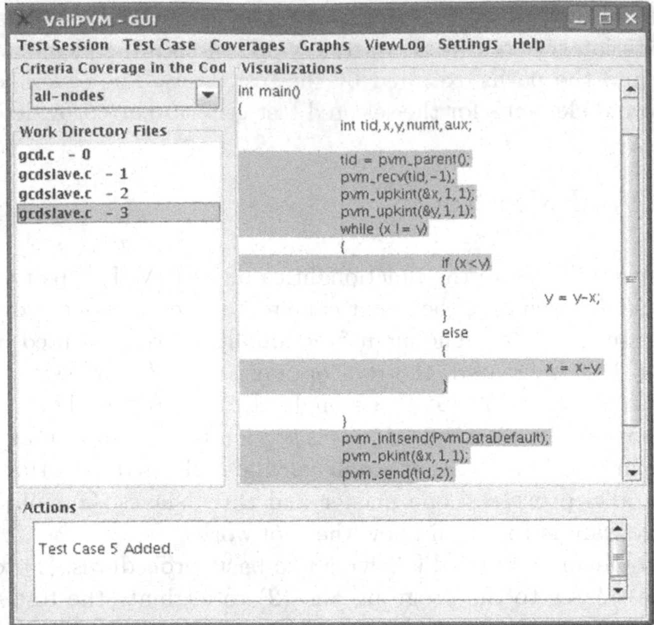

Fig. 4. ValiPVM - code visualization

In Case 1, there is no an initial test set. The following steps should be conducted: (a) choose a testing criterion to guide the test data selection; (b) identify test data that exercise the elements required by the testing criterion; (c) for each test case, analyse whether the output is correct; otherwise, the program must be corrected; (d) while uncovered required elements exist, identify new test cases that exercise each one of them; (e) proceeds with this method until the desired coverage is obtained (ideally 100%). In addition, other testing criteria may be selected to improve the quality of the generated test cases.

In Case 2, suppose that the tester has a test set T and he (or she) wants to know how good it is to test a parallel program. For this, the steps are: (a) execute the program with all test cases of T and analyse the output; (b) select a testing criterion and evaluate the coverage of T; (c) if the coverage is not the expected, the tester can improve this coverage by generating new test data.

Notice that these procedures are not mutually exclusive. If an *ad hoc* test set is available, it can be evaluated according to Case 2. If the obtained coverage is not adequate, this set can be improved by using Case 1. The use of such initial test set allows effort reduction in the application of the criteria. In this way, our criteria can be considered complementary to *ad hoc* approaches. They can improve the efficacy of the test cases generated by *ad hoc* strategies and offer a coverage measure to evaluate them. This measure can be used to know whether a program has been tested enough, as well as, to compare existing test sets.

During the insertion of a test case, the user is testing the parallel application. For each test case inserted, the parallel application is executed, the output is presented (Figure 3) and the user can visualize which program statements were executed (Figure 4 and 5). The highlighted lines in Figure 4 represent statements already exercised by test cases. The Figure 5 presents the PCFG for two slaves (processes 1 and 2) and master (process 0). In the same way, the filled nodes represent statements already exercised. In this graph, dashed edges represent synchronization edges (matches between a send and a receive primitive). So, it is possible to observe which synchronizations occur after each execution. These special visualizations contributes to the testing activity, by reducing testing effort for using the criteria and, by easing the debugging activity. The tester can easily to identify which statements were not executed yet and thus to improve the coverage of the parallel program under testing.

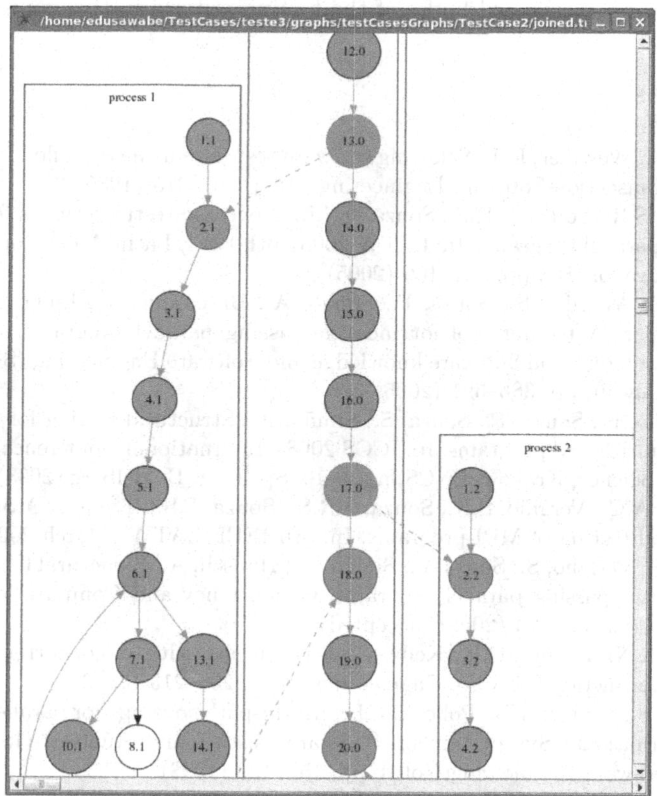

Fig. 5. ValiPVM - PCFG

5 Conclusion

This paper presented ValiPVM, a new graphical software tool for testing C/PVM parallel programs. It supports a family of testing criteria for control, data and communication flows [2]. ValiPVM automates the required elements generation and it provides facilities for coverage analysis. These features are the major contributions of this tool, allowing the application of the structural testing in PVM programs. ValiPVM can also be used to support test data selection and to offer coverage measures that can be used to evaluate quality of the test sets.

ValiPVM can be used through a graphical interface or from command line. The first one has facilities to help the test execution and the analysis of results. One important resource is the visualization of the PCFG with information about required element covered by test execution. This information also can be visualized in source code, faciliting on-the-fly the coverage analysis.

Future work includes the development of the empirical studies to analyse the applicability of the coverage criteria in the context of real parallel programs in PVM. In these studies we also intend to evaluate the application cost and the efficacy in terms of revealed faults of the testing criteria in this context.

References

[1] Rapps, S., Weyuker, E.J.: Selecting software test data using data flow information. IEEE Transaction Software Engineering 11(4), 367–375 (1985)

[2] Vergilio, S.R., Souza, S.R.S., Souza, P.S.L.: Coverage testing criteria for message-passing parallel programs. In: LATW 2005 - 6th IEEE Latin-American Test Workshop, Salvador, Ba, pp. 161–166 (2005)

[3] Souza, S., Vergilio, S., Souza, P., Simão, A., Bliscosque, T., Lima, A., Hausen, A.: Valipar: A testing tool for message-passing parallel programs. In: International Conference on Software knowledge and Software Engineering (SEKE 2005), Taipei-Taiwan, pp. 386–391 (2005)

[4] Sarmanho, F., Souza, P., Souza, S., Simão, A.: Structural testing for semaphore-based multithread programs. In: ICCS 2008 - International Conference on Computational Science, Krakow. LNCS, pp. 1–10. Springer, Heidelberg (2008) (accepted)

[5] Hausen, A.C., Vergilio, S.R., Souza, S.R.S., Souza, P.S.L., Simao, A.S.: A tool for structural testing of MPI programs. In: 8th IEEE LATW (March 2007)

[6] Souza, S., Vergilio, S., Souza, P., Simão, A., Hausen, A.: Structural testing criteria for message-passing parallel programs. Concurrency and Computation: Practice and Experience, 1–24 (2008) (accepted)

[7] Taylor, R.N., Levine, D.L., Kelly, C.: Structural testing of concurrent programs. IEEE Transaction Software Engineering 18(3), 206–215 (1992)

[8] Yang, C.S., Souter, A.L., Pollock, L.L.: All-du-path coverage for parallel programs. In: International Symposium on Software Testing and Analysis (ISSTA 1998), ACM-Software Engineering Notes, pp. 153–162 (1998)

[9] Yang, R.D., Chung, C.G.: Path analysis testing of concurrent programs. Information and Software Technology 34(1) (January 1992)

[10] Simão, A.S., Vincenzi, A.M.R., Maldonado, J.C., Santana, A.C.L.: A language for the description of program instrumentation and the automatic generation of instrumenters. CLEI Electronic Journal 6(1) (2003)

A Formal Approach to Detect Functionally Irrelevant Barriers in MPI Programs*

Subodh Sharma[1], Sarvani Vakkalanka[1], Ganesh Gopalakrishnan[1], Robert M. Kirby[1], Rajeev Thakur[2], and William Gropp[3]

[1] School of Computing, Univ. of Utah, Salt Lake City, UT 84112, USA
[2] Math. and Comp. Sci. Div., Argonne Nat. Lab., Argonne, IL 60439, USA
[3] Dept. of Computer Sci., Univ. of Illinois, Urbana, Illinois, 61801, USA

Abstract. We examine the unsolved problem of automatically and efficiently detecting functionally irrelevant barriers in MPI programs. A functionally irrelevant barrier is a set of `MPI_Barrier` calls, one per MPI process, such that their removal does not alter the overall MPI communication structure of the program. Static analysis methods are incapable of solving this problem, as MPI programs can compute many quantities at runtime, including send targets, receive sources, tags, and communicators, and also can have data-dependent control flows. We offer an algorithm called Fib to solve this problem based on dynamic (runtime) analysis. Fib applies to MPI programs that employ 24 widely used two-sided MPI operations. We show that it is sufficient to detect barrier calls whose removal causes a *wildcard receive statement* placed before or after a barrier to now begin matching a send statement with which it did not match before. Fib determines whether a barrier becomes relevant in *any* interleaving of the MPI processes of a given MPI program. Since the number of interleavings can grow exponentially with the number of processes, Fib employs a sound method to drastically reduce this number, by computing only the *relevant interleavings.* We show that many MPI programs do not have data dependent control flows, thus making the results of Fib applicable to all the input data the program can accept.

1 Introduction

The barrier construct (`MPI_Barrier`) is an important function in the MPI library. It is a *collective* call, meaning that all processes in the communicator must call the barrier. We define such a collective call defined by a set of barrier calls (one from each process) to be a *collective barrier.* A collective barrier is *functionally irrelevant* ("*irrelevant*" for short) if its removal does not alter the overall MPI communication structure of the program in terms of correctness and matching of operations. To the best of our knowledge, this problem has not

* Supported in part by NSF CNS-00509379, Microsoft HPC Institutes Program, and the Mathematical, Information, and Computational Sciences Division subprogram of the Office of Advanced Scientific Computing Research, Office of Science, U.S. Department of Energy, under Contract DE-AC02-06CH11357.

A. Lastovetsky et al. (Eds.): EuroPVM/MPI 2008, LNCS 5205, pp. 265–273, 2008.

been solved before. In this paper, we present an algorithm called Fib to solve this problem based on dynamic (runtime) analysis for MPI programs employing 24 widely used two-sided MPI operations (detailed on our web page [1]).

The importance of detecting irrelevant barriers comes from a number of perspectives. Many MPI users are known to employ collective barriers for "good measure;" they are unsure whether it is necessary. The authors of [2] narrate the example of an MPI program where a barrier was considered irrelevant, and removed. A year later, they were proven wrong, as a race condition was introduced by its removal. In [3], it is shown that barriers can consume a significant fraction of the total application time. Of course, users wanting to control performance by avoiding network or I/O contention may *insert* collective barriers. In this case, they are employing *functionally irrelevant* barriers for controlling the *non-functional* aspects of their program. The Fib algorithm can help these users by checking that these barriers are indeed functionally irrelevant.

Detecting irrelevant barriers by inspection is not straightforward, as we show through a number of small examples in Section 2. While each example seems to warrant a different justification, a nice feature of the Fib algorithm is that it reduces all these justifications to a single mathematical relation called the *completes-before relation.* This relation has two aspects: intra completes-before (IntraCB), and inter completes-before (InterCB). In a nutshell, the Fib algorithm detects a change in the set of communication possibilities by computing the InterCB relation in the presence of a barrier, and checks whether the barrier plays a role in ordering a send and a wildcard receive.

The examples given in Section 2 do not reflect the following additional difficulties. In realistic MPI programs, a user may forget to use a collective barrier (*i.e.* forget to place a barrier within a process), thus introducing a deadlock. Also, realistic programs may compute many quantities at run time, including send targets, receive sources, tags, and communicators. They also have data-dependent control flows which can determine the actual sends and receives issued. The Fib algorithm works in the presence of all these realities:

• Since Fib is implemented as an extension to the dynamic formal verification methodology employed in our tool ISP([9,13,4]), it is capable of detecting deadlocks, and then aborting its analysis. Here are some example deadlock scenarios that ISP can detect: (i) deadlocks due to a collective barrier being incorrectly placed, (ii) those introduced when the user forgets to issue the (supposed) collective call from within some of the processes, (iii) the user employing the wrong communicator for one of the barrier calls, or (iv) MPI messages not matching.

• Since Fib employs dynamic (runtime) analysis, all computed quantities would be fully resolved, and become known. For the same reason, data-dependent control flows are also not an issue for Fib, *in so far as path coverage goes.* It is clear that in general, the behavior of an MPI program can change in response to the input data being analyzed (addressing this issue is considered future research). However, a preliminary static analyzer that we have implemented confirms that for many examples (e.g., all our examples in [1]), control flow does not depend on data; for such programs, the analysis results of Fib are good for *all* input data.

Fib flags a barrier as functionally irrelevant *if and only if* it is functionally irrelevant across *all possible executions* (process interleavings) of the program *for the given input data.* Clearly, we cannot hope to examine all the interleavings of any realistic MPI program naïvely, because this number grows exponentially with the number of processes. Fortunately, the ISP tool actually generates only a *miniscule fraction* of all possible interleavings, by computing only the *relevant interleavings* of an MPI program using a formal verification method called *partial order reduction* [11,12]. Without such a reduction algorithm, an algorithm similar to Fib would be difficult to build.

Related Work: Fib is a significant extension of our POE algorithm implemented in the ISP verification tool. The mathematical relation IntraCB is employed in POE (formally defined in [5], but summarized in this paper). The relation InterCB builds on IntraCB, and is brand new to the Fib algorithm, and this paper.

In [6], the authors provide a formal approach for arguing about the relevance of barriers in MPI programs that do not employ wild-card receives. They prove that for *wild-card receive free* MPI programs that are deadlock free, all barriers are irrelevant. This justifies our criterion for relevant barrier detection, which is: *In a deadlock-free program, the removal of a barrier causes a* wildcard receive statement *placed before or after a barrier to now begin matching* a *send statement with which it did not match before.* The examples in Section 2 provide added insights into our criterion.

The work in [8] uses vector clocks [7], and provides a method for identifying the racing messages in a single trace of an MPI program execution across "frontiers" or *consistent cuts* [7]. While these ideas are somewhat related, the classical vector clock formulation does not directly apply to MPI because of its out-of-order completion semantics and barrier semantics, pointed out in Section 2.

Roadmap: Section 2 provides the intuition behind our Fib algorithm through several examples. The Fib algorithm itself is detailed in Section 3, where we also include sufficient background on the POE algorithm and our ISP tool. Section 4 provides experimental results, and Section 5 provides concluding remarks.

2 Overview of Fib, and the Completes-Before Relation

In this section, we present a number of examples, introducing the concepts of IntraCB and InterCB in context. These relations can be assumed to be always maintained in a transitively closed manner. Please note that we omit the prefix `MPI_` in most cases, and also suppress irrelevant arguments of MPI calls. Also for immediate-mode operations, we show a corresponding `Wait` only in some cases.

Example 1: As our simplest example, consider the following single process (rank) MPI pseudo-code program:

```
P0: Irecv(from P0, x, &h); Wait(&h); Barrier; Isend(to P0, 22);
```

In this program, the collective barrier is a singleton set containing Barrier from P0. Curiously, P0 is trying to send to itself, which is allowed in MPI. In this case, Fib will report a deadlock whether there is a barrier or not. This is because of an IntraCB edge from Wait to any following instruction. An IntraCB edge implies the MPI guarantee of not issuing any instruction after a Wait until Wait has been issued. In our example, there is an Isend after Wait, and unfortunately Wait cannot finish unless Isend finishes—a circular dependency causing the deadlock.

In MPI, there is also an IntraCB edge from a Barrier to any following instruction. This means that instructions following the barrier cannot be issued until the collective barrier can be crossed. Now, suppose we *alter this example* by moving Wait to be *after* the Isend. In this altered example, Barrier can be crossed after issuing Irecv, and this leads to Isend being issued. Thus, for this altered example, the barrier is **irrelevant**.

Example 2: Here * indicates ANY_SOURCE (a wildcard receive)[1]:

```
P0: Irecv(*, x, &h);  Barrier; Isend(to P0, 22); Wait(&h);
P1: Isend(to P0, 33); Barrier;
```

In this example, it is possible for x to attain the value 22, whether the collective barrier is there or not! This is because even though there *is* an IntraCB edge from Barrier to Isend in P0, there is no IntraCB edge from Irecv to Barrier in P0, and similarly there is no IntraCB edge from Isend to Barrier in P1. Therefore, for this program, Fib will flag the collective barrier as **irrelevant**.

Example 3: Consider the program:

```
P0: Irecv(*, x, &handle); Barrier; Wait(&handle);
P1: Isend(to P0, 33);     Barrier; Isend(to P0, 22);
```

Here, the collective barrier is indeed **irrelevant**, and will be flagged as such by the Fib algorithm, following this line of reasoning: (i) the first Isend of P1 and the Irecv of P0 can be issued; (ii) the Barrier in the respective processes can be crossed, as there is no IntraCB edge to these Barriers; (iii) before Irecv occurs, Isend(to P0, 22); can also be issued; (iv) however, MPI's message-matching rules require process-to-process FIFO message ordering; in other words, there is an IntraCB edge from the first Isend to the second Isend in P1. Therefore, x can attain the value of 33 only.

Example 4: In contrast with Example 3, in this program, we move the second send to process P2:

```
P0: Irecv(*, x, &handle); Barrier; Wait(&handle);
P1: Isend(to P0, 33);     Barrier; ...rest of P1...
P2: ...some code...       Barrier; Isend(to P0, 22);
```

[1] Note all examples upto ex 5 are deadlock free hence assume count of sends and recvs match in the program. For full code please refer [1].

The `Isends` are in different processes. Therefore, there is no IntraCB ordering between them. However, the `Irecv` of `P0` as well as `Isend` of `P1` also do not have an IntraCB to their barriers. Therefore, the collective barrier is **irrelevant**.

Now consider an alternative example (call it **Example 4(a)**) in which the `Wait` in `P0` is moved to be *before* its `Barrier`. Now, the collective barrier becomes **relevant**. This is because there would be an IntraCB edge from `Wait` to `Barrier`. Hence, `Barrier` cannot be crossed until the `Irecv` finishes. Therefore the `Isend` from `P2` cannot issue. Therefore, `Irecv` has to finish based on the `Isend` from `P1`.

InterCB: The reasoning employed in this example highlights the need for the notion of *InterCB* edges. Basically, the `Isend` of `P2` "wishes to match" the `Irecv` of `P0`. The only thing that prevents this is that the collective barrier orders `Irecv` to be before it, and `Isend` to be after it. This is the ordering defined by InterCB (detailed in Section 3). Furthermore, there is no alternative ordering path starting from this `Irecv` to `P2`'s `Isend` that does not involve a barrier. Hence the barrier is **relevant**.

Example 5: In all previous examples, the wildcard receive statement appeared before a barrier. In this example, it appears afterwards:

```
P0:                  Barrier; Send(to P2);
P1: Send(to P2);     Barrier;
P2: Irecv(from P1);  Barrier; Recv(*, ..);
```

Here, the barrier is **irrelevant** because `P2`'s `Irecv(from P1)` is ordered before `Recv(*)`. The reasoning now relies on another fact about MPI. If there is a specific-source nonblocking receive followed by a wildcard receive in an MPI program, the wildcard receive can *trump* the specific receive (i.e. may match before it), *if there is no matching sender to the specific-source receive*! In Example **5**, however, there *is* a matching `Send(to P2)` in `P1`, and so trumping does not happen. Since there is no trumping, the IntraCB ordering is maintained between `Irecv(from P1)` and `Recv(*,..)`.

3 The Fib Algorithm

We now provide an overview of the POE algorithm used in our ISP tool (Section 3.1), and then describe the Fib algorithm (Section 3.2).

3.1 POE Overview

The crucial idea embodied in POE is the notion of exploring only *relevant* interleavings—a technique known in model checking as *partial order reduction* [11]. Without this idea, any exploration method for MPI will go out of hand. For instance, consider **Example 6** below (used to illustrate the POE algorithm, and not used to illustrate Fib) that begins with two sends in P0 and P2, and a wildcard receive in P1. The total number of interleavings of all these MPI calls is

210.[2] However, to trigger error1, we need to consider the interleavings in which the send of P2 matches the wildcard receive. Testing oriented tools may easily miss these interleavings. Thanks to partial order reduction, POE will: (i) pick an arbitrary order for executing P0's first send and P1's first receive, (ii) pick an arbitrary order to execute P2's first send and P1's second receive, and then (iii) consider *both* the Send matches with the wildcard receive (shown by *). In other words, POE will examine only two interleavings.

Example 6

```
P0: MPI_Send(to P1...); MPI_Send(to P1, data = 22);
P1: MPI_Recv(from P0...); MPI_Recv(from P2...);
    MPI_Recv(*, x); IF (x==22) THEN error1 ELSE MPI_Recv(*, x);
P2: MPI_Send(to P1...); MPI_Send(to P1, data = 33);
```

POE has built-in knowledge of the commuting properties of MPI functions. As another example, consider an MPI program in which MPI_Barrier is invoked by N processes. POE would, in general, explore only one of the $N!$ ways in which to have invoked the barrier calls. In our implementation of the 24 MPI functions, the cases where alternate interleavings are to be explored include wildcard receives, WAIT_ANY, and TEST_ANY.

Overview of POE's use of PMPI: POE uses the well-known PMPI mechanism. It introduces an extra process called the *verification scheduler*. POE provides its own version of "MPI_f" for each MPI function f. Within MPI_f, POE arranges for handshakes with the scheduler that realizes the POE algorithm. When the scheduler finally gives permission to fire f, POE invokes PMPI_f. The MPI runtime only sees the PMPI_f calls.

Match Sets: Consider Example **6** above. If one repeatedly runs this example under MPICH2 (for example), it is not guaranteed that error1 will be caught. In other words, the matching of Send from P0 with the wildcard receive in P1 may prove *elusive*, as the MPI runtime may never schedule this match!

POE solves the problem of elusive matches *without requiring changes to the MPI library* and *without adding padding delays*—these are expensive, and/or brittle solutions. Instead, it *dynamically rewrites* wildcard receives into specific receives, one for each actual sender that, it computes to be a *certain* match. In the context of **Example 6**, if we can rewrite Recv(*) to Recv(from P0) and Recv(from P1), in turn, and: (i) pair the first one with Send(to P1, data=22), and (ii) pair the second one with Send(to P1, data=33), in turn, and (iii) issue only these send/receive pairs into the MPI system, we can force these matches to occur. Such groupings of sends and receives are called *match sets* in POE's parlance. In POE, in addition to match sets obtained by grouping dynamically rewritten wildcard receives with their matching sends, (i) point to point sends and their receives also form match sets, and (ii) a collective barrier also form match sets.

[2] (7!)/((2!).(3!).(2!)).

3.2 Fib Algorithm

The Fib algorithm can be expressed through the following pseudocode:

```
List IBL initialized to Empty; // Irrelevant Barrier List
FIB () {
   for each execution interleaving I {
      AddInterCB (I);
      for each (⟨R, S⟩ in each Match Set of I
         where R is a Wildcard Recv and S is a Send)
         CheckRelevant (R, S, I); }
   Print IBL;
}
AddInterCB (Interleaving I) {
   for each (Match Set M as ⟨x1, x2, ..., xn⟩ in I)
      for (i = 1 to n)
         for (j = 1 to n, j ≠ i)
            Add InterCB edge from xj to IntraCBSuccessor (xi)
}
CheckRelevant (R, S, I) {
   P = SetofPaths (R, S, I); // Set of all paths from R to S in I
   if(∄ some barrier B in every pi in P)
      return;
   if(barrrier B in every pi in P && B in IBL){
      Remove B and its match set barriers from IBL;
      return; }
   Add B to IBL;
   Add all barriers that form Match Set with B to IBL;
}
```

Fig. 1. Pseudocode for Fib

Illustration: In Example 4, there is no InterCB ordering from `Irecv` to the `Isend` of `P2`. Now in the alternate example called **Example 4(a)** discussed earlier, the above procedure will end up creating an IntraCB path from `Irecv` to `Wait` to `Barrier` in `P0`. Also, all of `Barrier` form a match set. Furthermore, `Isend` of `P2` is ordered to be after the `Barrier`. There is no alternate ordering path – so the collective barrier is relevant. Figure 2 summarizes the above explanation. The IntraCB edges depicted in Figure 2 for process P0 are easy to reason. In process P1, `Isend(to P0)` has no IntraCB edge to the following `Barrier` since Isend being a non blocking call has no obligation to finish before the barrier. However, since Fib knows that `Irecv(*)` in P0 matches this Isend, we add InterCB edges from Isend to operations that are bound to complete *after* Irecv. This explains the InterCB edge between `Isend( to P0)` to `Wait(&handle)`. The same reasoning explains the InterCB edge from `Irecv(*)` to `Finalize` of P1. After adding InterCB edges, the only path that reaches to

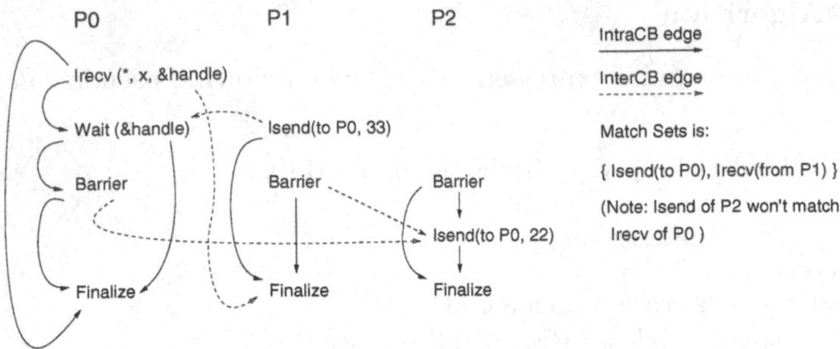

Fig. 2. Example 4(a) in Section 2 with InterCB and IntraCB edges

`Isend(to P0)` of P2 from `Irecv(*)` of P0 involves a barrier. Thus the barrier and all the barrier operations from other processes that formed the match set are flagged to be relevant.

4 Implementation and Experimental Results

We automatically instrument the MPI user code where all `MPI_Barrier(comm)` calls are replaced by `MPI_Barrier_new(comm,__LINE__, __FILE__)`. The two new arguments are system macros that keep the information of line number the function call and the file name that contains it. Our instrumentation tool is written using CIL [14] which offers a framework to create a custom source-to-source program-instrumentation pass. We have run our Fib tool on several MPI programs including: (i) the Monte-Carlo computation of *Pi*, (ii) 2D diffusion, and (iii) all 69 tests that came along with UMPIRE tool [10]. As for runtimes, the ISP algorithm introduces a slowdown because of its scheduler-mediated executions (in [13], we provide ideas for improving the execution time). The *added* overhead that Fib introduces over and above ISP is negligible. Our web page [1] provides detailed results; here is a summary:

- **Monte-Carlo:** The code of Monte-Carlo, did not have any barrier calls. To acid-test our implementation, we deliberately inserted an irrelevant collective barrier, which our implementation flagged as such. The run times of the Fib algorithm are as follows: (i) with 4 processes, it explored 6 interleavings in 0.2 seconds, and with 5 processes, it explored 24 interleavings in 1.52 seconds.
- **2D Diffusion:** This code had 2 irrelevant barriers which were caught by the tool. In fact, this example does not employ wildcard receives, and so all its barriers are irrelevant, and Fib finishes with one interleaving. The runtime of Fib on this example was less than a second. This reinforces that without wildcards we need only one interleaving.
- **Umpire test suite:** We ran our tool successfully on all the 69 tests that came along with Umpire tool [10]. Of the 36 tests that had barriers, all were flagged as **irrelevant**, with negligible runtimes.

5 Concluding Remarks

Removing unnecessary barriers is important, because they needlessly add to the program-execution time. This is particularly true for applications running on petascale machines with thousands of processors. We presented an algorithm, Fib, that is built as an extension to our verification tool ISP for MPI programs. Fib works by detecting, for each barrier, whether its removal causes a *wildcard receive statement* placed before or after a barrier to now begin matching a send statement with which it did not match before. We report success in detecting irrelevant barriers in a number of examples. Since all these examples have control that does not depend on data, the analysis is good for all input data. Our future plans include extending this analysis to cover interesting classes of data dependent control, as well as aiming to cover all of MPI 2.0.

References

1. `http://www.cs.utah.edu/formal_verification/europvm-mpi08/FIB`
2. Avrunin, G.S., Siegel, S.F., Siegel, A.R.: Finite-state Verification for High Performance Computing. In: Proc. Second Intl. Wkshp. on Soft. Eng. for High Perf. Computing Syst. Apps., pp. 68–72 (2005)
3. Rabenseifner, R.: Automatic MPI Counter Profiling. In: Proceedings of the 42nd Cray User Group Conference, CUG SUMMIT 2000, Noorwijk, The Netherlands, May 22-26 (2000)
4. Vakkalanka, S., DeLisi, M., Gopalakrishnan, G., Kirby, R.M., Thakur, R., Gropp, W.: Implementing Efficient Dynamic Formal Verification Methods for MPI Programs. In: Proceeding - EuroPVM/MPI 2008 (2008)
5. Vakkalanka, S., Gopalakrishnan, G., Kirby, R.M.: Dynamic verification of mpi programs with reductions in presence of split operations and relaxed orderings. In: Computer Aided Verification (2008) (accepted)
6. Siegel, S.F., Avrunin, G.S.: Modeling Wildcard-free MPI Programs for Verification. In: PPoPP, pp. 95–106 (2005)
7. Mattern, F.: Virtual Time and Global States of Distributed Systems. In: Parallel and Distributed Algorithms: Proc. Intl. Wkshp. Par. and Dist. Algo. (1989)
8. Netzer, R.H.B., Miller, B.P.: Optimal Tracing and Replay for Debugging Message Passing Parallel Programs. Supercomputing, 502–511 (1992)
9. Pervez, S., et al.: Practical model checking method for verifying correctness of MPI programs. In: EuroPVM/MPI, pp. 344–353 (2007)
10. Vetter, J.S., de Supinski, B.R.: Dynamic Software Testing of MPI Applications with Umpire. In: Proc. of SC 2000, pp. 70–79 (2000)
11. Clarke, E.M., Grumberg, O., Peled, D.A.: Model Checking. MIT Press, Cambridge (2000)
12. Flanagan, C., Godefroid, P.: Dynamic partial-order reduction for model checking software. In: POPL, pp. 110–121. ACM, New York (2005)
13. Vakkalanka, S., Sharma, S.V., Gopalakrishnan, G., Kirby, R.M.: ISP: A tool for model checking MPI programs. In: PPoPP 2008, pp. 285–286 (2008)
14. Necula, G.C.: CIL: Intermediate Language and Tools for Analysis and Transformation of C Programs. In: Horspool, R.N. (ed.) CC 2002. LNCS, vol. 2304, pp. 213–228. Springer, Heidelberg (2002)

Analyzing BlobFlow: A Case Study Using Model Checking to Verify Parallel Scientific Software

Stephen F. Siegel[1] and Louis F. Rossi[2]

[1] Verified Software Laboratory, Department of Computer and Information Sciences, University of Delaware, Newark, DE 19716, USA
siegel@cis.udel.edu

[2] Department of Mathematical Sciences, University of Delaware, Newark DE 19716, USA
rossi@math.udel.edu

Abstract. Model checking techniques are powerful tools for the analysis and verification of concurrent systems. This paper reports on a case study applying model checking techniques to a mature, MPI-based scientific program consisting of approximately 10K lines of code. The program, BLOBFLOW, implements a high order vortex method for solving the two-dimensional Navier-Stokes equations. Despite the complexity of the code, we verify properties including freedom from deadlock and the functional equivalence of sequential and parallel versions of the program. This has led to new insights into the technology that will be required to automate the modeling and verification process for complex scientific software.

1 Introduction

Over the past several years, there has been increasing interest in using *model checking* to debug and verify parallel scientific programs. Specific techniques vary, but usually involve three tasks: (1) an abstract model of the program is constructed, (2) correctness properties of the model are expressed in a formal language, and (3) automated algorithmic techniques are used to exhaustively explore all possible states of the model while checking that the properties hold. The model checking tool will either report that the properties always hold, or provide an explicit counterexample in the form of an execution trace of the model, greatly facilitating debugging.

This approach has several advantages over traditional testing and debugging techniques. The first is that it exhaustively explores all possible executions of the model, examining all allowable interleavings of statements from different processes, all choices available at a wildcard receive, and so on. These kinds of choices are difficult to explore at all by testing, let alone exhaustively.

One of the alleged shortcomings of the model checking approach is that it requires relatively small bounds on parameters such as the number of processes, the size of the input, and so on. This is not as serious a limitation as it first appears, because program defects almost always manifest themselves for relatively small values of these parameters. This point is often misunderstood by

A. Lastovetsky et al. (Eds.): EuroPVM/MPI 2008, LNCS 5205, pp. 274–282, 2008.

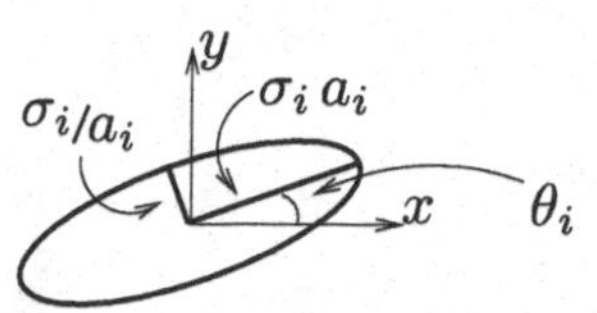

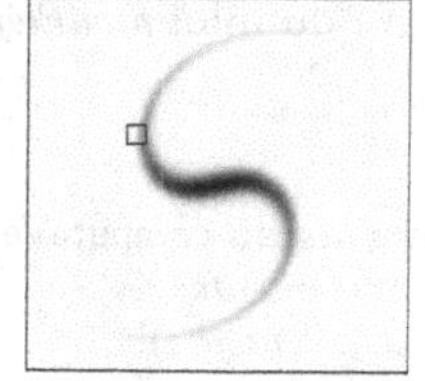
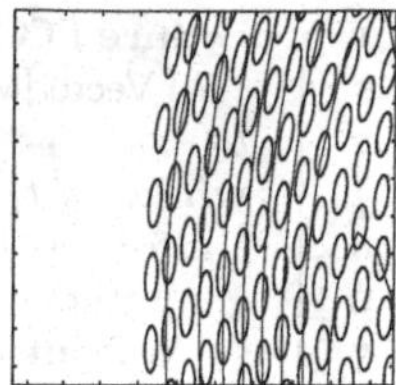

Fig. 1. Left: A schematic diagram of the geometry of an elliptical Gaussian basis function. Middle: a sample ECCSVM computation. Right: blowup of particles in the boxed area.

developers, who have seen many problems that only occur for very large process counts or inputs. A common example is an MPI program with the potential to deadlock: the program may run normally on small configurations if the MPI implementation chooses to buffer certain messages, but when the process count or memory requirements exceed some threshold, the implementation may choose to synchronize message delivery, revealing the deadlock. A model checker, however, will explore the possibility of forced synchronization in all states where it is permitted by the MPI Standard, and the failure will be detected in even the smallest configurations. Another example is an MPI program that may fail when the number of outstanding communication requests generated by the program exceeds some large bound. With model checking, the bound on the number of allowable requests is a parameter which can be given a small value, so the same failure will be detected for a small configuration of the program.

In fact, the ability of model checking to discover defects in small configurations is one of its most significant advantages over traditional debugging techniques. If one can only reproduce a failure using 1000 processes, and the resulting trace involves thousands of execution steps, analyzing the trace with a standard debugger can be difficult or impossible. The ability of model checkers to find small (even minimal) counterexamples is an immense advantage for debugging.

To date, most of the applications of model checking to scientific computing have involved only small example programs (e.g., [1, 2, 3]). The goal of the project reported on here was to determine whether model checking techniques could be successfully applied to a production code used in actual scientific research. The code we chose is L. Rossi's computational fluid dynamics program BlobFlow [4, 5, 6]. BlobFlow has been actively developed and used over the past 7 years to explore fluid flow phenomena and novel simulation algorithms. It consists of approximately 10K lines of C code and includes both a parallel (MPI-based) version and a sequential version. To the best of our knowledge, it is the largest parallel scientific application to have been successfully verified with model checking techniques.

The model checker used in this study is Mpi-Spin [7, 8], an extension to the standard model checker Spin [9]. Mpi-Spin adds to Spin's input language a large number of primitives corresponding to the types, constants, and functions comprising MPI. It also incorporates a precise model of the semantics of the

```
procedure ECCSVM(t0, tf, Δt: double; numElements: int; elements: Element[]) is
    vel : Vector[MaxElt];
    t ← t0;
    while t ≤ tf do
        for j ← 0 to numElements do computeVelocity(j, elements, vel);
        integrate(numElements, elements, vel);
        output(t, numElements, elements);
        t ← t + Δt;
procedure computeVelocity(numElements: int; elements: Element[]; vel: out Vector[]) is
    mpcoef : double[MpSize]; i, j, p : int;
    for i ← 0 to MpSize − 1 do mpcoef[i] ← 0;
    for i ← 0 to numElements − 1 do
        for p ← 0 to PMax − 1 do
            j ← f1(i, p, elements);
            mpcoef[j] ← mpcoef[j] + f2(i, p, elements);
    for i ← 0 to numElements − 1 do vel[i] ← f3(i, elements, mpcoef);
```

Fig. 2. ECCSVM algorithm, sequential version

MPI operations, based on the MPI Standards. Though there are no tools to automatically translate MPI programs into the input language for MPI-SPIN, this language support makes manual translation much more straightforward. We used MPI-SPIN to verify two important properties of BLOBFLOW: (1) freedom from deadlock, and (2) the functional equivalence of the sequential and parallel versions.

2 ECCSVM and BlobFlow

The Elliptical Corrected Core Spreading Vortex Method (ECCSVM) [4, 5] is an algorithm for computing the motions of incompressible gases and liquids in two dimensions. The algorithm falls under the general category of *vortex methods* [10]. Vortex methods represent vorticity as a sum of localized, moving basis functions, referred to as *elements* or *blobs*. Each element is characterized by the position of its center and other parameters (Fig. 1).

The high-level structure of the ECCSVM algorithm is shown in Fig. 2. The algorithm takes as input the initial values of the elements as well as the initial and final times and the length of each time step. At each iteration, the velocities (*vel*) are computed from the current vorticity data. This is the most computationally expensive part of the algorithm, but is necessary because the *integration step*, which updates the positions and parameters of the elements, requires explicit knowledge of the velocities.

The velocity is computed by evaluating a Biot-Savart integral [11, §2.3] over the entire spatial domain of the fluid. This means that the computation of the velocity at one element center requires knowledge of the vorticity at every element. A naïve approach would thus require $O(N^2)$ operations, where N is the

```
1  procedure slave (elements : Element[]; mpcoef : double[]; packsize : int) is
2      packbuf : byte[PBSize]; databuf : double[DataSize × WorkSize];
3      indexbuf : int[WorkSize]; i, pos : int;
4      sendreq, recvreq : MPI_Request; status : MPI_Status;
5      MPI_RECV_INIT(indexbuf, WorkSize, MPI_INT, 0, MPI_ANY_TAG, recvreq);
6      MPI_SEND_INIT(packbuf, packsize, MPI_PACKED, 0, 0, sendreq);
7      while true do
8          MPI_START(recvreq);
9          MPI_WAIT(recvreq, status);
10         if status.tag = Done then break;
11         for i ← 0 to WorkSize − 1 do
12             if indexbuf[i] ≠ −1 then
               databuf[i] ← f3(indexbuf[i], elements, mpcoef);
13         MPI_WAIT(sendreq, MPI_STATUS_IGNORE);
14         pos ← 0;
15         MPI_PACK(indexbuf, WorkSize, MPI_INT, packbuf, packsize, pos);
16         MPI_PACK(databuf, DataSize × WorkSize,
               MPI_DOUBLE, packbuf, packsize, pos);
17         MPI_START(sendreq);
18     MPI_REQUEST_FREE(recvreq);
19     MPI_WAIT(sendreq, MPI_STATUS_IGNORE);
20     MPI_REQUEST_FREE(sendreq);
```

Fig. 3. Slave

number of elements, in order to compute the velocity field at all element centers. The ECCSVM uses a more technique known as the *fast multipole method* (FMM) to approximate the integral in $O(N)$ operations. The FMM requires a decomposition of the domain into near and far elements, and therefore computational resources for evaluating the velocity will vary substantially from element to element, so dynamic load balancing is crucial in the parallel version.

The parallel version of the algorithm changes only computeVelocity. The computation of the FMM coefficients is distributed equally among the processes and the results are summed onto each process with an MPI_ALLREDUCE. The computation of the velocities from the FMM coefficients (the invocation of f_3 on line 16), on the other hand, uses a variation of the master-slave pattern. The master sends to a slave a list of element indices and the slave performs the f_3 computations on the specified elements and packs and sends the results back to the master. The master unpacks the results and updates *vel* appropriately. When all the velocity calculations are complete, the master broadcasts *vel* to all processes. In the actual BlobFlow code, there are many variations on the standard patterns, however. For example, the master initially sends two tasks to every slave, persistent requests are used in sometimes complicated ways, there is somewhat complex packing and unpacking of data. The slave code (which is a small portion of the parallel computeVelocity code) is summarized in Fig. 3.

3 Verification

The first goal was to verify freedom from deadlock, and we constructed an MPI-SPIN model of the parallel version of BLOBFLOW for this purpose. We found that a naïve translation, in which each variable in the original program is mapped to a unique variable in the model, would never scale, due to the sheer number of variables in the program. The key idea to making a tractable model is *conservative abstraction.* Conservative abstraction is a process by which we leave out some information from the model, but we ensure that when there is a decision that requires that information, the model checker will explore all possible outcomes (and perhaps some that are not possible).

To be precise, if M is a model of a program P, then every execution e of P maps to an execution $f(e)$ of M. Many distinct program executions may map to the same model execution; the extent to which this happens is a measure of the *abstraction* of the model. There may also be model executions that are not in the image of f; these are called *spurious* executions. The model is *conservative* for a property π if the following is true: for all e, π holds for $f(e)$ iff π holds for e. If the model checker verifies that π holds on all executions of a conservative model, one can conclude π holds for P. If, on the other hand, the model checker produces a counterexample e', it is possible that π still holds for P because e' is spurious. Hence one must examine e', and if it is determined to be spurious, the model is *refined* by adding sufficient information to eliminate the spurious execution. One continues to refine the model in this way until either an actual counterexample is produced or the property is verified successfully [12].

The natural starting point is a very abstract but conservative model that kept just enough information to capture the rank and request arguments occurring in the MPI function calls. An excerpt of this model is shown in Fig. 4(a) and corresponds to the main slave loop of Fig. 3. The conditional statement of Fig. 3, line 10, which controls when the slave breaks out of the loop, has been replaced by a nondeterministic choice, which means that execution may or may not break out of the loop—both possibilities are explored at each iteration. For this model, MPI-SPIN found a counterexample, and examination quickly revealed that it was spurious: one slave breaks out of the loop in its first iteration and the master ends

(a)
```
do :: MPI_Start(Pslave,&Pslave->recvreq);
      MPI_Wait(Pslave,&Pslave->recvreq, MPI_STATUS_IGNORE);
      if :: 1 -> break :: 1 fi;
      MPI_Wait(Pslave,&Pslave->sendreq, MPI_STATUS_IGNORE);
      MPI_Start(Pslave,&Pslave->sendreq) od
```

(b)
```
do :: MPI_Start(Pslave,&Pslave->recvreq);
      MPI_Wait(Pslave,&Pslave->recvreq, &Pslave->status);
      if :: status.tag == DONE -> break :: else fi;
      MPI_Wait(Pslave,&Pslave->sendreq, MPI_STATUS_IGNORE);
      MPI_Start(Pslave,&Pslave->sendreq) od
```

Fig. 4. (a) Abstract model of main slave loop. (b) Refined to include *status*.

```
do :: MPI_Start(Pslave, &Pslave->recvreq);
      MPI_Wait(Pslave, &Pslave->recvreq, &Pslave->status);
      if :: status.tag == DONE -> break :: else fi;
      do :: i < WorkSize -> c_code {
            if (Pslave->indexbuf[Pslave->i] != (uchar)(-1))
               Pslave->databuf[Pslave->i] =
                  SYM_cons(f_3_id,
                  SYM_cons(SYM_intConstant(Pslave->indexbuf[Pslave->i]),
                  SYM_cons(Pslave->elements,
                  SYM_cons(Pslave->mpcoef, SYM_NULL))));
         }; i++
         :: else -> i = 0; break od;
      MPI_Wait(Pslave, &Pslave->sendreq, MPI_STATUS_IGNORE);
      pos = 0;
      MPI_Pack(Pslave->indexbuf, WorkSize, MPI_BYTE, Pslave->packbuf,
        Pslave->packsize,&Pslave->pos);
      MPI_Pack(Pslave->databuf,DataSize*WorkSize,
        MPI_SYMBOLIC,Pslave->packbuf,Pslave->packsize,&Pslave->pos);
      MPI_Start(Pslave,&Pslave->sendreq) od
```

Fig. 5. Symbolic model of main slave loop

up waiting forever for a result from that slave. We refined the model (manually) by adding the *status* variable (Fig. 4(b)). This provided sufficient precision to verify deadlock-freedom.

With the resulting model, which we call the *communication skeleton*, we were able to verify deadlock-freedom for up to 5 processes. (Mpi-Spin actually checks several other correctness properties automatically, such as (1) there are no allocated request objects for a process when MPI_FINALIZE is called; and (2) MPI_START is never invoked on an active request.) The execution time for the 5-process verification was 41 minutes on a Sun Ray with two dual-core 2.6 GHz AMD Opteron processors. The verification consumed 7.7 GB of RAM and explored 3.5×10^7 states. The numbers for the 4-process run were 45 seconds, 163 MB, 1.5×10^6 states; for 3 processes, 1 second, 56 MB, 62,798 states.

Verifying the functional correctness of BlobFlow is a much harder problem. The goal is to verify that the sequential and parallel versions are *functionally equivalent*, i.e., they produce the same output on any given input. A method for this combining model checking and symbolic execution is described in [3] and we review it briefly here.

To perform this verification, we must model the program data *symbolically*: the input is represented as symbolic constants $X_1, X_2, \ldots$ and the output is represented as symbolic expressions in the X_i. Floating-point operations are replaced by corresponding symbolic operations, which simply build new symbolic expressions from their operands. These constructs are supported in Mpi-Spin through a symbolic type and a set of symbolic operations. A model of this type is made for both the sequential and parallel programs. Mpi-Spin is then used to explore all possible executions of the models and verify that in each case, the symbolic expressions output by the two versions agree. Fig. 5 shows the symbolic model of the main slave loop, which refines the abstract model used in the communication skeleton.

Because floating-point arithmetic is only an approximation to real arithmetic, we must clarify what is meant by *functionally equivalent.* Two programs that are equivalent when the arithmetic operations are interpreted as taking place in the set of real numbers may not be equivalent if those operations are implemented using IEEE754 floating-point arithmetic. Two programs that are equivalent with IEEE754 arithmetic may not be equivalent if some other floating-point arithmetic is used. Different notions of equivalence may be appropriate in different circumstances. For example, a parallel program containing an MPI reduction operation using floating-point addition may obtain different results when run twice on the same input, because the sum may be computed in different orders. Such a program cannot be floating-point equivalent to any sequential program, but it may be real-equivalent to one. MPI-SPIN deals with this situation by offering the user a choice of three successively stronger equivalence relations: real, IEEE, and Herbrand, the last holding only if the two programs produce exactly the same symbolic expressions.

A straightforward application of this method used in previous work would not scale to BLOBFLOW. In the earlier work, the symbolic variables corresponded one-to-one with the floating-point variables in the original program; the sheer number of such variables in BLOBFLOW meant that the memory required to store one state of the model would be prohibitive.

Our solution to this problem involved two related ideas. The first is that groups of variables that tend to be manipulated together can often be represented by a single symbolic variable in the model. The second is that sections of code that are shared by the sequential and parallel versions can be represented by a single abstract symbolic operation. An example is the function f_3 that is invoked in both Fig. 2, line 16, and Fig. 3, line 12. Since we are only trying to prove the equivalence of the two versions, there is no need to know exactly what f_3 computes. Instead, we can just introduce a new symbolic operation for f_3, and use it wherever this function is invoked in the codes. We used both of these techniques extensively to design a reasonably small, conservative symbolic model.

Space does not permit us to describe the model in detail, but the source for the model and all other artifacts used in this study are available at `http://vsl.cis.udel.edu`. Using this model, we were able to verify functional equivalence for up to 4 processes. Since the parallel version uses a floating-point addition reduction operation (in the FMM computation), we used the real-equivalence mode. The 4-process run lasted 21 minutes, consumed 1.2 GB, explored 2.0×10^7 states, and generated 988 distinct symbolic expressions. The numbers for the 3-process run were 19 seconds, 96 MB, 691, 837 states, and 731 expressions.

4 Conclusion

In this paper, we have reported on our investigation using the model checker MPI-SPIN to verify correctness properties of a nontrivial parallel scientific program. We were able to verify generic concurrency properties, such as freedom from deadlock, for models of the program with up to 5 processes. We also verified the functional equivalence of the parallel and sequential versions of the program, for up to 4 processes and within specific bounds on certain parameters.

In carrying out this study, we have developed—at least in outline—a methodical way to construct the models. The method begins with a very abstract but conservative model that encodes only the data necessary to represent the rank and request arguments occurring in the MPI function calls of the program. The model is then progressively refined until sufficient precision is achieved to either verify or produce a valid counterexample to freedom from deadlock. This "communication skeleton" model is then augmented by representing the program data symbolically, and the resulting model is used to verify functional equivalence.

Several abstraction techniques proved useful in the construction of the symbolic model. The most important requires one to locate units of code common to the sequential and parallel versions of the program and abstract these segments using uninterpreted symbolic operations. This process is made easier if the sequential and parallel versions share code. It also requires partitioning the data of the programs and assigning a symbolic variable to each partition.

Further progress will require the development of a formal basis for these abstraction techniques. This would lay the groundwork for automating (at least in part) the model construction process. Indeed, there are clearly some established static analysis techniques that could be brought to bear on our method. These include standard techniques to estimate the set of variables read or written to by a program unit, and dependence analysis, which could be used to determine that certain variables must be incorporated into the models. These techniques, however, must be made aware of certain aspects of the semantics of the MPI functions used in the programs. In other contexts, the counterexample-driven refinement loop has been automated using theorem-proving techniques, and it is possible that similar techniques could be adopted for MPI-based programs.

Finally, more work is needed to address the state-explosion problem for master-slave style programs. In this study, we observed a very steep blowup in the number of states with the process count. The reason for this appears to be the combinatorial explosion inherent in the master-slave architecture, due to the large number of ways tasks can be partitioned among slaves and the differing orders in which results can be received by the master. Various reduction techniques have been devised for programs that avoid the nondeterministic constructs used in master-slave programs, such as MPI_ANY_SOURCE and MPI_WAITANY (e.g., [13]). These techniques allow model checking to scale effectively for many types of programs, such as standard discrete grid simulations. Other reduction strategies have been proposed to deal with wildcard receives (e.g., [14, 15]). These have not been incorporated into MPI-SPIN, but it is not clear they would make a substantial difference for master-slave programs in any case. It appears that some fundamental algorithmic advance must be made in this area if model checking is to become practical for this important class of parallel scientific programs.

These are some of the many avenues of future research.

Acknowledgments. We are grateful to the U.S. National Science Foundation for funding under grant CCF-0733035 and to Samuel Moelius for assistance in constructing the MPI-SPIN models.

References

1. Siegel, S.F., Avrunin, G.S.: Verification of MPI-based software for scientific computation. In: Graf, S., Mounier, L. (eds.) SPIN 2004. LNCS, vol. 2989, pp. 286–303. Springer, Heidelberg (2004)
2. Palmer, R., Gopalakrishnan, G., Kirby, R.M.: Semantics driven partial-order reduction of MPI-based parallel programs. In: Parallel and Distributed Systems: Testing and Debugging (PADTAD V), London (July 2007)
3. Siegel, S.F., Mironova, A., Avrunin, G.S., Clarke, L.A.: Combining symbolic execution with model checking to verify parallel numerical programs. Transactions on Software Engineering and Methodology 17(2), 1–34 (2008)
4. Rossi, L.F.: Resurrecting core spreading methods: A new scheme that is both deterministic and convergent. SIAM J. Sci. Comp. 17(2), 370–397 (1996)
5. Rossi, L.F.: Achieving high-order convergence rates with deforming basis functions. SIAM J. Sci. Comput. 26(3), 885–906 (2005)
6. Rossi, L.F.: Evaluation of the Biot-Savart integral for deformable elliptical gaussian vortex elements. SIAM J. Sci. Comput. 28(4), 1509–1532 (2006)
7. Siegel, S.F.: The MPI-Spin web page (2007), `http://vsl.cis.udel.edu/mpi-spin`
8. Siegel, S.F.: Model checking nonblocking MPI programs. In: Cook, B., Podelski, A. (eds.) VMCAI 2007. LNCS, vol. 4349, pp. 44–58. Springer, Heidelberg (2007)
9. Holzmann, G.J.: The SPIN Model Checker. Addison-Wesley, Boston (2004)
10. Cottet, G.H., Koumoutsakos, P.D.: Vortex methods: Theory and Practice. Cambridge University Press, Cambridge (2000)
11. Saffman, P.G.: Vortex Dynamics. Cambridge University Press, Cambridge (1992)
12. Ball, T., Rajamani, S.K.: Automatically validating temporal safety properties of interfaces. In: Model Checking of Software: 8th Intl. SPIN Workshop, pp. 103–122. Springer, Heidelberg (2001)
13. Siegel, S.F., Avrunin, G.S.: Modeling wildcard-free MPI programs for verification. In: Proceedings of the 2005 ACM SIGPLAN Symposium on Principles and Practice of Parallel Programming (PPoPP 2005), pp. 95–106. ACM Press, New York (2005)
14. Siegel, S.F.: Efficient verification of halting properties for MPI programs with wildcard receives. In: Cousot, R. (ed.) VMCAI 2005. LNCS, vol. 3385, pp. 413–429. Springer, Heidelberg (2005)
15. Vakkalanka, S., Gopalakrishnan, G., Kirby, R.M.: Dynamic verification of MPI programs with reductions in presence of split operations and relaxed orderings. In: Computer Aided Verification: 20th International Conference, CAV 2008, Princeton, USA, July 7–14, (to appear, 2008)

7th International Special Session on Current Trends in Numerical Simulation for Parallel Engineering Environments: New Directions and Work-in-Progress* (ParSim 2008)

Carsten Trinitis[1] and Martin Schulz[2]

[1] Lehrstuhl für Rechnertechnik und Rechnerorganisation (LRR)
Institut für Informatik
Technische Universität München, Germany
Carsten.Trinitis@in.tum.de
[2] Center for Applied Scientific Computing
Lawrence Livermore National Laboratory
Livermore, CA, USA
schulzm@llnl.gov

In today's world, the use of parallel programming and architectures is essential for simulating practical problems in engineering and related disciplines. Remarkable progress in CPU architecture (multi- and manycore, SMT, transactional memory, virtualization support, etc.), system scalability, and interconnect technology continues to provide new opportunities, as well as new challenges for both system architects and software developers. These trends are paralleled by progress in parallel algorithms, simulation techniques, and software integration from multiple disciplines.

In its 7th year ParSim continues to build a bridge between computer science and the application disciplines and to help with fostering cooperations between the different fields. In contrast to traditional conferences, emphasis is put on the presentation of up-to-date results with a shorter turn-around time. This offers the unique opportunity to present new aspects in this dynamic field and discuss them with a wide, interdisciplinary audience. The EuroPVM/MPI conference series, as one of the prime events in parallel computation, serves as an ideal surrounding for ParSim. This combination enables the participants to present and discuss their work within the scope of both the session and the host conference.

After a quick turn-around, yet thorough review process we again picked three papers for publication and presentation during the ParSim session. These papers cover a diverse set of topics in parallel simulation and their support infrastructure: the first paper describes a generic and high-level class hierarchy for the parallelization of structured grid codes; the second paper demonstrates the scaling behavior of an earthquake simulation code on up to 4096 processors; and the

* Part of this work was performed under the auspices of the U.S. Department of Energy by Lawrence Livermore National Laboratory under Contract DE-AC52-07NA27344 (LLNL-CONF-404491).

A. Lastovetsky et al. (Eds.): EuroPVM/MPI 2008, LNCS 5205, pp. 283–284, 2008.

final paper addresses the efficient use of GPGPUs by providing fast data transfer mechanisms to and from the accelerator hardware. We are confident that these selections resulted in an attractive program and that ParSim will yet again be an informal setting for lively discussions and for fostering new collaborations.

Several people contributed to this event. Thanks go to Jack Dongarra, the EuroPVM/MPI general chair, and to Alexey Lastovetsky and Tahar Kechadi, the PC chairs, for their support to continue the ParSim series at EuroPVM/MPI 2008. We would also like to thank the numerous reviewers, who provided us with their reviews in such a short amount of time (in most cases in just a few days) and thereby helped us to maintain the tight schedule. Last, but certainly not least, we would like to thank all those who took the time to submit papers and hence made this event possible in the first place.

We are confident that this session will fulfill its purpose to provide new insights from both the engineering and the computer science side and encourages interdisciplinary exchange of ideas and collaborations. We hope that this will continue ParSim's tradition at EuroPVM/MPI.

LibGeoDecomp: A Grid-Enabled Library for Geometric Decomposition Codes

Andreas Schäfer and Dietmar Fey

Lehrstuhl für Rechnerarchitektur und -kommunikation, Institut für Informatik, Friedrich-Schiller-Universität, 07737 Jena, Germany
{gentryx,fey}@cs.uni-jena.de

Abstract. In this paper we present first results obtained with LibGeoDecomp, a work in progress library for scientific and engineering simulations on structured grids, geared at multi-cluster and grid systems. Today's parallel computers range from multi-core PCs to highly scaled, heterogeneous grids. With the growing complexity of grid resources on the one hand, and the increasing importance of computer based simulations on the other, the agile development of highly efficient and adaptable parallel applications is imperative. LibGeoDecomp is to our knowledge the first library to support all state of the art features from dynamic load balancing and exchangeable domain decomposition techniques to ghost zones with arbitrary width and parallel IO, along with a hierarchical parallelization whose layers can be adapted to reflect the underlying hierarchy of the grid system.

Keywords: Grid computing, self-adaptation, hierarchical parallelization.

1 Introduction

Fueled by the rise of computer based simulations and stagnating clock rates, the trend towards highly parallel supercomputers puts scientists in the awkward position of having to be experts in both, their actual subjects and parallel computing. This problem is aggravated by the hierarchical structure of grid computing resources, being comprised of heterogeneous networks and multi-socket or multi-core machines. Achievement relies on the efficient usage of such systems, but capable applications usually represent the outcome of expensive, multidisciplinary projects. Thus scientists and engineers require tools to reduce application development time and complexity.

During the MuCluDent [1] (Multi-Cluster DendriTe) project, a simulation software for cooling molten metal alloys based on a combination of cellular automaton and finite difference method, we were facing a number of challenges. First, the program had to run on a variety of machines, ranging from notebooks for model tests to multi-clusters for larger simulations. Manually adapting the code to run most efficiently on each setup soon proved to be tedious and error prone. Second, the model exhibited strong computational hot spots, making some regions four times more expensive to update than others. And third, previous parallelizations did not scale well on the multi-cluster setup we were using.

A. Lastovetsky et al. (Eds.): EuroPVM/MPI 2008, LNCS 5205, pp. 285–294, 2008.

Similarly, a number of simulations can be modeled as time-discrete evolutions on structured grids and can be parallelized by geometric decomposition. This means that the simulation domain can be broken down into atomic cells, arranged in a grid. While the geometry of the cells may vary, the grid's topology has to be equivalent to a cube of appropriate dimension. The cells encapsulate the data on which the simulation operates, as well as how they are updated to the next time step. For instance, a cell for Conway's game of life will contain only one bit as data (dead or alive) and its update method will set this bit depending on the cell's old state and the number of living neighbors.

Typically, implementations will store two grids at any given time: one containing the cells from the past time step, and another one to store the newly updated cells. This is to prevent cells referencing already updated neighbors during the update process. A typical approach for parallelization is to decompose the simulation grid into smaller regions and distribute them among the participating processors. Since the cells require the states of their neighbors during update, the nodes need to synchronize the rims of their regions. These rims are also referred to as ghost zones. Section 5 deals with the crucial question of how to subdivide the grid optimally. This process is known as domain decomposition, and the optimum depends on numerous factors, such as load balance and fluctuation or communication costs.

Many frameworks aim to support developers with varying degrees of genericness, ranging from domain specific libraries like COOLFluiD to fully fledged problem solving environments like Cactus, but most of them target homogeneous clusters and support only limited domain decomposition types. As an alternative, LibGeoDecomp focuses on the parallelization. While it does not provide domain specific libraries (e.g. physical modules or numerical methods), the main design goals were adaptability and scalability. Its hierarchical parallelization can be tuned to reflect the characteristics of the underlying systems. On each level the domain decomposition, the load balancing scheme and the ghost zone width can be varied. This way wide ghost zones can be used on WAN connections to hide latency while intra-cluster links may use smaller widths, thus increasing efficiency. Small subsystems may use remapped load balancing while larger systems may use diffusive algorithms. An additional layer can be added where necessary to reflect multi-core CPUs. Using threads on that level can speed up the simulation, for instance by reducing the number of MPI processes taking part in collective operations.

The rest of the paper is organized as follows: Section 2 gives a brief overview of the current state of the art. Section 3 outlines our design and Sections 4 and 5 yield in depth explanations of our parallelization and the domain decomposition. Section 6 closes with initial benchmark results.

2 Related Work

Cactus [2] is a widely used framework for three-dimensional physical simulations, written mainly in C. Cactus consists of two types of components: the Thorns provide functionality like IO or certain numerical methods. They communicate via the Flesh. Cactus' emphasis lies on a large library of ready to use modules.A special type of Thorns, the Drivers, are used for parallelization. PUGH is Cactus' default driver for regular grids, but it can only subdivide the dimensions of the simulation grid in parts who are divisors of the number of processes. If the number of processes is for instance not a power of two, this may yield a sub-optimal surface to volume ratio. Carpet, the Driver for mesh-refinement simulations, handles this case more gracefully, but cannot yet perform dynamic load balancing. COOLFluiD [3] is a toolkit for computational fluid dynamics. It can be extended for multiple numerical methods and complex geometries (unstructured grids, hybrid meshes etc.), but is not targeted for grid systems. It cannot yet perform dynamic load balancing.

3 LibGeoDecomp Overview

This sections contains a brief overview of LibGeoDecomp's structure and user interface. On the basis of our experiences with MuCluDent we started to develop LibGeoDecomp as a generic library for time discrete simulation codes on structured grids that can be parallelized by geometric decomposition, written in C++. Figure 1(a) illustrates the basic structure of LibGeoDecomp: Objects of the `Simulator` class conduct the parallel evolution, based on the user supplied model `Cell`. At the begin of the simulation the grid is set up by the `Initializer`, while `Writer` objects perform output for various formats (e.g. the `PPMWriter` for basic graphical output). The MPIIO Writer and Initializer can be used for application level checkpoint/restart functionality. LibGeoDecomp is based on template classes, so that user supplied classes for initialization and cells can be integrated with minimal overhead.

To solve the adaptation problem, we built LibGeoDecomp with the Pollarder [4] framework. Pollarder is a library that can perform an environment discovery at application start-up time and can then select the most suitable user provided components automatically, thereby freeing the user from doing this on each system manually.

The following listing is a simple example how to use LibGeoDecomp with a custom cellular automaton. All the user has to do is to specify his evolution code (in this case it is Conway's game of Life) and how the grid is initialized (via the `CellInitializer` class). He can then request a `Simulator` object from Pollarder's factory. For simplicity, this example does not use output. Usually the user would add output objects to the factory by calling the factory's `addWriter()` method. The user does not have to worry about the environment, Pollarder will automatically chose a suitable parallelization along with a load balancer and compatible IO objects.

```
#include <libgeodecomp.h>
using namespace LibGeoDecomp;

class Cell {
public:
    static inline unsigned nanoSteps() { return 1; }
```

```
    Cell(const bool& _alive = false) : alive(_alive) {}
    void update(CoordMap<Cell>& neighborhood, unsigned&) {
        int livingNeigbors = countLivingNeighbors(neighborhood);
        if (neighborhood[Coord(0, 0)].alive);
            alive = (2 <= livingNeigbors) && (livingNeigbors <= 3);
        else
            alive = (livingNeigbors == 3);
    }
    int countLivingNeighbors(const CoordMap<Cell>& nhood) { ... }
    bool alive;
};

class CellInit : public SimpleInitializer<Cell> { ... };

int main(int argc, char *argv[])
{
    MPI::Init(argc, argv);
    Simulator<Cell> *sim =
        Pollarder::Factory<Simulator>().get<Cell>(new CellInit());
    sim->run();
    MPI::Finalize();
    return 0;
}
```

4 Parallelization Architecture

One of our goals was to develop a hierarchical parallelization which could be adapted level-wise to the characteristics of a grid system. Thereby we could handle the aspects of its subsystems individually, instead of having to worry about every participating system simultaneously.

Pollarder's Hierarchical Adaptive Parallelization (HAP) Pattern is a solution intended for this use case. If a algorithm can be reformulated in a recursive manner, then HAP enables the developer to create a number of parallelization classes, each for a single type of system (e.g. SMP machines or for slow inter-cluster links). These classes have to be registered with Pollarder.

Pollarder will then perform an environment discovery whose outcome is a tree-shaped representation of the grid system. Depending on the system the top level node might represent a multi-cluster, its children could be the head nodes of the subordinate clusters and the leaves would be actual cluster nodes. SMP machines would contain an extra level to reflect the parallelism offered by their processors. All tree nodes will then be mapped path-wise to the processing cores. Figure 1(b) illustrates this mapping for the example of a a dual-core head node in a multi-cluster setup. The first core would receive a parallelization for inter-cluster synchronization, one for intra-cluster communication and one OpenMP parallelization. The second core would only get another one for OpenMP. Each level would get its own balancer to drive the parallelizations parameters (e.g. the ghost zone width).

The following pseudo code sketches out our hierarchical parallel algorithm. Ghost zones are updated first and sent asynchronously before updating the inner kernel in order to let communication and computation overlap. Upper level parallelizations shield their lower levels by aggregating outer communication. To allow for high latency on outwards connections, wider ghost zones on higher

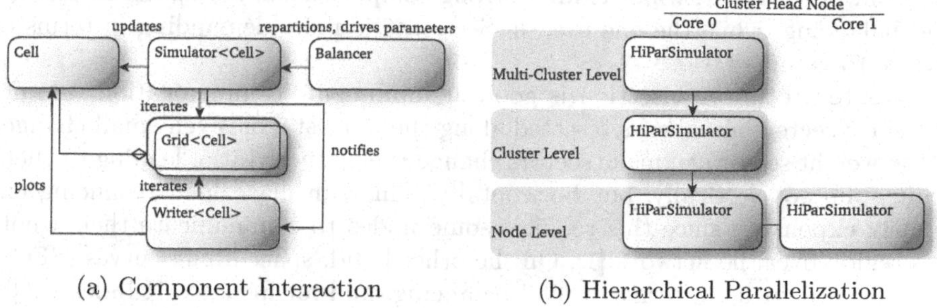

(a) Component Interaction (b) Hierarchical Parallelization

Fig. 1. Component Architecture. HiParSimulator is a subclass of Simulator. Associated Balancers have been omitted in Fig. 1(b) for clarity.

levels are advantageous. The grid itself and the ghost zones are stored decentralized. Upper level parallelizations delegate updates to their children. To perform synchronization on their level, they collect the updated ghost zone fragments from their children and send them back the parts they have received from their peers. A major challenge during the implementation has been to manage the different ghost zone widths, which influence the rhythm of the updates in subsidiary components (see Section 5). Since nodes in each sub-system are shielded by their head nodes from other sub-systems, they can perform load-balancing independently, thus reducing the over all cost of local balancing.

```
def update
  async_recv(@outer_ghost_region)
  if (children) then children.update(@inner_ghost_region)
  else do_update(@inner_ghost_region)
  async_send(@inner_ghost_region)
  if (children) children.update(@kernel)
  else do_update(@kernel)
  wait_for_communication
end

def main
  @outer_ghost_region, @inner_ghost_region, @kernel =
    initialize_simulation
  loop { update }
end
```

5 Domain Decomposition

The domain decomposition is one of the crucial parts of any parallel simulation. It splits up the simulation grid into smaller chunks which are then assigned to the processors. While its ultimate goal is to maximize performance, this can only be achieved by a compromise between its sub-goals: minimum surface to volume ratio (to minimize communication volume), high locality in neighborhood relationship (to minimize inter cluster communications), maximum overlap for two decompositions with similar weight vectors (to minimize communication costs for load balancing) and minimal computational overhead. The actual performance of a technique depends on the simulation model, too. Models like the one from MuCluDent, which exhibit strong computational hot-spots, emphasize load balancing, while the one used in Section 6 is more demanding in terms of surface to volume ratio.

Weighted recursive bisection is good at minimizing communication volume. But since sectors are always bisected along the longest axis, even small changes to the weight sector can make sectors change their aspect ratio, leading to them being split not vertically, but horizontally. This can make load balancing extremely expensive, since this requires some nodes to communicate their whole grid region over the network [1]. On the other hand, space filling curves (SFCs) are typically docile in terms of load balancing, and for instance H-Indexing [5] is near optimal in terms of neighborhood locality, but less good in respect to minimum node surfaces. All in all, SFCs are associated with high computational overhead and for a given point on the curve it is hard find the curve coordinates in spacial neighborhood. Basically this requires one to invert the curve, mapping space coordinates back to curve coordinates.

To avoid both, the problems of SFCs and to keep our parallelization independent from fixed decomposition schemes, we have devised an alternative representation for the coordinate set allocated on a node. Figure 2(b) illustrates how the `Region` data structure performs a run length compression of a stream of coordinates provided for instance by the H-Indexing curve. Instead of having to store all coordinates, we just store the starting point and length of consecutive coordinate streaks. These streaks are then stored in an associative map. This allows a much faster iteration in the stored section of the curve, but the original order is lost.

`Region` objects support Boolean operators like `and`, `or` and `and not`. Additionally they can be modified to include all neighboring coordinates up to a given distance n by the method `expand(n)`. If two given nodes are to simulate the parts of the grid described by the regions `a` and `b` and the ghost zone has the width n, then the ghost zone to be sent by one node to the other can be obtained by `a.expand(n) & b`.

The actual mesh data is stored in a `DisplacedGrid` which is basically a `boost::multi_array`[1], which is additionally able to transform absolute coordinates into local ones by using a displacement and intercept out of bounds accesses to route them to dedicated boundary cells. This way each node can use the same coordinate system despite storing only a fraction of the whole simulation grid (namely the bounding box of its region). These abstractions allow us to test virtually any domain decomposition, ranging from simple striping to SFCs. Even an adapter for ParMETIS [6] could be built.

Figures 2(c) to 2(e) illustrate a hierarchical decomposition with two levels in a multi-cluster setup with the Z curve. Figure 2(c) depicts the cluster level decomposition, which is further broken down on the node level and exemplary documented for node 7 in Fig. 2(d). The inter-cluster ghost zone from Fig. 2(c) is stored decentralized with each node on the boundary hosting the fragment relevant to him (Fig. 2(e)). Besides his actual region from Fig. 2(d), node 7 has

[1] `http://www.boost.org`

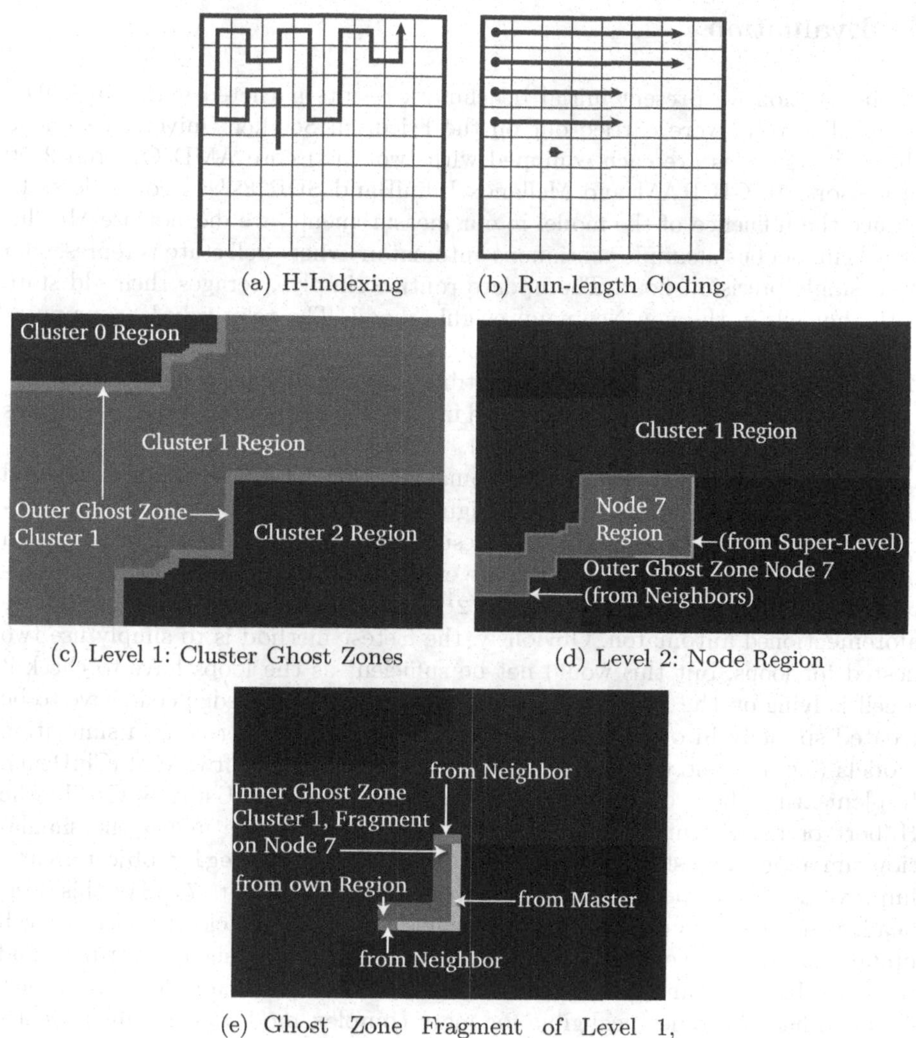

(a) H-Indexing (b) Run-length Coding

(c) Level 1: Cluster Ghost Zones (d) Level 2: Node Region

(e) Ghost Zone Fragment of Level 1, Hosted on Node 7.

Fig. 2. Coordinate compression by the Region data structure and two-level hierarchical domain decomposition

to update parts of its cluster's inter cluster ghost zone. These update regions in turn have ghost zones themselves which have to be filled from multiple sources. During the update of the cluster ghost zone in Fig. 2(e), the node has to save patches of certain time steps in order to paste them into the actual region's ghost zone (see *Super-Level* region in Fig 2(d)). On the other hand, the update for the inter cluster ghost zone fragment requires patches from its own region, from its neighbors and additionally from its master (in this case the head node which aggregates inter-cluster communication).

6 Evaluation

In this section we present initial benchmark results gathered with LibGeoDecomp. The tests were carried out on the Friedrich Schiller University's `omega` cluster[2]. Its nodes are each equipped with two quad core AMD Opteron 2350 processors, 16 GB RAM and Mellanox InfiniBand SDR 8Gbit/s controllers. To reduce the influence of the model in our measurements, we did not use MuCluDent's model but a simple continuous automaton, whose cell state is represented by a single precision float. The update routine simply averages their old state with the cells in the von Neumann neighborhood. The relatively low computational overhead of five floating-point operations per update at a cell size of four bytes makes the parallelization susceptible to communication delays (for comparison: the continuous automaton used in [2] takes 780 floating-point operations per update at a cell size of eight bytes).

One could argue that the complex domain decomposition techniques proposed in the previous section would inflict significant overhead, making their advantages moot. Table 1 summarizes the results of a sequential benchmark we ran to test the speed of different coordinate enumeration techniques. The times correspond to the time taken to simulate 2^{16} iterations on a 128×128 grid of the aforementioned automaton. Obviously, the fastest method is to simply use two nested for loops, but this would not be sufficient as the loops have to check if a cell is lying on the border of the simulation grid. Those edge cells have to be treated specially in order to satisfy boundary conditions for certain simulation models (e.g. the one used for our metal alloy simulation software MuCluDent). Implementing these checks costs about 3% performance. Using SFCs like the Hilbert or the Z Curve [7] to directly traverse the square makes the simulation run about 50% slower. However, packing them into a `Region` object greatly improves iteration speed, resulting in an overhead of not quite 7%. For this overhead, which equates on the testing machine to about five clock cycles per cell update, we can move away from cuboid decompositions (such as striping and recursive bisection and get all the benefits of arbitrary domain decompositions. The overhead becomes negligible for more complex models, e.g. MuCluDent's cells take on average about 135 clock cycles on the test machine.

To evaluate the scalability of LibGeoDecomp, we ran weak and strong scaling tests. All tests used the Z curve for domain decomposition (since it yields a good surface to volume ratio) and a ghost zone width or 4. Figure 3(a) contains the result of two speedup test. The test on a grid of 2048×2048 cells leads to a speedup of 62.44 on 80 cores. This equates to a overhead of 21.95%, which is quite high, but about half of it (11.46%) is caused by the additional updates required by the employed ghost zone width of 4. The results are better for the larger grid of 8192×8192 cells (speedup of 34.86, as opposed to 31.82, for 40 cores). The efficiency test in Fig. 3(b) evaluates how well LibGeoDecomp scales for growing problem sizes. In this case we have used for n cores a grid with the edge length $2048 \cdot \sqrt{n}$. For 32 cores we have achieved an efficiency of more than 95%.

[2] `http://www.uni-jena.de/omega.html`

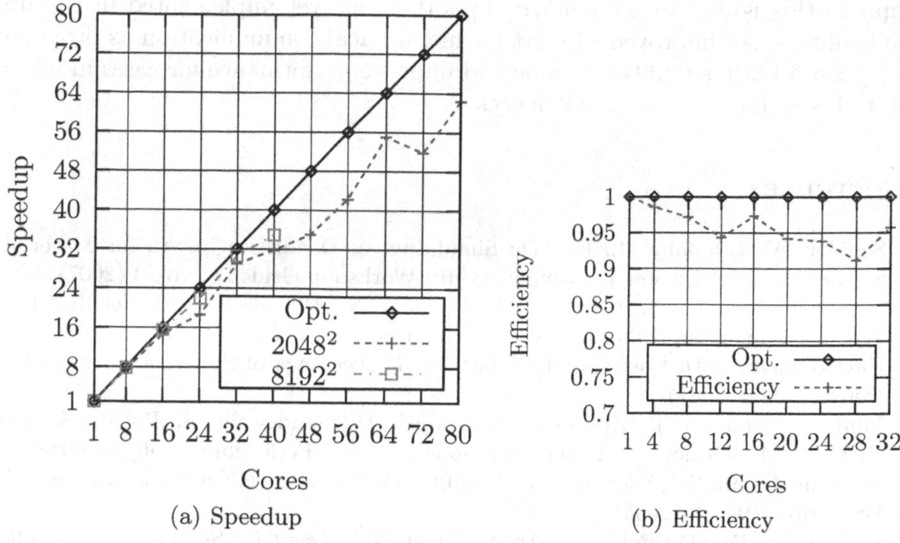

(a) Speedup

(b) Efficiency

Fig. 3. Speedup and weak scaling results

Table 1. Coordinate Enumeration Overhead for Domain Decomposition

Type	Time (s)	Overhead (%)
Nested loop	35.5	
Nested loop with bounds checking	36.6	2.97
Hilbert curve enumeration	54.3	52.79
Z curve enumeration	56.8	59.76
Region	38.0	6.76

7 Summary and Future Plans

We have presented LibGeoDecomp, a library for scientific simulations in grid and multi cluster environments. An environment discovery facility enables static adaptation, while balancer objects can drive the parallelizations parameters to achieve run time adaptation. A hierarchical parallelization can be automatically tuned to match the supercomputer's structure. Overlapping communication and computation, ghost zones of arbitrary width and exchangeable domain decomposition techniques improve performance and adaptability. Initial benchmarks prove its scalability, although still numerous tests are missing and a lot optimization has to be done. Especially the various domain decomposition techniques have to be evaluated for different network setups and simulation models.

Currently LibGeoDecomp only has weak support for three-dimensional simulations, as all decompositions can only work on two spacial dimensions, which can lead to sub-optimal surfaces. An extension of the `Region` class is meant to improve this issue. Also, OpenMP support is not yet implemented in the update routine. An improved support for hierarchical communication as proposed in [8] could both, simplify our code and improve performance for cases in which inter-cluster links are not a bottleneck.

References

1. Schäfer, A., Erdmann, J., Fey, D.: Simulation of Dendritic Growth for Materials Science in Multi-Cluster Environments. In: Workshop Grid4TS, vol. 3 (2007)
2. Allen, G., Dramlitsch, T., Foster, I., Karonis, N., Ripeanu, M., Seidel, E., Toonen, B.: Supporting Efficient Execution in Heterogeneous Distributed Computing Environments with Cactus and Globus. In: Proceedings of Supercomputing 2001, Denver, USA (2001)
3. Lani, A., Quintino, T., Kimpe, D., Deconinck, H., Vandewalle, S., Poedts, S.: The COOLFluiD Framework: Design Solutions for High Performance Object Oriented Scientific Computing Software. In: Computational Science, ICCS 2005, Atlanta, GA, USA, pp. 279–286 (2005)
4. Schäfer, A., Fey, D.: Pollarder: An Architecture Concept for Self-adapting Parallel Applications in Computational Science. In: Computational Science - ICCS. LNCS, vol. 5101, pp. 174–183. Springer, Heidelberg (2008)
5. Niedermeier, R., Reinhardt, K., Sanders, P.: Towards Optimal Locality in Mesh-Indexings. In: FCT 1997. LNCS, vol. 1279, pp. 364–375. Springer, Heidelberg (1997)
6. Schloegel, K., Karypis, G., Kumar, V.: Parallel Multilevel Diffusion Algorithms for Repartitioning of Adaptive Meshes. Technical Report TR 97-014 (1997)
7. Morton, G.M.: A computer Oriented Geodetic Data Base and a New Technique in File Sequencing. Technical report, IBM, Ottawa, Canada (1966)
8. López, R., Pérez, C.: Improving mpi support for applications on hierarchically distributed resources. In: Cappello, F., Herault, T., Dongarra, J. (eds.) PVM/MPI 2007. LNCS, vol. 4757, pp. 187–194. Springer, Heidelberg (2007)

Using Arithmetic Coding for Reduction of Resulting Simulation Data Size on Massively Parallel GPGPUs

Ana Balevic, Lars Rockstroh, Marek Wroblewski, and Sven Simon

Institute for Parallel and Distributed Systems,
Universitaetsstr. 38, 70596 Stuttgart, Germany
ana.balevic@ipvs.uni-stuttgart.de

Abstract. The popularity of parallel platforms, such as general purpose graphics processing units (GPGPUs) for large-scale simulations is rapidly increasing, however the I/O bandwidth and storage capacity of these massively-parallel cards remain the major bottle necks. We propose a novel approach for post-processing of simulation data directly on GPGPUs by efficient data size reduction immediately after simulation that can considerably reduce the influence of these bottlenecks on the overall simulation performance, and present current performance results.

Keywords: HPC, GPGPU, I/O Bandwidth, Data Compression, Arithmetic Coding.

1 Introduction

In general, simulations require not only high amount of computing power but also the transfer and storage of large amounts of data. Due to high computational requirements, large-scale simulations are typically conducted on parallel systems comprising of multiple computing nodes. With recent advances in development of massively parallel graphics cards suitable for general purpose computations (GPGPUs) and their general affordability, with prices as low as $400 per unit (2008) featuring up to 128 streaming processors, the computers used in simulations are increasingly equipped with one or more GPGPUs integrated as arithmetic co-processors that enable hundreds of GFLOPs of raw processing power in addition to the CPU.

One of the major bottlenecks of such parallel computing systems, besides the storage of large amount of data, is the I/O bandwidth required for run-time communication and synchronization of numerous processing elements, as well as the transfer of the resulting data from the arithmetic co-processors to the central processing unit. As the time spent in data transfers between computational nodes can significantly reduce observed speedups and thus severely influence the performance benefit of using a parallel system for computations, there is a demand for novel approaches to the storage and transfer of data.

A. Lastovetsky et al. (Eds.): EuroPVM/MPI 2008, LNCS 5205, pp. 295–302, 2008.

The study of data compression algorithms in the computer science has resulted in efficient coding algorithms such as Huffman coding, Arithmetic/Range coding, Lempel-Ziv family of dictionary compression methods and various transforms such as Burrows-Wheeler Transform (BWT), that are now a part of widely used compression utilities, such as Zip, RAR, etc as well as image and video codecs. In this paper we explore use of entropy coding algorithms on high performance computing systems containing massively parallel GPGPUs, such as NVidia GeForce 8800 GT, for efficient stream reduction by processing of the resulting simulation data in between the simulation steps, and prior to the transfer and storage on the host computer.

The paper is structured as follows. In Section 2, the current approaches to data size reduction on GPUs are reviewed. An overview of fundamental compression methods is given in Sections 3 and 4. Section 5 presents design of a block-parallel entropy coding algorithm. Sections 6 and 7 give current performance results in compression of floating-point data from a light scattering simulation on a GPGPU, and an overview of the future research.

2 Related Work

The popularity of parallel platforms, such as GPGPUs for large-scale simulations is rapidly increasing, however the I/O bandwidth and storage capacity of GPGPUs remain a bottle neck. In simulations of large systems, a variety of approaches has been used to reduce run-time size of simulation data set. Some common approaches include different methods for the storage of sparse matrices, use of reduced precision for calculations, etc. Fast lossless compression approaches to floating-point data, including the overviews of older approaches can be found in [1,2]. As GPGPUs impose numerous constraints on the data types that could be efficiently used for storage of the simulation data, it is worth exploring which other approaches to the data size reduction are available and could be efficiently used on GPGPUs. The most notable approaches for data reduction that used on GPUs are stream reduction and texture compression of computer graphics:

Texture compression is driven by the need for reducing the amount of physical memory required for the storage of texture images that enhance gaming experience. A distinctive characteristic of the texture compression is that it provides a fixed ratio compression coupled with single-memory data access, which makes it ideally suited for computer graphics. Texture compression is a lossy data compression scheme, with common implementations being S3TC family of algorithms (DXT1-DXT5), and DXTC. For gaming purposes, the loss of fidelity is acceptable and can even account for a perceptually better experience as the decrease in data size enables the storage of higher-resolution textures in the memory of a graphics card.

The second notable approach on GPUs is stream reduction, which is the process of removing elements that are not necessary from the output stream. Stream reduction is frequently used in multi-pass GPU algorithms, where the stream output of the first pass is used as the input for the next pass. An efficient implementation of the stream reduction on GPUs is given in [3], and achieves a linear performance by using divide-and-conquer approach that is well applicable to GPGPU block-oriented architectures.

3 Data Compression

Data compression deals with the data size reduction by removing redundancy from the input data. The theoretical bound of the compression, i.e. the maximum theoretical compression ratio, is given by Claude Shannon's Source Coding Theorem, which establishes that the average number of bits required to represent an uncertain event is given by its entropy (H). Data compression methods are classified according to the information preservation to lossless and lossy. Lossless compression algorithms are used in areas where absolutely accurate reconstruction of the original data is necessary, such as in compression of text, medical, scientific data, etc. In the applications targeted toward human end-users, lossy compression is applied to audio, video and image data in order to provide perceptively (near) lossless or acceptably distorted representation of data by using perceptual models of the human audio-visual system. We consider two fundamental lossless algorithms for the compression of the simulation data, which could be easily combined with intermediate lossy steps, e.g. quantization, if further increase of the compression ratio at the expense of accuracy is desired:

Huffman Coding: As a statistical lossless data compression algorithm, Huffamn coding gives a reduction in the average code length used to represent the symbols of an alphabet by assigning shorter codewords to more frequent symbols and vice versa. The Huffman code is an optimal prefix code in the case where exact symbols probabilities are known in advance and are integral powers of 1/2 [4]. In real-world scenarios, the exact distribution of symbol probabilities is rarely known in advance, so this means either acceptance of lower compression rates or use of adaptive Huffman algorithms that provide one-pass encoding and adaptation to changing statistics of the input data. The major disadvantage of the adaptive Huffman coding is relatively high cost of tree maintenance operations, especially in GPU environments, where non-aligned memory access are penalized in terms of performance. When the symbol probabilities are highly skewed, which is often in the case of the simulation data, Huffman coding does not provide good compression rates as the generated codewords, being external nodes of a binary tree, are always represented by an integral number of bits.

Arithmetic Coding: Arithmetic coding treats the whole input data stream as a single unit that can be represented by one *real* number in the interval $[0, 1)$. As the input data stream becomes longer, the interval required to represent it becomes smaller and smaller, and the number of bits needed to specify the final interval increases. Successive symbols in the input data stream reduce this interval in accordance with their probabilities. The more likely symbols reduce the range by less, and thus add fewer bits to the coded data stream.

By allowing fractional bit codeword length, arithmetic coding attains the theoretical entropy bound to compression efficiency, and thus provides better compression ratios than Huffman coding on input data with highly skewed symbol probabilities. The arithmetic coding gives greater compression, is faster for adaptive models, and clearly separates the model from the channel encoding [5]. As simulation data is usually biased to certain values (or could be transformed into a set of biased data e.g. by some sort of predictive coding), we chose to further experiment with arithmetic coding for simulation data compression on GPGPUs.

4 Fundamental Principles of Arithmetic Coding

The central concept behind arithmetic coding with integer arithmetic is that given a large-enough range of integers, and frequency estimates for the input stream symbols, the initial range can be divided into sub-ranges whose sizes are proportional to the probability of the symbol they represent[4,5]. Symbols are encoded by reducing the current range of the coder to the sub-range that corresponds to the symbol to be encoded. Finally, after all the symbols of the input data stream have been encoded, transmitting the information on the final sub-range is enough for completely accurate reconstruction of the input data stream at the decoder. The fundamental sub-range computation equations are given recursively as:

$$low^n = low^{n-1} + (high^{n-1} - low^{n-1})P_l(x_n) \tag{1}$$

$$high^n = low^{n-1} + (high^{n-1} - low^{n-1})P_h(x_n) \tag{2}$$

where P_l and P_h are the lower and higher cumulative probabilities of a given symbol (or cumulative frequencies) respectively, and low and high represent the sub-range boundaries after encoding of the n-th symbol from the input data stream. As an illustration of the arithmetic coding concepts, a basic encoding to a real number, for the input sequence 'bac', with the given symbol distribution is depicted in Fig. 1. The decoding algorithm works in an analogous way, and must be synchronized with the encoder. The practical integer-implementation of the arithmetic coder function according to the same principle as illustrated in Fig. 1, but uses frequencies of occurrence instead of symbol probabilities and a range of $[0, N)$, where typically N is an integer value $N >> 1$.

To avoid arithmetic overflows on 32-bit architectures, a maximally 31-bit integer range can be used to represent the full range of the coder. To avoid underflows, which would happen if the current sub-range would become too small to distinctively encode the symbol, i.e. when the upper and lower boundaries of the range converge, several methods have been proposed for range renormalization[4,5,6].

For the range renormalization and generation of the compressed data bit stream, we use a method of dividing the initial range into quarters described in detail in [6] that works as follows: After the coder detects that the current sub-range falls into a certain quarter, it is ensured that the leading bits of the numbers representing the sub-range boundaries are set, and cannot be changed

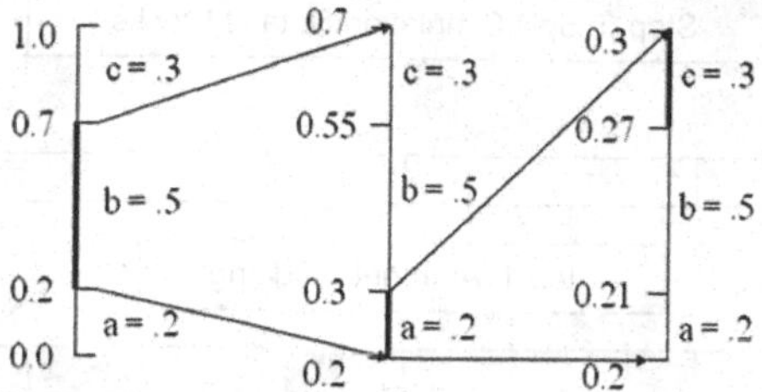

Fig. 1. Example of arithmetic encoding of the input sequence 'bac'. Symbols of the alphabet $A = a, b, c$ have probabilities of occurrence $P = .2, .5, .3$ Final range is $[0.27, 0.3)$. The sequence 'bac' can be thus coded with 0.27.

by subsequent symbols. A series of scaling operations is conducted, and the set bits are output one after the other, thus generating the compressed data output stream. These operations result in the renormalization of the coder range back to the full supported range, thus eliminating possibility of incorrect en/decoding due to the range underflow.

5 Block-Parallel GPGPU Arithmetic Encoder

Simulations run on general purpose graphics hardware often produce large amount of data that after a number of iterations hardly fits into the memory of a graphics card, thus imposing a need for a memory transfer so that free space is made available for subsequent iterations. As the frequent data transfers from the memory of a GPGPU to the host PC reduce the overall performance of the simulation, it is our goal to lessen the frequency of these data transfers. We propose processing of simulation data directly on the GPGPUs after each simulation step to reduce the resulting data size, and thus resources required for the storage and transfer of results.

First, the simulation data is partitioned into the data blocks as in [7,3], which are then processed by a number of replicated coders running in parallel, as depicted in Fig 2. Each block of simulation data is processed by an instance of the encoder running in a separate thread. In the CUDA computational model threads are executed in the thread blocks, each of which is scheduled and run on a single multi-processor. Our block-parallel encoder implementation can be executed by multiple blocks containing multitude of threads, where each thread executes the CUDA-specific code that implements the arithmetic encoding process described in Sect.4 (Fig.2,Step1). The data block size, as well as the number of blocks and threads, is configurable as the compression kernel execution parameter. Based on different block sizes, different compression ratios are obtained - typically resulting in higher compression ratio for larger data block sizes.

After the complete input stream is encoded, the coded data blocks are prepared for the storage at the adequate global memory locations, prior to the transfer to the host computer. The first preparation step for storage in the parallel

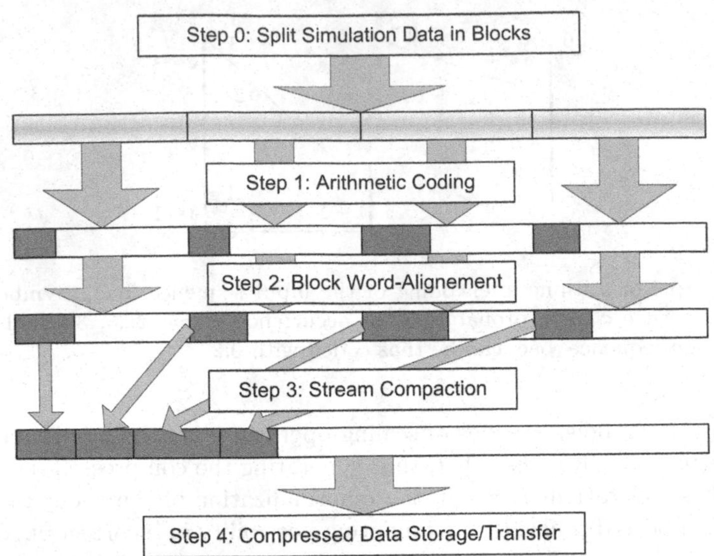

Fig. 2. Block-Parallel GPGPU Simulation Data Compression Process

implementation of encoder on GPGPU is alignment of the coded data bitstream to the byte or word boundary (Fig.2,Step2). The padding to the boundary increases the compressed data length. The decrease of the performance ratio due to this operation is dependent on the initial data block length and its entropy - the larger the resulting coded block is, the smaller difference those couple of padded bits make.

To obtain highly biased data model, the floating-point numbers from the simulation are processed on the byte level. Each of 256 possible byte values is assigned a corresponding symbol. The model constructed in this manner exploits the statistical properties of data much better than if we would assign each different floating-point value a single symbol, typically resulting in probabilities highly biased to some symbol e.g. 0x00. Another advantage of this modeling approach is that it can be without any modification applied to other data types, without a loss of generality. After byte-level arithmetic encoding, the coded data is aligned to the word boundary, e.g. 8-bit before transferring the results into the global memory of device.

The compacted output stream is obtained by the concatenation of the codewords at the block-level by stream compaction (Fig.2,Step3) that produces a single continuous array containing the coded data. The concatenation process is executed fully in parallel on the GPGPU, by creating an array of the coded data block lengths from the resulting data of encoding process. After generation of the codewords and alignment to desired word boundary length (i.e. 8-bits or 32 bits), the information on the coded block lengths is used to generate the array of the pointers to the starting positions of the coded data coming from parallel coders by using parallel prefix sum primitives.

For correct functioning of the method, the stream compaction is not necessary, as the data from each coded block can be transferred separately to the host computer followed by storage into a continuous array. However, it is worth examining, as the burst mode for data transfer generally achieves better performance than the iterative data transfer.

6 Performance Results and Discussion

The block-parallel implementation of integer-based arithmetic coder for GPGPU was tested on data from the simulation of light scattering [8]. As the test data for the compression were taken the results of finite-difference time-domain simulation iterations on the grid of 512x512 cells. The distinctive characteristics of the test data set were low values with highly biased symbol probability distribution, resulting in very low entropy when using the model described in Sect. 5. The output of parallel arithmetic encoder running on the GPGPU is decompressed by sequential decoder runing on the host PC, and the results are verified by byte comparison functions, as well as the external file comparison tool WinDiff.

Table 1. Performance results on test configuration: AMD Athlon 2.41GHz, 2GB RAM, nVidia GeForce 8800GT 128SPs, 768MB RAM. CUDA 1.0. Total time corresponds to the time required for compression and transfer of data including the overheads, such as. alignment, stream compaction and block sizes array transfer.

	Test Data Set	Block Size [B]	Parallel Coders	CR	GPU Encode T [ms]	Direct Transf. T[ms]	Comp. Transf. [ms]	Total Time [ms]	
								Iterat.	Burst
1	Size: 1MB	512	2048	438	5.33	1.69	0.038	44.15	5.89
	H=0.00289908 [b/B]	1024	1024	765	5.25	1.61	0.037	25.20	5.80
	CPU Encode T=0.5s	4096	256	1727	8.98	1.59	0.06	14.54	9.46
2	Size: 4 MB	1024	4096	979	11.39	4.44	0.04	87.86	11.97
	H=0.00037884 [b/B]	4096	1024	1288	13.50	4.49	0.059	33.52	14.08
	CPU Encode T=1.6s	8192	512	1528	15.60	4.40	0.23	26.14	16.67
3	Size: 16 MB	1024	16384	242	51.22	16.13	0.23	354.4	52.05
	H=0.01047371 [b/B]	4096	4096	293	59.28	15.59	0.37	136	60.19
	CPU Encode T=6.5s	8192	2048	304	76.76	16.90	0.56	115	77.86

The performance results in Table 1. show that the parallel implementation of arithmetic encoder achieves compression ratios (CR) competitive with a sequential coder, but in a considerably shorter time, with the compression ratio approaching the lower entropy bound as the data block size increases. The transfer times for the compressed data (Col. 7) are significantly lower than those for the direct transfer of data (Col. 6) without any compression; however as the compression process inevitably introduces an overhead, the gains achieved so far are mostly in the required space on the GPGPU for the storage of the temporary results, with more work on the speed optimization of the codec required for making it a competitive method for reduction of the I/O bandwidth requirements. The storage savings are a significant achievement, as the frequency with which the simulation data needs to be transferred considerably influences overall simulation speed-up. If the storage of simulation results requires less space, there is a more room for the new data, resulting in a lower number of required memory transfers from the GPGPU to the host computer, and thus a better overall simulation performance.

7 Conclusions and Future Work

The implementation of the block-parallel arithmetic encoder proved that use of statistical coding methods for the compression of simulation data directly on GPGPUs has a potential for the efficient reduction of simulation data size. The compression ratios of the parallel coder approach entropy as the theoretical boundary of compression ration with the increasing block sizes. Furthermore, the parallel implementation exhibits a significant speed-up over the sequential data compression algorithm, thus showing high potential to reduce influence of the limited resources for storage and transfer on the simulation performance on parallel systems. Our ongoing work focuses on optimization of computational performance of entropy coders. Further work will examine strategies for preprocessing of simulation data that could account for high compression efficiency coupled with high processing speed.

References

1. Isenburg, M.: Fast and efficient compression of floating-point data. IEEE Transactions on Visualization and Computer Graphics 12(5), 1245–1250 (2006)
2. Ratanaworabhan, P., Ke, J., Burtscher, M.: Fast lossless compression of scientific floating-point data. In: DCC 2006: Proceedings of the Data Compression Conference, pp. 133–142. IEEE Computer Society Press, Washington (2006)
3. Roger, D., Assarsson, U., Holzschuch, N.: Efficient stream reduction on the gpu. In: Kaeli, D., Leeser, M. (eds.) Workshop on General Purpose Processing on Graphics Processing Units (October 2007)
4. Sayood, K. (ed.): Lossless Compression Handbook. Academic Press, London (2003)
5. Howard, P.G., Vitter, J.S.: Arithmetic coding for data compression. Technical Report Technical report DUKE–TR–1994–09 (1994)
6. Bodden, E.: Arithmetic coding revealed - a guided tour from theory to praxis. Technical Report 2007-5, Sable (2007)
7. Boliek, M.P., Allen, J.D., Schwartz, E.L., Gormish, M.J.: Very high speed entropy coding, pp. 625–629 (1994)
8. Balevic, A., Rockstroh, L., Tausendfreund, A., Patzelt, S., Goch, G., Simon, S.: Accelerating simulations of light scattering based on finite-difference time-domain method with general purpose gpus. In: Proceedings of 2008 IEEE 11th International Conference on Computational Science and Engineering (2008)

Benchmark Study of a 3d Parallel Code for the Propagation of Large Subduction Earthquakes

Mario Chavez[1,2], Eduardo Cabrera[3], Raúl Madariaga[2], Narciso Perea[1], Charles Moulinec[4], David Emerson[4], Mike Ashworth[4], and Alejandro Salazar[3]

[1] Institute of Engineering, UNAM, C.U., 04510, Mexico DF, Mexico
chavez@servidor.unam.mx, narpere@iingen.unam.mx
[2] Laboratoire de Géologie CNRS-ENS, 24 Rue Lhomond, Paris, France
raul.madariaga@ens.fr
[3] DGSCA, UNAM, C.U., 04510, Mexico DF, Mexico
alejandro@labvis.unam.mx, eccf@super.unam.mx
[4] STFC Daresbury Laboratory, Warrington WA4 4AD, UK
c.moulinec@dl.ac.uk, d.r.emerson@dl.ac.uk,
m.ashworth@dl.ac.uk

Abstract. Benchmark studies were carried out on a recently optimized parallel 3D seismic wave propagation code that uses finite differences on a staggered grid with 2nd order operators in time and 4th order in space. Three dual-core supercomputer platforms were used to run the parallel program using MPI. Efficiencies of 0.91 and 0.48 with 1024 cores were obtained on HECToR (UK) and KanBalam (Mexico), and 0.66 with 8192 cores on HECToR. The 3D velocity field pattern from a simulation of the 1985 Mexico earthquake (that caused the loss of up to 30000 people and about 7 billion US dollars) which has reasonable agreement with the available observations, shows coherent, well developed surface waves propagating towards Mexico City.

Keywords: Benchmark, modeling, finite difference, earthquakes, parallel computing.

1 Introduction

Realistic 3D modeling of the propagation of large subduction earthquakes, such as the 1985 Mexico earthquake (Fig. 1), poses both a numerical and a computational challenge, particularly because it requires enormous amounts of memory and storage, as well as an intensive use of computing resources. As the recurrence time estimated for this highly destructive type of event in Mexico is only a few decades, there is a seismological, engineering and socio-economical interest in modeling them by using parallel computing [1].

In this paper, we present the results from benchmark studies performed on a recently optimized parallel 3D wave propagation staggered-grid finite difference code, using the Message Passing Interface (MPI). The code was run on three dual-core platforms, i.e.: KanBalam (KB, Mexico, [2]), HPCx (UK, [3]) and HECToR (UK, [4]). Characteristics

A. Lastovetsky et al. (Eds.): EuroPVM/MPI 2008, LNCS 5205, pp. 303–310, 2008.

of the three systems are shown in Table 1. In section 2, a synthesis of the 3D wave propagation problem and the code are presented; a description of the strategy followed for the data parallelism of the problem and the MPI implementation are discussed in section 3. The benchmark experiment performed on the code and its main conclusions are addressed in section 4 and in section 5, the results obtained for the modeling of the seismic wave propagation of the Mexico 1985 Ms 8.1 subduction earthquake are given.

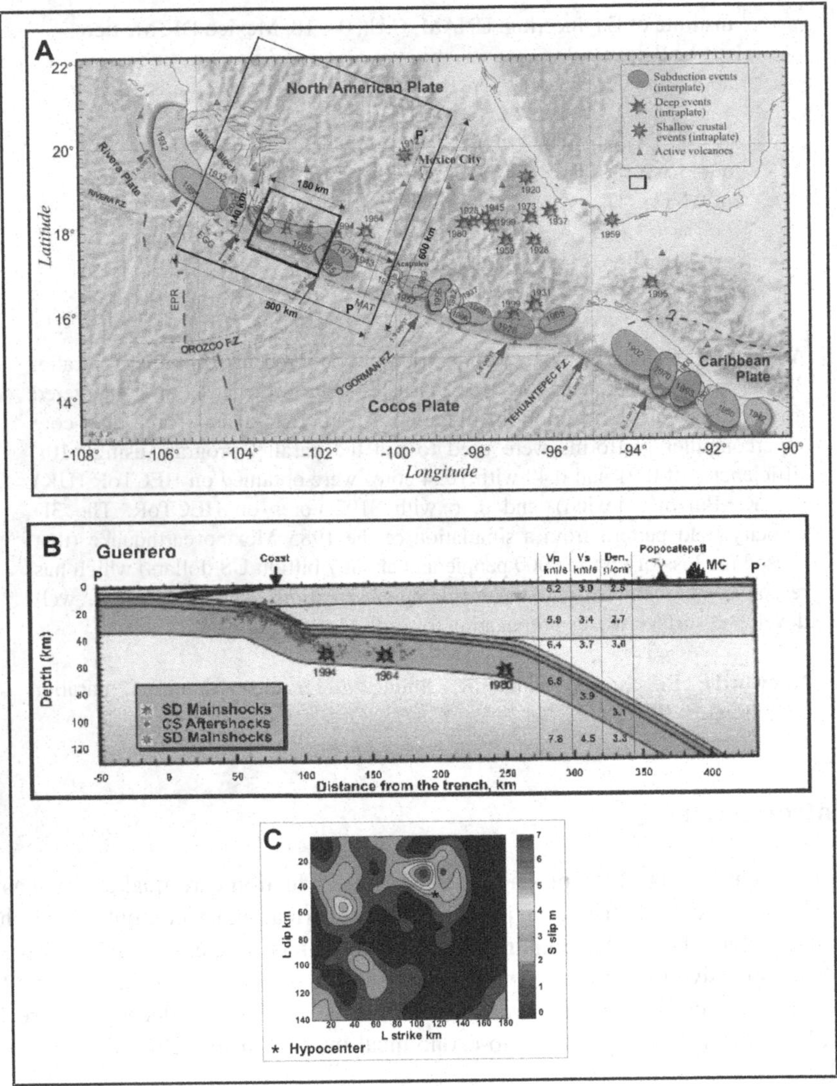

Fig. 1. A) Inner rectangle is the rupture area of the 19/09/1985 Ms 8.1 earthquake on surface projection of the 500x600x124 km earth crust volume 3DFD discretization; B) profile P-P´; C) Kinematic slip distribution of the rupture of the 1985 earthquake [1]

Table 1. Characteristics of the 3 Supercomputer platforms used in the benchmark study

Platform	HPCx	KB	HECToR
Processor	IBM PowerPC 5 1.5GHz dual core	AMD Opteron 2.6GHz dual core	AMD Opteron 2.8GHz dual core
Cache	L1 data 32KB and L1 instr 64KB per core L2 1.9MB shared L3 128 MB shared	L1 instr and data 64KB per core L2 1MB shared	L1 instr and data 64KB per core L2 1MB shared
FPU's	2 FMA	1Mult, 1Add	1Mult, 1Add
Peak performance/core	6 GFlop/s	5.2 GFlop/s	5.6 GFlop/s
Cores	2560	1368	11328
Peak Perf	15.4 TFLOP/s	7.12 TFLOP/s	63.4 TFLOP/s
Linpack	12.9 TFLOP/s	5.2 TFLOP/s	54.6 TFLOP/s
Interconnect	IBM High performance switch	Infiniband Voltaire switch 4x, fat tree topology	Cray SeaStar2 3D toroidal topology
Bandwidth	4GB/s	1.25 GB/s	7.6 GB/s
latency	5.5 μs	13 μs	5.5 μs
File system	GPFS	Lustre	Lustre

2 3D Wave Propagation Modeling and Its Algorithm

The 3D velocity-stress form of the elastic wave equation, consists of nine coupled, first order partial differential hyperbolic equations for the three particle velocity vector components and the six independent stress tensor components [1, 5].

The finite difference staggered algorithm applied to the mentioned equations is an explicit scheme which is second-order accurate in time and fourth-order accurate in space. Staggered grid storage allows the partial derivatives to be approximated by centered finite differences without doubling the spatial extent of the operators, thus providing more accuracy. The discretization of the 3D spatial grid is such that $x_i=x_0+(i-1)h_x$, $y_j=y_0+(j-1)h_y$, and $z_k=z_0+(k-1)h_z$ for i=1, 2, 3, … I, j= 1, 2, 3, … ,J, and k=1, 2, 3, … ,K, respectively. Here x_0, y_0, z_0 are the minimum grid values and h_x, h_y, h_z give the distance between points in the three coordinate directions. The time discretization is defined by $t_l=t_0+(l-1)h_t$ for l=1, 2, 3,…,L. Here t_0 is the minimum time and h_t is the time increment.

3 Parallel Implementation of the 3DFD Algorithm

We use 3D data parallelism for efficiency. The domain was decomposed into small subdomains and distributed among a number of processors, using simple partitioning to give an equal number of grid points to each processor [1]. This approach is appropriate for the 3DFD wave propagation code, as large problems are too big to fit on a single processor [1].

The Message Passing Interface (MPI) was used to parallelize the 3DFD code [1]. In particular, MPI_Bcast, MPI_Cart_Shift and MPI_SendRecv instructions were used; the first two to communicate the geometry and physical properties of the problem, before starting the wave propagation loop, and the last to update the velocities and stresses calculated at each time step. The nature of the chosen 3DFD staggered scheme precluded the efficient application of overlapping MPI_Cart_Shift, MPI_SendRecv operations with computations.

Parallel I/O from MPI-2 was used in the code to read the earth model data by all processors and to write the velocity seismograms by the processors corresponding to the free surface of the physical domain [1], which is only a small percentage of the total number of processors. As this type of parallel I/O is machine independent, it fitted the benchmark experiment performed on the three platforms.

4 Benchmark Experiment

As mentioned above the code was run on three dual-core platforms, i.e.: KanBalam (KB, Mexico, [2]), HPCx (UK, [3]) and HECToR (UK, [4]).

The actual size of the problem is 500 x 600 x 124 km (Fig 1), and its physical properties are also shown in the Fig. We used spatial discretizations hx = hy = hz, of: 1.0, 0.500, 0.250 and 0.125 km (to include thinner surficial geologic layers in the Z direction) and the associated time discretizations were 0.03, 0.02, 0.01 and 0.005 s, respectively (to comply with the Courant-Friedrich-Lewy condition). Therefore, *Nx*=500, 1000, 2000, 4000; *Ny*=600, 1200, 2400, 4800 and *Nz*=124, 248, 496, 992 are, the model size in the X, Y and Z directions, respectively (notice that *Nz* is about 0.25 of *Nx* and *Ny*). The number of time steps, *Nt*, used for the experiment was 4000.

Speedup, *Sp*, and efficiency, *E*, among others, are the most important metrics to characterize the performance of parallel programs. Theoretically, speedup is limited by Amdahl's law [6], however there are other factors to be taken into account, such as: communications costs, type of decomposition and its resultant load balance, I/O and others [1]. *Sp* and *E*, disregarding those factors, can be expressed by:

$$Sp \equiv mT_1(n/m)/T_m(n/m), E \equiv T_1(n/m)/T_m(n) \tag{1}$$

for a scaled-size problem *n* (weak scaling), and for a fixed-size problem (strong scaling)

$$Sp \equiv T_1/T_m \quad , E \equiv T_1/T_m m \tag{2}$$

where T_1 is the serial time execution and T_m is the parallel time execution on *m* processors.

If the communications costs and the 3D decomposition are taken into account, the expression for *Sp* is:

$$Sp \equiv \frac{A\Gamma R^3}{A\Gamma R^3 / m + 24\left(\iota + 4\beta R^2 / m^{2/3}\right)} \tag{3}$$

where the cost of performing a finite difference calculation on $mx \times my \times mz$, m, processors is $A\Gamma R^3 / m$; A is the number of floating operations in the finite difference scheme (velocity-stress consists of nine coupled variables); Γ is the computation time per flop; R is equal to Nx x Ny x Nz ; ι is the latency and β is the inverse of bandwidth [1]. This scheme requires the communication of two neighbouring planes in the 3D decomposition [1].

This benchmark study consisted of both scaled-size (weak scaling) and fixed-size (strong scaling) problems. In the former, the number of processors (m) utilized for KB and HECToR varied from 1 - 8192 and for the latter, 64 and 128 processors were used on the three platforms. For both type of problems, and whenever it was possible, experiments with one or two cores were performed, for KB, HECToR, and HPCx platforms.

The results of the two studies are synthesized in Table 2 and Fig. 2. From the results of the weak-scaling problems, it can be concluded that when large amounts of cores (1024 for KB) and (8192 for HECToR), with respect to the total number available in the tested platform, *Sp* and *E* decrease considerably, to 492 and 0.48 and 5375 and 0.66, for KB and HECToR, respectively. We think that this behavior is due to the very large number of communications demanded among the processors by the 3DFD algorithm [1]. This observation is more noticeable for the dual-core results, due to, among other factors, the fact that they are competing for the cache memory available and the links to the interconnect, and that this is stressed when thousands of them are used. The opposite behavior of *Sp* and *E* is observed when only tens, hundreds (for KB) or up to 1024 cores are used for HECToR, Table 2, Fig 2.

From the results for the strong-scaling problem shown in Table 2, it can be concluded that for the three platforms, the observed *Sp* and *E* are very poor, particularly when the two cores were used,. The "best" results were obtained for HECToR, followed by KB and HPCx. Given that the mentioned observation is valid for the three platforms, we can conclude that the 3DFD code tested is ill suited for strong-scaling problems.

5 Seismological Results for the 19/09/1985 Mexico's Ms 8.1 Subduction Earthquake

Herewith we present examples of the type of results that for the 1985 Mexico earthquake (Fig. 1) were obtained on the KB system with the parallel MPI implementation of the 3DFD code. At the top of Fig 3, the 3D low frequency velocity field patterns in the X direction, and the seismograms obtained at observational points, in the so-called near (Caleta) and far fields (Mexico City), of the wave propagation pattern for times equal to 49.2 and 136.8 s. The complexity of the propagation pattern at t = 49.2 s, when the seismic source is still rupturing, is contrasted by the one for t = 136.8 s, in which packages of coherent, well developed surface waves, are propagating towards Mexico City. Finally, at the bottom of Fig. 3 we show the observed and synthetic (for a spatial discretization dh = 0.125 km) low

Table 2. Scaled and fixed –size* models: *mi* (i = x, y, z) processors used in each axis (*mz* was fixed to 4 because *Nz* is about one fourth of *Nx* and *Ny*), timings, speedup, efficiency and memory per subdomain (Mps) obtained for KB, HECToR and HPCx. The total run time of KB of 37600 s was used to compute *Sp* and *E* for the (*) cases

Size model and spatial step (dh km)	m	mx	my	mz	Cores per chip used	Total run time (s)	Speedup (Sp)	Efficiency (E)	Mps (GB)
500x600x124 (1) **KB**	1	1	1	1	1	13002	1	1	1.9
1000x1200x248 (0.5) **KB**	16	1	4	4	1	6920	30	1.9	0.97
1000x1200x248 (0.5) **KB**	16	1	4	4	2	11362	18	1.14	0.97
2000x2400x496 (0.25) **KB**	128	4	8	4	2	15439	108	0.84	0.97
4000x4800x992 (0.125) **KB**	1024	16	16	4	2	27033	492	0.48	0.97
500x600x124 (1) **HECToR**	1	1	1	1	1	11022	1	1	1.9
1000x1200x248 (0.5) **HECToR**	16	1	4	4	1	6404	28	1.7	0.97
1000x1200x248 (0.5) **HECToR**	16	1	4	4	2	10583	17	1.04	0.97
2000x2400x496 (0.25) **HECToR**	128	4	8	4	1	6840	207	1.6	0.97
2000x2400x496 (0.25) **HECToR**	128	4	8	4	2	11083	127	0.99	0.97
4000x4800x992 (0.125) **HECToR**	1024	16	16	4	1	7200	1568	1.53	0.97
4000x4800x992 (0.125) **HECToR**	1024	16	16	4	2	12160	928	0.91	0.97
8000x9600x1984 (0.0625) **HECToR**	8192	32	32	8	2	16800	5375	0.66	0.97
1000x1200x248 (0.5) **KB***	1	1	1	1	1	37600	1	1	14.3
1000x1200x248 (0.5) **KB***	64	4	4	4	1	2699	13.9	0.22	0.242
1000x1200x248 (0.5) **KB***	64	4	4	4	2	3597	10.5	0.16	0.242
1000x1200x248 (0.5) **KB***	128	4	8	4	1	1681	22.4	0.18	0.121
1000x1200x248 (0.5) **KB***	128	4	8	4	2	2236	16.8	0.13	0.121
1000x1200x248 (0.5) **HECToR***	64	4	4	4	1	1898	19.8	0.31	0.242
1000x1200x248 (0.5) **HECToR***	64	4	4	4	2	2910	12.9	0.20	0.242
1000x1200x248 (0.5) **HECToR***	128	4	8	4	1	878	42.8	0.33	0.121
1000x1200x248 (0.5) **HECToR***	128	4	8	4	2	1420	26.5	0.21	0.121
1000x1200x248 (0.5) **HPCx***	64	4	4	4	2	4080	9.2	0.14	0.242
1000x1200x248 (0.5) **HPCx***	128	4	8	4	2	2100	17.9	0.14	0.121

frequency, north-south velocity seismograms of the 19/09/1985 Ms 8.1 Mexico earthquake, and their corresponding Fourier amplitude spectra for the firm soil Tacubaya site in Mexico City, i.e. at a far field observational site. Notice in Fig. 3, that the agreement between the observed and the synthetic velocity seismogram is reasonable both in the time and in the frequency domain.

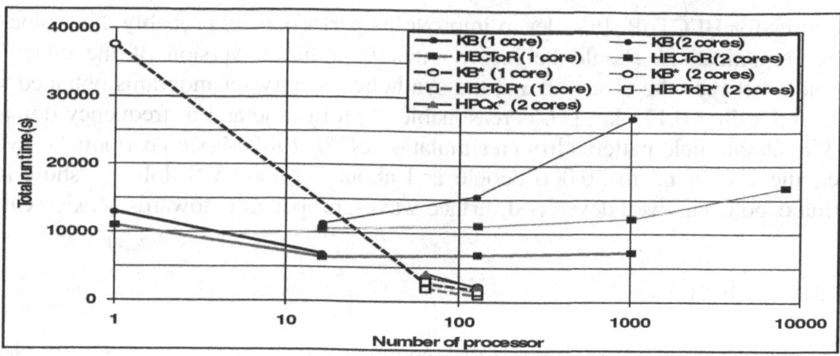

Fig. 2. Execution time vs number of processors for the three platforms for Scaled and fixed – size* models of Table 2

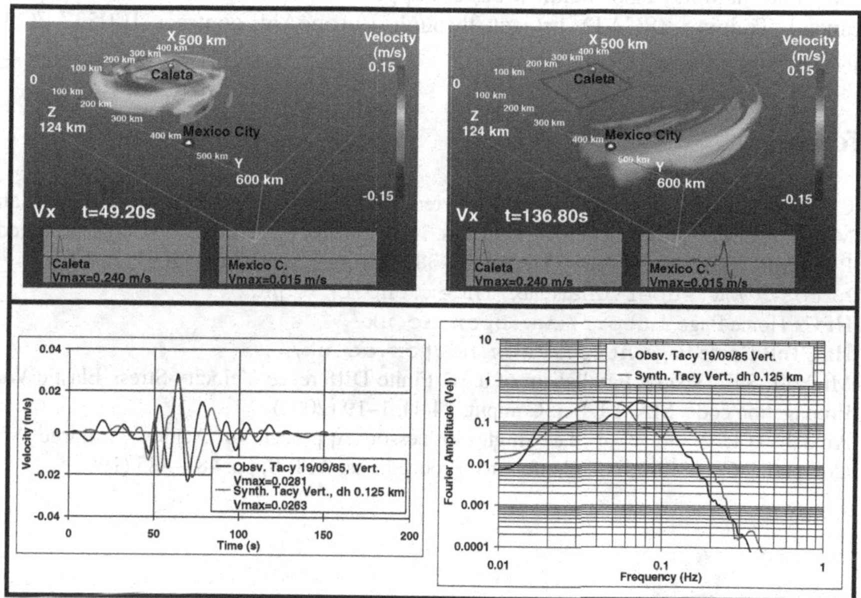

Fig. 3. Top) 3D Snapshots of the velocity wavefield in the X direction of propagation for t = 49.2 and 136.8 s for the 1985 Mexico earthquake; Bottom) Left side observed and synthetic seismograms at Mexico City, right side Fourier amplitude spectra

Conclusions

Benchmark studies were carried out on a recently optimized seismic wave propa-gation 3D, parallel MPI finite difference code that uses 2nd order operators in time and 4th order in space on a staggered grid, 3DFD. Three dual-core supercomputer platforms were used to test the program. Efficiencies of 0.91 and 0.48 with 1024 cores were obtained for the HECToR (UK) and KanBalam (Mexico) machines, respectively, and of 0.66 for

8192 cores for HECToR. In order to improve its performance, probably, Non-blocking MPI communications should be incorporated in a future version of the code. The agreement between the observed and the synthetic velocity seismograms obtained with 3DFD and a dh = 0.125 km [1], is reasonable, both in time and in frequency domains. The 3D velocity field patterns from a simulation of the 1985 Mexico earthquake (which caused the loss of up to 30,000 people and about 7 billion US dollars), show large amplitude, coherent, well developed surface waves, propagating towards Mexico City.

Acknowledgments

We would like to thank the support of Genevieve Lucet, José Luis Gordillo, Hector Cuevas, the supercomputing staff and Marco Ambriz, of DGSCA, and the Institute of Engineering, UNAM, respectively. We acknowledge DGSCA, UNAM for the support to use KanBalam, as well as the STFC Daresbury Laboratory to use HECToR and HPCx. The authors also acknowledge support from the Scientific Computing Advanced Training (SCAT) project through EuropeAid contract II-0537-FC-FA (http://www.scat-alfa.eu).

References

[1] Cabrera, E., Chavez, M., Madariaga, R., Perea, N., Frisenda, M.: 3D Parallel Elastodynamic Modeling of Large Subduction Earthquakes. In: Cappello, F., Herault, T., Dongarra, J. (eds.) PVM/MPI 2007. LNCS, vol. 4757, pp. 373–380. Springer, Heidelberg (2007)

[2] http://www.super.unam.mx/index.php?op=eqhw

[3] HPCx Home Page, http://www.hpcx.ac.uk/

[4] HECToR Home Page, http://www.hector.ac.uk/

[5] Minkoff, S.E.: Spatial Parallelism of a 3D Finite Difference Velocity-Stress Elastic Wave Propagation code. SIAM J. Sci. Comput. 24(1), 1–19 (2002)

[6] Amdahl, G.: Validity of the Single Processor Approach to Achieving Large Scale Computing Capabilities. In: Conference Proceedings, AFIPS, pp. 483–485 (1967)

Vis-OOMPI: Visual Tool for Automatic Code Generation Based on C++/OOMPI

Chantana Phongpensri (Chantrapornchai) and Thanarat Rungthong

Department of Computing, Faculty of Science
Silpakorn University, Thailand
ctana@su.ac.th

Abstract. This work proposes a visual tool called , Vis-OOMPI, which targets to help programmers to write parallel programs. The platform OOMPI is considered in the tool. Since MPI is one of popular APIs in C which are used in parallel programs for message passing, the object concept is incorporated and then becomes OOMPI [5]. The tool helps the beginners in message passing programming and includes some additional concepts with OO supports which is exhibited in OOMPI.

Keywords: visual parallel programming, OOMPI, MPI, message-passing programming interface.

1 Introduction

Parallel programs are becoming popular in many computational sciences. Many parallel programs are developed based on C/Fortran and using MPI. However, it always takes time to learn the concept of parallel program development.

In this paper, we are interested in the object paradigms. We focus on the visual tool that encapsulates object message passing scheme. The library is written based on MPI and uses a class concept from C++. It has been shown in the literatures that the performance of OOMPI is comparable to others. The proposed tool will be suitable for beginners in OOMPI/C++ and to demonstrate the object message passing concept.

There are many existing works in visual parallel tools. Brown et. al. also developed a framework called HeNCE [2]. The framework is graph-based and generated the code based on HeNCE compiler and PVM. MPI-Delphi [1] is another variation of MPI. It is a framework which is developed to imitate the object Pascal programming in Delphi. Also, there are many existing works on object-based message passing interface such as JOPI, jmpi, etc[3,4]. These works rely on Java language.

2 Vis-OOMPI IDE and Code Generation

In Vis-OOMPI, the user specifies the processes and messages sent using a convention of UML (sequence diagram). A user specifies a model of communication on their owns. There is a support for group communication and communicator in the tool.

A. Lastovetsky et al. (Eds.): EuroPVM/MPI 2008, LNCS 5205, pp. 311–312, 2008.

There are two types of communications: peer-to-peer message and group message. The peer-to-peer message is generated by Send/Receive APIs among processes. Processes with the matching send and receive models are both constrained by object message type, communicator, and tag. In this way, the programmer errors due to the wrong send and receive parameters are reduced. For the group communication, the examples are Bcast,Gather/Scatter,Reduce APIs. For these, the users need to specify the communicator/group to use with. The generated code inserted some necessary code in MPI for building communicator/group and then in OOMPI. Several complexities in the tool interface are : 1) to build a new data type as class. 2) to create communicator and group. Since OOMPI sends a message as object, our object type is according to the OOMPI, primitive type, array, user-defined object. Since OOMPI uses a communicator as a port to communicate between processes, a communicator is important. The default one is always the built-in, OOMPI_COMM_WORLD. Since the rank of a process is based on a communicator, when a process uses many communicators, it may have many ranks. Then the user has an option to specify which communicator is used for each communication. Also, the case becomes complicated when the process is in many groups which use many communicators. For code generation, we insert an initialization part which is about OOMPI initialization as well as defining user variables and temporary variables from the tool. Then the code for building communicator/group is automatically generated. After that, the portion of the code is generated according to each process. For each process, we generate the code according to the timeline based on the data property for each model for each communicator. Then the finalization of the code is generated. The code can be saved as text files for other modification. If the user changes the communication model, the code can be regenerated.

Acknowledgments. This work is partially supported by Faculty of Science Funding, Silpakorn University and Silpakorn University Research and Development Institute, Thailand.

References

1. Acacio, M., López-de-Teruel, P.E., García, J.M., Cánovas, O.: MPI-Delphi: an MPI implementation for visual programming environments and heterogeneous computing. Journal of Future Generation Computer Systems 18(3), 317–333 (2002)
2. Browne, J.C., et al.: Visual programming and debugging for parallel computing. IEEE Paralell and Distributed Computing 3(1), 75–83 (1995)
3. Nader, M., et al.: JOPI: a Java object-passing interface. In: Proceedings of 2002 joint ACM-ISCOPE conference on Java Grande, pp. 37–45 (2002)
4. Kivanc, D.: Jmpi and a Performance Instrumentation Analysis and Visualization Tool for jmpi. In: First UK Workshop on Java for High Performance Network Computing, EUROPAR 1998, Southampton, UK, September 2-3 (1998)
5. Squyres Jeffrey, M., Willcock, J., McCandless, B.C., Rijks, P.W., Lumsdaine, A.: Object Oriented MPI (OOMPI):A C++ Class Library for MPI, User Guide

A Framework for Deploying Self-predefined MPI Communicators and Attributes

Carsten Clauss, Boris Bierbaum, Stefan Lankes, and Thomas Bemmerl

Chair for Operating Systems, RWTH Aachen University, Germany
{clauss,lankes,bemmerl}@lfbs.rwth-aachen.de
http://www.lfbs.rwth-aachen.de/content/research

Abstract. In this contribution, we present a new library (that can be used in addition to any MPI library) which allows writing optimized applications for heterogeneous environments without the need of relying on intrinsic adaptation features provided by a particular MPI implementation. This is achieved by giving the application programmer the ability to introduce new self-predefined attributes and communicators whose semantics can be adjusted to each respective environment. Furthermore, we have also developed an additional tool set that form a framework around this library in order to simplify such an adjustment for the user.

Keywords: MPI communicators/attributes, heterogeneity-aware MPI, optimization features, topology analyzer, graphical user interface, XML.

1 Introduction

An important feature of the Message Passing Interface is the communicator concept. This concept allows the application programmer to group the parallel processes by assigning them to abstract objects called *communicators*. For that purpose, the programmer can split the group of initial started processes into subgroups, each forming a new self-contained communication domain. This concept usually follows a *top-down* approach where the process groups are built according to the communication patterns required by the parallelized algorithm.

In contrast to common MPI implementations, heterogeneity-aware MPI libraries often provide special adaptation features which help the application programmer to adapt the algorithms' communication patterns to the respective heterogeneity of the physical communication topology (see, e.g., [1,2]). The implementation of such optimization features normally follows a *bottom-up* method where the topology information must be passed from the MPI runtime system to the application in a more or less unconventional way. One possible way is to provide this information to the application programmer in terms of additional predefined MPI communicators with group affiliations that try to reflect the respective heterogeneity. Another possible way is to pass the needed environmental information to the application level by means of communicator-attached *attributes*. In fact, even the standard defines a set of such predefined attributes that describe the actual execution environment of an MPI job. Furthermore, the

A. Lastovetsky et al. (Eds.): EuroPVM/MPI 2008, LNCS 5205, pp. 313–315, 2008.

standard also suggests that vendors may add their own implementation specific attributes to this set [3]. However, if the symbol names of those additional attributes and/or communicators are already set at compile time of the respective MPI implementation, an application gets tied to this library when using those symbols and hence makes its source code less portable.

2 An Additional Library

For that reason, we have developed a new library (that can be used in addition to any MPI library) which allows writing optimized applications without the need of relying on intrinsic adaptation features provided by a particular MPI implementation. This is achieved by giving the application programmer the ability to introduce new self-predefined attributes and communicators whose semantics can be adjusted to each respective environment. For that purpose, the library provides a special function that expect a reference to an uninitialized MPI communicator. By means of the respective communicator name, the function can then look up a table containing the descriptions of all desired predefined communicators. If such an inquired communicator can be found in the table, each calling process checks whether it takes part in this predefined communicator or not. This identification can be done by means of comparing the own *processor name* against a list of names associated with that communicator. Afterwards, the included processes will build the inquired communicator and will attach all related attributes stated in the description table, too. That way, the actual hardware topology can become visible also at application level.

In order to supply the building function with the needed information stated in the lookup table, an additional initialization function must firstly read the desired communicator and attribute configurations from an XML file. For illustration, consider the following example which may be written for a coupled-code simulation on a hierarchical system consisting of two linked clusters:

```
<comm name="MPI_COMM_OCEAN">
     <processor name="clusterA:[0-7]>
          <attribute key="CPU_SPEED" value="2.2" />
     </processor>
</comm>
<comm name="MPI_COMM_ATMOSPHERE">
     <processor name="clusterB:[0-3]>
          <attribute key="CPU_SPEED" value="2.0" />
     </processor>
     <processor name="clusterB:[4-7]>
          <attribute key="CPU_SPEED" value="1.8" />
     </processor>
</comm>
<intercomm name="MPI_COMM_INTER">
      <first>MPI_COMM_OCEAN</first>
     <second>MPI_COMM_ATMOSPHERE</second>
</intercomm>
```

This example assumes that the node names on the two cluster sites are composed in a simple `cluster_name:node_number` manner, so that the site affiliation can then be determined by using regular expressions. As one can see, attribute keys and values are stated as strings in the configuration file, whereas MPI uses integers and void-pointers for that purpose. Therefore, it is a task of the presented library to assure a proper key allocation and assignment, while the return value of MPI's *attribute-get* function can then just be interpreted as a pointer to a string. The displacement of these configurations into an external file offers several advantages and opportunities: For example, the application does not need to be recompiled if the desired grouping scheme or the attribute values have changed. In addition, also the user can easily employ this mechanism in order to pass job- or environment-dependent parameters to the application. Furthermore, the configuration needs not necessarily be written by a user or an application programmer. In fact, the XML file containing the communicator and attribute definitions can rather be generated, for example, by a process scheduler, by a topology analyzing tool or even by the MPI runtime environment itself.

3 The Framework

Actually, we have already developed such a topology analyzer being capable of exploring a heterogeneous system and storing the required information in XML-coded files. Although such an analyzer can gather information representing a system's hierarchy, the gap between this hardware topology description and the logical communication patterns governed by the respective application must still be closed by the user. Therefore, we have also developed a graphical frontend tool that helps the user to adapt such automatic generated configurations to the actual needs of an optimized application. For that purpose, the tool visualizes the initially explored hardware topology and allows the user to map groups of processes in terms of new MPI communicators onto the given hardware structure. This mapping and the attachment of additional attributes is supported by the graphical frontend in a simple *drag-and-drop* manner. Afterwards, the tool generates a suitable configuration file for the presented library. That way, this tool chain can facilitate the deployment of such self-predefined MPI communicators and attributes within optimized applications for heterogeneous systems.

References

1. Karonis, N., Toonen, B., Foster, I.: MPICH-G2: A Grid-Enabled Implementation of the Message Passing Interface. Journal of Parallel and Distributed Computing 63(5), 551–563 (2003)
2. Pöppe, M., Schuch, S., Bemmerl, T.: A Message Passing Interface Library for Inhomogeneous Coupled Clusters. In: Proc. of the 17th IEEE International Parallel and Distributed Processing Symposium, Nice, France, April 2003, p. 199 (2003)
3. Snir, M., Otto, W., Huss-Lederman, S., Walker, D.W., Dongarra, J.: MPI: The Complete Reference, Scientific and Engineering Computation Series, p. 287. MIT Press, Cambridge (1996)

A Framework for Proving Correctness of Adjoint Message-Passing Programs

Uwe Naumann[1], Laurent Hascoët[2], Chris Hill[3], Paul Hovland[4], Jan Riehme[5], and Jean Utke[4]

[1] Corresponding Author: LuFG Informatik 12 (Software and Tools for Computational Engineering), Department of Computer Science RWTH Aachen University, 52056 Aachen, Germany
naumann@stce.rwth-aachen.de
http://www.stce.rwth-aachen.de

[2] Projet TROPICS, INRIA Sophia-Antipolis, France

[3] Department of Earth, Atmospheric, and Planetary Sciences, Massachusetts Institute of Technology, Cambridge, MA, USA

[4] Mathematics and Computer Science Division, Argonne National Laboratory, Argonne, IL, USA

[5] Department of Computer Science, University of Hertfordshire, Hatfield, UK

Abstract. We propose a technique for proving correctness of adjoint message passing programs that relies on data dependences in partitioned global address space. As an example we discuss asynchronous unbuffered send/receive using MPI.

1 Adjoint Numerical Programs

Numerical simulation and optimization in computational science and engineering have gained significant importance over the past few decades. For example, our ability to understand physical, chemical, and biological processes has improved with the increased power of computational resources as well as with the deepened insight into mathematical and algorithmic issues. Numerical simulation programs map n *independent* inputs onto m *dependent* outputs (also referred to as the objectives). Often n is very large in comparison to m. The classical numerical approach to quantifying the sensitivities of those objectives with respect to the inputs through finite-difference quotients yields a computational complexity of $O(n)$. Note that certain high-end applications such as the simulation of ocean circulation [11] may have a runtime of several days to produce physically relevant results on the latest high-performance computing platforms. The number of independent inputs may reach values of the order of $n = 10^9$. Hence, forward sensitivity analysis requiring n runs of the simulation program is simply not feasible.

Adjoint methods and corresponding program transformation techniques have been developed to replace the dependence on n with that on the number of objectives m. If $m = 1$, then adjoint programs deliver the sensitivities of the objective with respect to all independent inputs at $O(1)$. Adjoint programs can

A. Lastovetsky et al. (Eds.): EuroPVM/MPI 2008, LNCS 5205, pp. 316–321, 2008.

be generated from a given numerical simulation program by a semantic program transformation technique known as *automatic differentiation* (AD) [8]. A large number of successful applications of AD to real-world problems in science and engineering have been reported in the literature. Refer, for example, to [2].

Adjoint numerical programs consist of two parts. The *augmented forward section* is an instrumented version of the original program containing statements to memorize certain intermediate values that are required for the correct (and efficient) evaluation of the adjoint program variables. The *reverse section* propagates values of adjoint program variables in the opposite direction of the original data flow. Optimal data-flow reversal is NP-complete [12,13]. It involves the reversing of the flow of control (which implies reversing the order of the statements within basic blocks) and generating the corresponding adjoint statements. Proofs of correctness of sequential adjoint programs are based on the chain rule of differential calculus and, in particular, on its associativity. Refer to [8] for a comprehensive discussion of the mathematical foundations of adjoint programs.

This paper is motivated by the need for automatically generated adjoint versions of parallel programs that use message passing. Related work comprises [3,4,7,9,10,16]. We describe a proof technique that allows us to show the correctness equivalent adjoint versions can be generated for a given message-passing program. As developers of adjoint code compilers, we consider the scenario of a given transformation algorithm that needs to be proved right or wrong in the sense that correct adjoints are computed for arbitrary inputs.

2 Correctness of Adjoint Communication Patterns

We consider the *partitioned global address space (PGAS)* [5] version P_s of a message-passing program P involving n processes $p_1,\ldots,p_n$. In order for P_s to operate on the union of the n memory spaces all program variables are augmented with an additional dimension of length n. Communications are translated into *x-assignments* between augmented program variables belonging to disjoint address spaces. Auxiliary variables are introduced for buffered communication. Barriers in asynchronous communication yield a set of PGAS versions for a given message-passing program.

2.1 Example

The program

s_0

if $(myrank == 1)$ isend(a, r); s_1; if $(myrank == 2)$ irecv(b, r); s_2; wait(r)

s_3

with unspecified sequences of statements s_i for $i = 0,\ldots,3$ yields the following set of constraints for the placement of the x-assignment χ:

$$s_0^1 < \chi; \quad s_1^2 < \chi; \quad \chi < s_3^1; \quad \chi < s_3^2.$$

These constraints lead to the following six PGAS codes:

$$s_0;\ s_1^2;\ b^2 = a^1;\ s_1^1;\ s_2;\ s_3$$
$$s_0;\ s_1;\ b^2 = a^1;\ s_2;\ s_3$$
$$s_0;\ s_1;\ s_2^2;\ b^2 = a^1;\ s_2^1;\ s_3$$
$$s_0;\ s_1;\ s_2^1;\ b^2 = a^1;\ s_2^2;\ s_3$$
$$s_0;\ s_1^2;\ s_2^2;\ b^2 = a^1;\ s_1^1;\ s_2^1;\ s_3$$
$$s_0;\ s_1;\ s_2;\ b^2 = a^1;\ s_3$$

The statements executed in section i by processor j are denoted by s_i^j. In this example we assume two processors. Note that $(s_i^1;\ s_i^2) = (s_i^2;\ s_i^1)$ as a result of the disjoint address spaces. Hence, the PGAS code $s_i;\ s_{i+1}$ yields the following six semantically equivalent sequential codes:

$$s_i^1;\ s_{i+1}^1;\ s_i^2;\ s_{i+1}^2 \qquad s_i^2;\ s_{i+1}^2;\ s_i^1;\ s_{i+1}^1$$
$$s_i^1;\ s_i^2;\ s_{i+1}^1;\ s_{i+1}^2 \qquad s_i^1;\ s_i^2;\ s_{i+1}^2;\ s_{i+1}^1$$
$$s_i^2;\ s_i^1;\ s_{i+1}^1;\ s_{i+1}^2 \qquad s_i^2;\ s_i^1;\ s_{i+1}^2;\ s_{i+1}^1$$

The partial order of the statements is induced by $s_i^j < s_{i+1}^j$. Any two statements from s_i^j and s_{i+1}^k can be executed in arbitrary order for $j \neq k$. Further combinations resulting from feasible (wrt. data dependence) switches of the x-assignment and statements in certain s_i^j lead to an exponential number of possible actual execution orders that need to be taken into account when proving properties of PGAS programs. For this example we observe that the original program must satisfy the restriction for *isend* that a^1 is not written by s_1 nor s_2.[1] Similarly, for *irecv* it must satisfy that b^2 is neither read nor written by s_2.

To prove the correctness of an adjoint of a message-passing program, we need to show that its adjoint PGAS versions are semantically equivalent to the PGAS versions of its adjoint. We do so by looking at all possible actual execution orders.

2.2 Case Study: Asynchronous Unbuffered Send/Receive

We present here a case study to illustrate the use of the proposed formalism. Similar proofs are required for a large number of communication patterns. We are analyzing all communication patterns used by our main target applications, including MITgcm (`mitgcm.org`) as well as ICON (`www.icon.enes.org`).

Proposition: *Let P be a message-passing program involving processes p_1 and p_2, and let the integer variable myrank contain the respective process identifiers. The communication pattern*

[1] a^1 must not be written by s_1^1 nor s_2^1. It is not written by s_1^2 nor s_2^2 due to the separate address spaces.

$$s_{i-1};\ \text{if } (myrank == 1)\ \text{isend}(a, r);\ s_{i+1}$$
$$\ldots$$
$$s_{j-1};\ \text{if } (myrank == 2)\ \text{irecv}(b, r);\ s_{j+1}$$
$$\ldots$$
$$s_{k-1};\ \text{wait}(r);\ s_{k+1}$$

in the forward section of the adjoint code yields

$$\begin{array}{l} \bar{s}_{k+1} \\ \text{if } (myrank == 2)\ \text{isend}(\bar{b}, r) \\ \text{if } (myrank == 1)\ \text{irecv}(t, r) \\ \bar{s}_{k-1} \\ \ldots \\ \bar{s}_{j+1};\ \text{if } (myrank == 2)\ \text{wait}(r);\ \bar{b} = 0;\ \bar{s}_{j-1} \\ \ldots \\ \bar{s}_{i+1};\ \text{if } (myrank == 1)\ \text{wait}(r);\ \bar{a}\mathrel{+}=t;\ \bar{s}_{i-1} \end{array}$$

in the reverse section, where $\bar{s}_k$ are the adjoint statements corresponding to s_k.

Proof. The forward PGAS codes are given as

$$\begin{array}{l} s_{i-1};\ s_{i+1};\ \ldots\ s_{j-1};\ s^2_{j+1};\ b^2 = a^1;\ s^1_{j+1};\ \ldots\ s_{k-1};\ s_{k+1} \\ s_{i-1};\ s_{i+1};\ \ldots\ s_{j-1};\ s_{j+1};\ b^2 = a^1;\ \ldots\ s_{k-1};\ s_{k+1} \\ \ldots \\ s_{i-1};\ s_{i+1};\ \ldots\ s_{j-1};\ s_{j+1};\ \ldots\ s_{k-1};\ b^2 = a^1;\ s_{k+1} \end{array}$$

The reverse sections of the adjoint PGAS codes become

$$\begin{array}{l} \bar{s}_{k+1};\ \bar{s}_{k-1};\ \ldots\ \bar{s}^1_{j+1};\ \bar{a}^1\mathrel{+}=\bar{b}^2;\ \bar{b}^2 = 0;\ \bar{s}^2_{j+1};\ \bar{s}_{j-1};\ \ldots\ \bar{s}_{i+1};\ \bar{s}_{i-1} \\ \bar{s}_{k+1};\ \bar{s}_{k-1};\ \ldots\ \bar{a}^1\mathrel{+}=\bar{b}^2;\ \bar{b}^2 = 0;\ \bar{s}_{j+1};\ \bar{s}_{j-1};\ \ldots\ \bar{s}_{i+1};\ \bar{s}_{i-1} \\ \ldots \\ \bar{s}_{k+1};\ \bar{a}^1\mathrel{+}=\bar{b}^2;\ \bar{b}^2 = 0;\ \bar{s}_{k-1};\ \ldots\ \bar{s}_{j+1};\ \bar{s}_{j-1};\ \ldots\ \bar{s}_{i+1};\ \bar{s}_{i-1} \end{array}$$

The variable a^1 is not written by any of the statements in $s_{i+1};\ \ldots\ s_{k-1}$ because the original message-passing program is assumed to satisfy the restrictions on *isend*. Similarly, b^2 is neither read nor written by $s_{j+1};\ \ldots\ s_{k-1}$. However, the value of a^1 may be read by statements in $s_{i+1};\ \ldots\ s_{k-1}$, implying that while $\bar{a}^1$ may be incremented by $\bar{s}_{k-1};\ \ldots\ \bar{s}_{i+1}$, it is not read or written otherwise. The order of two successive increment operations can be switched if the incremented variable is neither read nor written in between the two increment operations.[2]

[2] For a given use of a variable we distinguish between reads, writes, and increment operations as a special case of a read-write combination.

Moreover, the placement of these increment operations is arbitrary as long as the values of the increments do not change. The value of $\bar{b}^2$ is neither read nor written by $\bar{s}_{k-1}; \ldots \bar{s}_{j+1}$. Hence, the statement $\bar{a}^1$+=$\bar{b}^2$ can be inserted at any position between $\bar{s}_{k+1}$ and $\bar{s}_{j-1}$. In other words, the adjoints of all PGAS versions of the given message-passing program are equivalent.

The adjoint message passing program yields the following set of constraints for the placement of the adjoint x-assignment $\bar{\chi} \equiv$ "$t = \bar{b}^2$" :

$$\bar{s}^1_{k+1} < \bar{\chi}; \quad \bar{s}^2_{k+1} < \bar{\chi}; \quad \bar{s}^2_{j-1} > \bar{\chi}; \quad \bar{s}^1_{i-1} > \bar{\chi} \quad .$$

Hence, the PGAS versions of the adjoint message-passing program are the following:

$$\bar{s}_{k+1};\ \bar{s}_{k-1};\ \ldots\ \bar{s}_{j+1};\ t = \bar{b}^2;\ \bar{b}^2 = 0;\ \bar{s}_{j-1};\ \ldots\ \bar{s}_{i+1};\ \bar{a}^1\text{+=}t;\ \bar{s}_{i-1}$$

$$\ldots$$

$$\bar{s}_{k+1};\ t = \bar{b}^2;\ \bar{s}_{k-1};\ \ldots\ \bar{s}_{j+1};\ \bar{b}^2 = 0;\ \bar{s}_{j-1};\ \ldots\ \bar{s}_{i+1};\ \bar{a}^1\text{+=}t;\ \bar{s}_{i-1}$$

As a compiler-generated auxiliary variable, t can be guaranteed not to be read or written by any of the statements $\bar{s}_{k-1}; \ldots \bar{s}_{i+1}$. From our previous argument we recall that $\bar{a}^1$ may be incremented by $\bar{s}_{k-1}; \ldots \bar{s}_{i+1}$ but it is not read or written otherwise. Hence, the increment operation of $\bar{a}^1$ with t can be placed in between $\bar{s}_{i+1}$ and $\bar{s}_{i-1}$. As the value of $\bar{b}^2$ is neither read nor written by $\bar{s}_{k-1}; \ldots \bar{s}_{j+1}$, the fixed placement of $\bar{b}^2 = 0$ in between $\bar{s}_{j+1}$ and $\bar{s}_{j-1}$ does not change the program's semantics either. The auxiliary variable t can be removed as the result of copy-propagation [1], yielding a possible replacement of the first assignment in $t = \bar{b}^2; \ldots \bar{a}^1$+=$t$ with $\bar{a}^1$+=$\bar{b}^2$. Consequently, the adjoint PGAS versions of the message-passing program are semantically equivalent to the PGAS versions of the adjoint message-passing program. ■

3 Conclusion and Outlook

A formalism for proving the correctness of adjoint message-passing programs has been illustrated by means of an asynchronous unbuffered send/receive communication between two processes. This method is applied to a large number of transformation rules currently being implemented in OpenAD [17] and the differentiation-enabled NAGWare Fortran compiler [14]. It is based on analyzing the data dependences in the PGAS versions of the original message-passing program. Rigorous proofs can thus be constructed that rely only on program analysis techniques used in classical compiler construction. We intend to consider ideas presented in [15] in order to investigate a potential automatization of this proof technique.

One of our long-term goals is to build an adjoint message-passing library on top of MPI. Such an extension is desirable for achieving satisfactory efficiency. The ability to prove the correctness of given communication patterns is a fundamental ingredient of this ambitious research and development project.

References

1. Aho, A., Sethi, R., Ullman, J.: Compilers. Principles, Techniques, and Tools. Addison-Wesley, Reading (1986)
2. Bischof, C., Bücker, M., Hovland, P., Naumann, U., Utke, J. (eds.): Advances in Automatic Differentiation. LNCSE, Berlin. Springer, Heidelberg (2008)
3. Carle, A., Fagan, M.: Automatically differentiating MPI-1 datatypes: The complete story. In: [6] ch. 25, pp. 215–222. Springer, Heildberg (2002)
4. Faure, C., Dutto, P.: Extension of Odyssée to the MPI library – reverse mode. Rapport de recherche 3774, INRIA, Sophia Antipolis (October 1999)
5. Coarfa, C., Dotsenko, Y., Mellor-Crummey, J., Cantonnet, F., El-Ghazawi, T., Mohanti, A., Yao, Y., Chavarría-Miranda, D.: An evaluation of global address space languages: co-array fortran and unified parallel c. In: PPoPP 2005: Proceedings of the tenth ACM SIGPLAN symposium on Principles and practice of parallel programming, pp. 36–47. ACM, New York (2005)
6. Corliss, G., Faure, C., Griewank, A., Hascoët, L., Naumann, U. (eds.): Automatic Differentiation of Algorithms – From Simulation to Optimization. Springer, New York (2002)
7. Faure, C., Dutto, P., Fidanova, S.: Odysée and parallelism: Extension and validation. In: Procceedings of the 3rd European Conference on Numerical Mathematics and Advanced Applications, Jyväskylä, Finland, July 26-30, 1999, pp. 478–485. World Scientific, Singapore (2000)
8. Griewank, A.: Evaluating Derivatives: Principles and Techniques of Algorithmic Differentiation. SIAM, Philadelphia (2000)
9. Heimbach, P., Hill, C., Giering, R.: Automatic generation of efficient adjoint code for a parallel Navier-Stokes solver. In: Sloot, P.M.A., Tan, C.J.K., Dongarra, J., Hoekstra, A.G. (eds.) ICCS-ComputSci 2002. LNCS, vol. 2330, pp. 1019–1028. Springer, Heidelberg (2002)
10. Hovland, P., Bischof, C.: Automatic differentiation of message-passing parallel programs. In: Proceedings of the First Merged International Parallel Processing Symposium and Symposium on Parallel and Distributed Processing, pp. 98–104. IEEE Computer Society Press, Los Alamitos (1998)
11. Marotzke, J., Giering, R., Zhang, K., Stammer, D., Hill, C., Lee, T.: Construction of the adjoint MIT ocean general circulation model and application to Atlantic heat transport variability. J. Geophysical Research 104, C12:29, 529, 547 (1999)
12. Naumann, U.: Call tree reversal is NP-complete. In: [2] (to appear, 2008)
13. Naumann, U.: DAG reversal is NP-complete. J. Discr. Alg. (to appear, 2008)
14. Naumann, U., Riehme, J.: A differentiation-enabled Fortran 95 compiler. ACM Transactions on Mathematical Software 31(4), 458–474 (2005)
15. Shasha, D., Snir, M.: Efficient and correct execution of parallel programs that share memory. ACM Trans. Program. Lang. Syst. 10(2), 282–312 (1988)
16. Mills Strout, M., Kreaseck, B., Hovland, P.: Data-flow analysis for MPI programs. In: ICPP 2006: Proceedings of the 2006 International Conference on Parallel Processing, pp. 175–184. IEEE Computer Society Press, Washington (2006)
17. Utke, J., Naumann, U., Wunsch, C., Hill, C., Heimbach, P., Fagan, M., Tallent, N., Strout, M.: OpenAD/F: A modular, open-source tool for automatic differentiation of Fortran codes. ACM Transactions on Mathematical Software 34(4) (2008)

A Compact Computing Environment for a Windows Cluster: Giving Hints and Assisting Job Execution

Yuichi Tsujita, Takuya Maruyama*, and Yuhei Onishi**

Department of Electronic Engineering and Computer Science,
School of Engineering, Kinki University
1 Umenobe, Takaya, Higashi-Hiroshima, Hiroshima 739-2116, Japan
tsujita@hiro.kindai.ac.jp

Windows Compute Cluster Server 2003 [1] (hereafter Windows CCS) has been focused to build and utilize a PC cluster with high availability and a Windows graphical user interface (GUI). The Windows CCS consists of an operating system and an add-on software toolkit. The former is Windows Server 2003, Compute Cluster Edition and the latter is Compute Cluster Pack. The Compute Cluster Pack provides many useful software tools such as a well-tuned MPI [2] library named MS-MPI [1] and a job scheduler.

Even in remote accesses to the Windows CCS system, the software tools are available with a Windows GUI by using a Windows Remote Desktop. However, most of the application users do not expect to encompass the expertises for the Windows CCS system. As a result, such the users may face difficulties and make failures in operations of their programs because of complexity in selecting and utilizing essential functionalities of the system. It is also remarked that there is not any support to check standard output/error of a running job in a GUI menu named Compute Cluster Job Manager of the job scheduler. Furthermore, normal users can not utilize Compute Cluster Administrator of the scheduler to indicate, e.g., the number of available CPUs because it is available for only administrative users. Instead, a command line interface is available for such the users, however, it is too complicated for non-expert users.

To remove such the difficulties, we have focused into building a compact and user-friendly computing environment which is available on a client PC. The main objective of this system is to give hints for job creations and to assist job executions on a remote Windows CCS system through a simple interface. To provide a user-friendly interface for non-expert users, we consider that a GUI interface is essential. It is also remarked that portability is an important issue. We selected Java to develop the computing environment with regarding to these requirements. In this implementation, we have adopted Java SE Development Kit 6 [3]. This system hides complexity of job execution on a remote Windows CCS system for the usability issues.

* Present address: Fujitsu Business Systems Ltd.

** Present address: Hitachi Business Solution Co., Ltd.

A. Lastovetsky et al. (Eds.): EuroPVM/MPI 2008, LNCS 5205, pp. 322–323, 2008.

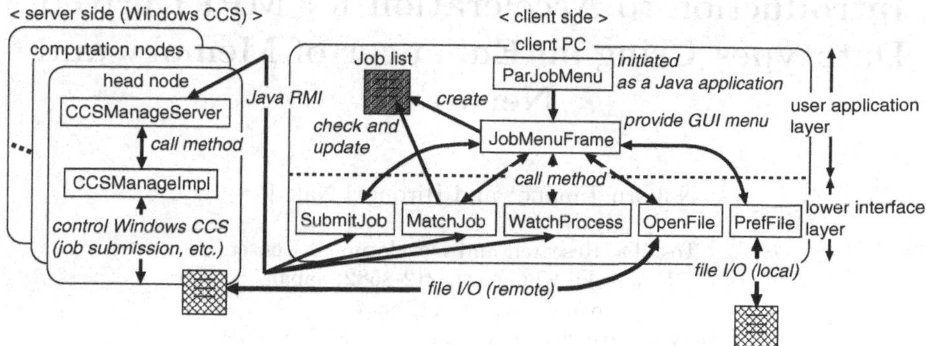

Fig. 1. Organization of developed Java classes

The system consists of two parts; one is a server side part which runs on a head node of a PC cluster and the other is a client side part which runs on a client PC as shown in Figure 1. Data communications between server and client parts are carried out by using Java's Remote Method Invocation (RMI) over a network in the current implementation. This system is available in a Windows domain provided by an Active Directory server. Users are expected to have their user accounts in the domain. Currently we do not have any user authentication to establish the connections.

The client side has seven Java classes to provide GUI menus and realize cooperations with the server side. The Java classes are separated into two layers; one is a user application layer and the other is a lower interface layer. The former has GUI menus for job creation, job submission, and monitoring a PC cluster and jobs. While the latter provides control methods which send Java RMI requests to the server side on the Windows CCS.

On the other hand, the server side part has a role to receive Java RMI requests for job submission and so on from a client side application. The RMI server carries out operations for the Windows CCS by using Windows CCS command line interface with the help of a Java's Runtime class.

With this computing environment, users can easily submit their jobs and monitor job status without deep understandings about the Windows CCS. As a future work, we would like to rebuild this system to be independent of a Windows CCS system in the client side by encapsulating its infrastructure-dependent parts in the server side program code. We would like to also prepare a Java's jar file by collecting commonly used Java classes.

References

1. Russel, C.: Overview of Microsoft Windows Compute Cluster Server 2003. White paper, Microsoft Corporation (November 2005)
2. MPI Forum, `http://www.mpi-forum.org/`
3. Java Technology, `http://java.sun.com/`

Introduction to Acceleration for MPI Derived Datatypes Using an Enhancer of Memory and Network

Noboru Tanabe[1] and Hironori Nakajo[2]

[1] Toshiba, Research and Development Center,
Kawasaki, Kanagawa 212-8582, Japan
noboru.tanabe@toshiba.co.jp

[2] Tokyo University of Agriculture and Technology,
Koganei, Tokyo 184-8588, Japan
nakajo@cc.tuat.ac.jp

Abstract. We present a support function for MPI derived datatypes on an Enhancer of Memory and Network named DIMMnet-3 which is under development. Semi-hardwired derived datatype communication based on RDMA with hardwired gather and scatter is proposed. This mechanism and MPI using it are implemented on DIMMnet-2 which is a former prototype. The performance of gather or scatter transfer of 8byte elements with large interval by using vector commands of DIMMnet-2 is 6.8 compared with software on a host. Proprietary benchmark of MPI derived datatype communication for transferring a submatrix corresponding to a narrow HALO area is executed. Observed bandwidth on DIMMnet-2 is far higher than that for similar condition with VAPI based MPI implementation on Infiniband, even though poorer CPU and motherboard are used.

MPI provides a powerful mechanism for describing non-contiguous memory locations: derived datatypes. Because derived datatype communications generate non-contiguous memory accesses, their performance has been far lower than that of burst communications.

Acceleration method for derived datatype communications on a vector supercomputer was proposed [1]. However, vector supercomputers are very expensive. Therefore, the acceleration of derived datatype communication on COTS PC cluster is important.

In research at Argonne National Laboratory [2], the performance of derived datatype communication has been improved by selecting optimal packing algorithms with access patterns. However, some big performance degradations remain in some communication patterns. Furthermore, when the packing and unpacking of a message are performed manually or derived datatype communication is carried out by MPI system, a lot of data arranged discontinuously are delivered via a cache on CPU, usually. Consequently, there is a danger that pollution of a cache will have an adverse influence on the performance of the following process.

A. Lastovetsky et al. (Eds.): EuroPVM/MPI 2008, LNCS 5205, pp. 324–325, 2008.

On the other hand, a team at Ohio State University implements MPI based on RDMA with Gather/Scatter functions of Infniband[3]. In this implementation, some processing is offloaded to firmware processing on the Network Interface Card (NIC). However, for the following three reasons, we think there is still a good possibility of accelerating by employing new hardware. (1) The frequency of CPU on the NIC is dozens of times slower than that of host CPU. (2) Since CPU on the NIC is usually based on cache, when carrying out discontinuous access, it wastes the bandwidth of a main memory and an I/O bus by the access in a cache line unit. (3) In the case that the NIC accesses a main memory through an I/O bus, access by short burst length cannot operate efficiently.

The goal of this research is acceleration of MPI derived datatype communication with hardware supports on the NIC. We investigate a hardware support function for MPI on DIMMnet-3 which is an enhancer of memory and network using DIMMnet-2 prototype[4].

In this poster presentation, we introduce overviews of the DIMMnet-2 and DIMMnet-3 which is under development. We show the hardware supports for MPI derived datatype communication implemented on these NICs. We propose a new communication protocol named SDDC (Semi-hardwired Derived Datatype Communication). MPI-2 over DIMMnet-2 is implemented with SDDC.

We also present performance evaluation with DIMMnet-2 employing our proposed mechanisms. The performance gain of gather or scatter transfer of 8byte elements with large interval by using vector commands of DIMMnet-2 is 6.8 compared with software on a host. Proprietary benchmark of MPI derived datatype communication for transferring a sub-matrix corresponding to a narrow HALO area is executed. Observed bandwidth on DIMMnet-2 is far higher than that for similar condition with VAPI based MPI implementation on Infniband, even though poorer CPU and motherboard are used.

Acknowledgment

This work is supported by the Ministry of Internal Affairs and Communications (Soumu-sho).

References

1. Träff, J.L., Hempel, R., Ritzdorf, H., Zimmermann, F.: Flattening on the Fly: Efficient Handling of MPI Derived Datatypes. In: Margalef, T., Dongarra, J., Luque, E. (eds.) PVM/MPI 1999. LNCS, vol. 1697, pp. 109–116. Springer, Heidelberg (1999)
2. Byna, S., Gropp, W., Sun, X., Thaku, R.: Improving the performance of mpi derived datatypes by optimizing memory-access cost. In: International Conference on Cluster Computing (CLUSTER 2003), pp. 412–419 (December 2003)
3. Wu, J., Wyckoff, P., Panda, D.K.: High Performance Implementation of MPI Derived Datatype Communication over InfiniBand. In: 18th International Parallel and Distributed Processing Symposium (IPDPS 2004), pp. 26–30 (April 2004)
4. Tanabe, N., Nakatake, M., Hakozaki, H., Dohi, Y., Nakajo, H., Amano, H.: A New Memory Module for COTS-Based Personal Supercomputing. Innovative Architecture for Future Generation High-Performance Processors and Systems, 40–48 (July 2004)

Efficient Collective Communication Paradigms for Hyperspectral Imaging Algorithms Using HeteroMPI

David Valencia[1,*], Antonio Plaza[1], Vladimir Rychkov[2], and Alexey Lastovetsky[2]

[1] Technology of Computers and Communications Dept.,Technical School of Cáceres University of Extremadura, Avda. de la Universidad S/N, E-10071 Cáceres (Spain)
{davaleco,aplaza}@unex.es

[2] Heterogeneous Computing Laboratory, School of Computer Science and Informatics University College Dublin, Belfield, Dublin 4, Ireland
{vrychkov,alastovetsky}@ucd.ie

Abstract. Most of the parallel strategies used for information extraction in remotely sensed hyperspectral imaging applications have been implemented in the form of parallel algorithms on both homogeneous and heterogeneous networks of computers. In this paper, we develop a study on efficient collective communications based on the usage of HeteroMPI for a parallel heterogeneous hyperspectral imaging algorithm which uses concepts of mathematical morphology.

Keywords: Hyperspectral Imaging Algorithms, HeteroMPI.

1 Introduction

Hyperspectral imaging identifies materials and objects in the air, land and water on the basis of the unique reflectance patterns that result from the interaction of solar energy with the molecular structure of the material[1]. Most applications of this technology require timely responses for swift decisions which depend upon high computing performance of algorithm analysis. Examples include target detection for military and defense/security deployment, urban planning and management, risk/hazard prevention and response including wild-land fire tracking, biological threat detection, monitoring of oil spills and other types of chemical contamination. These images are characterized by covering tens or even hundreds of kilometers long, having hundreds of MB in size. Few consolidated parallel techniques for analyzing this kind of data currently exist in the open literature, and mainly all of them implemented on homogeneous networks of computers using MPI. Although the standard MPI[3] has been widely used to implement parallel algorithms for Heterogeneous Networks of Computers (HNOCs), it does not provide specific means to address some additional challenges posed by these networks, including the distribution of computations and communications unevenly,

* Corresponding author.

A. Lastovetsky et al. (Eds.): EuroPVM/MPI 2008, LNCS 5205, pp. 326–331, 2008.

taking into account the computing power of the heterogeneous processors and the bandwidth of the communications links. To achieve these goals, HeteroMPI was developed as an extension of MPI which allows the programmer to describe the performance model of a parallel algorithm in generic fashion[4], a very useful feature for heterogeneous hyperspectral imaging applications to define distribution of workload and communications, which typically make intensive use of scatter/gather communication operations).

In this paper, our main goal is to study on several approximations for efficient collective communications adapted to the particularities of a heterogeneous hyperspectral image processing scenario already developed using HeteroMPI, basing our developments on the communication model by Lastovetsky et al.[9]. The paper is structured as follows. Section 2 first describes hyperspectral imaging algorithm considered in this study and main features of HeteroMPI. Section 3 explore the different paradigms studied. Finally, section 4 concludes with the experimental results obtained and some remarks and hints at plausible future research.

2 Related Work

Several hyperspectral imaging algorithms have been implemented using MPI as a standard development tool. Examples include the distributed spectral-screening principal component transform algorithm (S-PCT)[6], D-ISODATA[7], a computationally efficient recursive hierarchical image segmentation algorithm hybrid method (called RHSEG)[8], and a morphological approach for classification of hyperspectral images called automated morphological classification (AMC)[10], which takes into account both the spatial and the spectral information in the analysis in a combined fashion. An MPI-based parallel version of AMC has been developed and tested on NASA's Thunderhead cluster[12], showing parallel performance results superior to those achieved by other parallel hyperspectral algorithms in the literature[2]. In particular, this algorithm is the one used in our experiments because it is an exemplar algorithm with the main characteristics of the different hyperspectral imaging existing in the literature. An important limitation in the mentioned parallel techniques is that they assume that the number and location of processing nodes are known and relatively fixed, allowing the use of the standard MPI specification. This approach is feasible when the application is run on a homogeneous distributed-memory computer system. However, selection of a group for execution on HNOCs must take into account the computing power of the heterogeneous processors and the speed/bandwidth of communication links between each processor pair[5]. This feature is of particular importance in applications dominated by large data volumes such as hyperspectral image analysis, but is also quite difficult to accomplish from the viewpoint of the programmer. The main idea of HeteroMPI is to automate and optimize the selection of a group of processes that executes a heterogeneous algorithm faster than any other possible group.

Particularly, HeteroMPI has been used to measure the processing power of each processor in the moment the execution of the heterogeneous algorithm is to be made. To measure this, the directive **HeteroMPI_Recon** has been used along with a benchmark defined to reflect the most important features of the real algorithm in terms of computational cost and to stress and activate the whole memory hierarchy. Then, with directive **Hetero_Group_create** and performance model defined through mpC[11], the best heterogeneous executing group is created, and data is distributed based on the actual processing power available at each node.

3 Communication Patterns

Recently, Lastovetsky et al.[5][9] designed a new model for describing performance of all collective communications that generally take place in parallel MPI applications and, in particular, in those applications executed on heterogeneous clusters based on a switched Ethernet networks. The idea is to model a few simple parameters with point-to-point communication between each pair of nodes on the network, and then use these parameters to build an estimate for collective communications based on a one-to-many and many-to-one pattern. In particular, in this paper we have further studied different solutions to the problem of sending information with different sizes located on the limits of partitions between processes (see Fig. 1(b)), whose size is located on the congestion area predicted by the communication model. The communication paradigms considered are: Chaotic Non-Blocking (CNB), Divided Chaotic Non-Blocking (DCNB) and Subgroup-Based (SB) Communications. CNB is characterized as a naive approximation, with highly balanced computing phase (thanks to the benchmark and directives of HeteroMPI) and the use of non-blocking communication directives for overlapping. DCNB is developed with the idea of coping with the problem of having communications located on the congestion region. In order to evade the congestion region predicted by the model in the network, it is necessary to introduce very complex control code to correctly retrieve the data, also making it completely independent of the particular algorithm, thus only dependent on the parameters of the network and the size of the message passed to the communication framework, posing as a robust algorithm for subdivision of messages and ordered reconstruction upon reception that evades the congestion area. On the other hand, SB is developed with the idea of evade control code and make use of divided messages. Introducing an ordered communication pattern by means of subgroups of processes and collective communications we eliminate the need of control code.

4 Experimental Results

In the present section, we describe the images and heterogeneous cluster used in our studies, along with a comparison of the communication times obtained for the different communication frameworks mentioned before.

4.1 Heterogeneous Cluster and Hyperspectral Image

The heterogeneous cluster used is located in the Heterogeneous Computing Laboratory of the University College Dublin. It is formed by 16 different machines

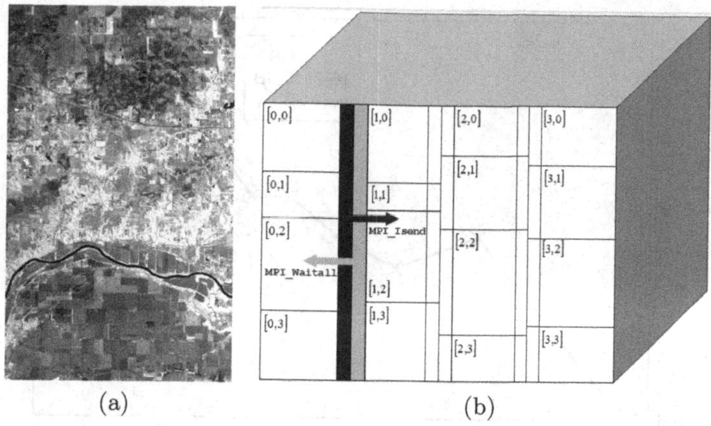

Fig. 1. (a) Spectral band at 587 nm wavelength of an AVIRIS scene comprising agricultural and forest features at Indian Pines, Indiana. (b) Communication of a shared part of the hyperspectral image between neighboring processes.

interconnected by two level 5 Cisco switchs that allows hardware reconfiguration of bandwidth between nodes. The processors are as follows: one IBM x306 3.0GHz AMD processor; two IBM x326 2.2GHz AMD processors; two Dell PowerEdge SC1425 Xeon processors at 3.0GHz and 2.2GHz; 6 Dell PE750 Pentium 3.4GHz processors; 3 HP DL140 Xeon Processors at 2.8GHz, 3.4GHz and 3.6GHz; two HP DL320 Celeron at 2.9GHz and 3.4GHz Pentium 4 Processors. The cluster is connected via an Ethernet switch with adjustable bandwidth (from few Kilobytes) on each link. In this research, we have only used 15 machines due to a problem of disk space in node 2 during experiments.

The image used in the experiments is characterized by very high spectral resolution (224 narrow spectral bands in the range 0.4-2.5 μm) and moderate spatial resolution (614 samples, 512 lines and 20-meter pixels). It was gathered over the Indian Pines test site in Northwestern Indiana, a mixed agricultural/forested area, early in the growing season. Fig. 1(a) shows the Indian Pines AVIRIS hyperspectral data set considered in experiments. The data set represents a very challenging classification problem and it is a scene universal and extensively used as benchmark to validate classification accuracy of hyperspectral imaging algorithms.

4.2 Communication Times

Our experiments have focused on the measurements of the communication times for each paradigm used on communicating the data located on the borders of each partition assigned to the different processor on a processing power basis, producing thus different number of messages and sizes. Each execution has been made with the same group of processors, only varying the data assigned due to particular processing load at each node, except in the case of SB, where additional subgroups are created to scatter the data from the borders.

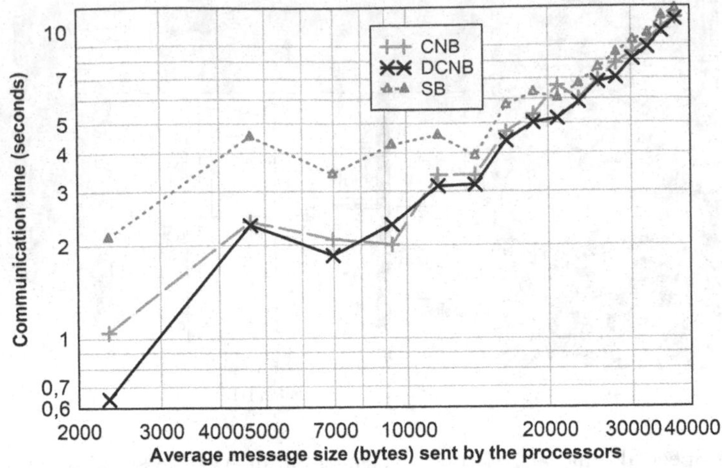

Fig. 2. Mean communication time for each particular communication paradigm

In Fig. 2, we show the mean communication time of the 15 machines using each one of the communication paradigms before mentioned. As can be seen, before reaching 6972-9296 bytes the CNB method is similar to DCNB, as expected from the model (we are still before the congestion area in most of the processors which is located around 3-4KB), thanks to small messages and no overhead for control in this implementation. Once we reach 9296 bytes, all the processes enter the congestion region, occurring then the effects of non-linearity in the communications[9]. Now, the best results are obtained by the DCNB. This is due to the use of division of the original message into several smaller messages that will fall out of the congestion area. Even though the overhead introduced with the control code, this implementation gives the best results, showing that the division of messages poses as a key solution to the problem itself. Also from the figure, we can see that the times of the SB are worst than those of the DCNB, but still very close, specially when the size of messages reach the congestion area, due to elimination of control overhead and ordered nature imposed by groups and Scatter operations. This is a very promising solution to the communication problem studied in this paper, upon the inclusion of non-blocking divided collective communications and overlapping groups.

In general, the best results are those of the DCNB, but all the paradigms show a logarithmic scaling behavior and approximation between values due to higher message sizes and overhead of the network, until the linearity is regained when reaching 65KB (as predicted by the model).

5 Conclusion

The aim of this paper has been the study of different collective communication paradigms for its use on the implementation of parallel hyperspectral imaging algorithms on heterogeneous networks of computers, using for it HeteroMPI library and communication models. The results obtained are very promising and reveal different solutions and approaches varying in complexity. As future work, we plan to integrate subgroups and collective nonblocking scatter/gather operations which may allow us to resolve the problem of excessive communications in the congestion area

References

1. Chang, C.-I.: Hyperspectral imaging: Techniques for spectral detection and classification. Kluwer, NY (2003)
2. Plaza, A., Valencia, D., Plaza, J., Martinez, P.: Commodity cluster-based parallel processing of hyperspectral imagery. Journal of Parallel and Distributed Computing 66, 345–358 (2006)
3. Dongarra, J., Huss-Lederman, S., Otto, S., Snir, M., Walker, D.: MPI: The complete reference. MIT Press, Cambridge (1996)
4. Lastovetsky, A., Reddy, R.: HeteroMPI: Towards a message-passing library for heterogeneous networks of computers. Journal of Parallel and Distributed Computing 66, 197–220 (2006)
5. Lastovetsky, A., Mkwawa, I., O'Flynn, M.: An accurate communication model for a heterogeneous cluster based on a switched-enabled Ethernet network. In: The 12th International Conference on Parallel and Distributed Systems (ICPADS) (2006)
6. Achalakul, T., Taylor, S.: A distributed spectral-screening PCT algorithm. Journal of Parallel and Distributed Computing 63, 373–384 (2003)
7. Dhodhi, M.K., Saghri, J.A., Ahmad, I., Ul-Mustafa, R.: D-ISODATA: A distributed algorithm for unsupervised classification of remotely sensed data on network of workstations. Journal of Parallel and Distributed Computing 59, 280–301 (1999)
8. Tilton, J.C.: Parallel implementation of the recursive approximation of an unsupervised hierarchical segmentation algorithm. In: Plaza, A., Chang, C.-I. (eds.) High-performance computing in remote sensing. Chapman & Hall/CRC Press, Boca Raton (2007)
9. Lastovetsky, A., O'Flynn, M., Rychkov, V.: Optimization of Collective Communications in HeteroMPI. In: Cappello, F., Herault, T., Dongarra, J. (eds.) PVM/MPI 2007. LNCS, vol. 4757, pp. 135–143. Springer, Heidelberg (2007)
10. Plaza, A., Martinez, P., Plaza, J., Perez, R.: Dimensionality reduction and classification of hyperspectral image data using sequences of extended morphological transformations. IEEE Transactions on Geoscience and Remote Sensing 43(3), 466–479 (2005)
11. Valencia, D., Lastovetsky, A., Plaza, A.: Design and implementation of a parallel heterogeneous algorithm for hyperspectral image analysis using HeteroMPI. In: Proceedings of the Fifth International Symposium on Parallel and Distributed Computing (ISPDC 2006), Timisoara, Romania (2006)
12. NASA's Thunderhead Cluster webpage, `http://thunderhead.gsfc.nasa.gov/`

An MPI-Based System for Testing Multiprocessor and Cluster Communications

Alexey N. Salnikov and Dmitry Y. Andreev

Faculty of Computational Mathematics and Cybernetics,
Lomonosov Moscow State University, Russia
salnikov@cs.msu.su, andreev@angel.cs.msu.su

Abstract. This article describes a system of MPI-tests that collects statistics on delays during message transmitting through communications of multiprocessor or cluster and Graphic User Interface (GUI) to visualize it. The system is capable to imitate communication activity level and to count delays for targeted messages passing. A cluster with 940 multicore processors has been tested.

Communications in modern multiprocessors and clusters are rather complex. It is difficult to predict duration of message transfer between two processors for a message with a given length and an expected level of communications loading. Difficulty is a result of many components that form multiprocessor or cluster communications. The most popular technology of message passing in multiprocessors and clusters is MPI. We expect several intelligent MPI-based testing applications that will particularly solve the problem of time prediction of interaction through communications to be developed. NetPIPE [2] is an example of MPI-based test which performs testing by means of point-to-point messages. There is another point of view on communication testing process. Often clusters have thousands of processors. This forces the tests to generate a great amount of data; it's impossible to comprehend this data. So it is necessary to develop visualization tools for it.

The goal of our research is to develop several MPI-based tests for extracting statistics on communication delays and to develop an application for visualizing information collected after testing MPI-application work.

The authors introduce MPI-application, a part of PARUS [1] project called *network_tests*, which implements six modes of communication testing. User defines several parameters for testing application: interval of message length, step of message length and number of repeats for each message length. The MPI-application begins its work with the lowest message length in the interval, upon each step it increases the message length by value of step parameter that has been defined by the user. The algorithm finishes its work with the last message length located in the interval. For each message length application performs data transmissions trough communications. Type and volume of transmit tings depends on mode that is chosen by user in application parameters. The number of repeats parameter determines the number of independent iterations of data

A. Lastovetsky et al. (Eds.): EuroPVM/MPI 2008, LNCS 5205, pp. 332–333, 2008.

transmission. The testing results for all iterations constitute a sample which will be used for searching minimum value, median and counting average and standard deviation. All found values form a set of matrices, each matrix corresponds one of the messages length. The i, j matrix cell corresponds transfer duration from MPI-process with number i to process with number j. Duration is estimated by `MPI_Wtime` function. Data for median, average and so on is stored in different text files. Let's describe several available test modes:

- *one_to_one* – there is only one pair of MPI-processes that are involved simultaneously in message transmitting with `MPI_Send` and `MPI_Recv` functions. MPI_Recv duration is stored in the matrix.
- *all_to_all* – all MPI-processes are involved in message transmitting. An application uses `MPI_Isend`, `MPI_Irecv` where time intervals from `MPI_Irecv` to `MPI_Waitany` are stored in the matrix.
- *async_one_to_one* – is similar to one_to_one mode but it uses unblocked functions.
- *test_noise_blocking* – MPI-processes are divided in three groups: pair of target processes, noise and idle processes. Noise processes are chosen randomly from idle processes where number of them is defined by user. Target processes are involved in message passing similar to one_to_one and durations of them are stored in the matrix. Noise processes perform all_to_all transmissions with fixed by application parameters message length, but their durations are not stored in the matrix.

There is a Sun Java 1.5 GUI application developed to visualize results of communications testing with three modes of data visualisation. In the first mode a matrix of delays for fixed message length is drawn. This mode has two internal modes of data normalization: local in one matrix and global in all results. The intensity of black corresponds normalized duration of transmitting. Min value is converted to white color and max value is converted to black color. In the second mode the user chooses one row or column in matrix and the program draws this for all messages length. This mode highlights delays for one fixed MPI-process. In the third mode a plot for chosen pair of MPI-processes is built.

Both applications have been tested on mvs100k (cluster of 470 nodes with four Intel Xeon 5160 processors which are connected through Infiniband network) and IBM pSeries 690 (SMP system with 16 processors in our configuration). The code for both applications is available from SourceForge PARUS project page.

References

1. Salnikov, A.N.: PARUS: A Parallel Programming Framework for Heterogeneous Multiprocessor Systems. In: Mohr, B., Träff, J.L., Worringen, J., Dongarra, J. (eds.) PVM/MPI 2006. LNCS, vol. 4192, pp. 408–409. Springer, Heidelberg (2006)
2. Turner, D., Chen, X.: Protocol-Dependent Message-Passing Performance on Linux Clusters. In: IEEE International Conference on Cluster Computing (CLUSTER 2002), p. 187 (2002) (ISBN: 0-7695-1745-5)

MPI in Wireless Sensor Networks*

Yannis Mazzer and Bernard Tourancheau

LIP
UMR 5668 CNRS-ENS-INRIA-University of Lyon
ENS-Lyon, 69364 Lyon - France
firstname.lastname@ens-lyon.fr

Abstract. This paper presents the possible usage of MPI in our Wireless Sensor Networking software stack project.

1 Introduction

Wireless Sensor Networks (WSN) are built around systems on a board or on a chip that gathers a micro-controller coupled with a radio chip and some memory chips. These systems can embed a bunch of environmental sensors and relays through there ADC/DAC interfaces. This is a booming technology [1,2] that can be applied to numerous fields of our industrial societies that need fine grain control and ubiquitous field knowledge.

We intentionally reduced our scope to home automation. And more precisely in that particular domain, our primary interest is energy savings in buildings with the help of WSN infrastructures. With this research in WSN platform, our aim is to participate to the reduction of energy wastes: in OCDE societies, energy spent in buildings represent more than 43% of the raw energy spent per year.

WSN allows for monitoring and control of appliances. Sensors monitor the physical parameters of the building in several location of importance. From this observation, actions can be computed to control the building equipment like the heater, ventilation, cooler, shades, ... and this can help to simulate and predict the building behavior [3] from the energy spending point of view. In the monitoring phase, the micro-controller acts as a network connection for the sensors and in the reaction phase, as a network connection for the actuators relays.

Thus, the WSN is a networked communication platform and each node loaded with a lightweight OS is able to do basic IO operations and data computation. We are investigating MPI as an application communication middleware paradigm for these wireless micro-controller networks embedding sensors.

2 Our WSN Model

Most of WSN communications are based on the IEEE 802.15.4 radio and MAC layers with dedicated radio chips coupled to the micro-controllers. Usually, PC-based

* This work was made possible thanks to a BQR grant from ENS-Lyon and our reception in the GRAAL team.

A. Lastovetsky et al. (Eds.): EuroPVM/MPI 2008, LNCS 5205, pp. 334–339, 2008.

gateways interface the WSN with the Internet world. Thus, the resulting networking schema is hierarchical and, for the WSN layer, has the following properties :

- it is build with "small" autonomous communication nodes
- it is based on ad-hoc wireless meshed network
- the position of nodes and gateways is fixed, moreover, the location of each node will be important to the user
- the network connectivity is dense because of radio broadcast mean
- the network is faulty because of the radio mean, the battery life and other field constraints
- the energy saving is the primary goal because the lifetime is expected in years with as small as possible battery powered nodes
- in this micro-controller world, data communication cost is of the order of a thousand times the local computation on this same nodes

Thus, in such an architecture, the communication system and middleware should try to reduce the communication volume and choose the best communication paths to preserve and equilibrate battery loads. Moreover, distributed operations may be preferred against data treatment in a remote server if this reduces data communication number and volume, thus preserves the battery life.

3 WSN and Communication

3.1 Monitoring Measurements Requirements

Up to now, WSNs simply collect data through their meshed network. The monitoring data is obtained through mostly one sided communications from a remote server acting as a database sending requests to the nodes when needed :

- interactive requests
- timed regular reporting

However, one can imagine from the WSN side that regular checking can be done in order to provide alarms, local or global pre-computed function of the whole WSN sensor data leading to :

- computed thresholds events
- locally computed transfer functions
- distributed functions compute in the WSN

Hence, from the communication middleware point of view, the needed communication functions include all the MPI spectrum[4] from send and receive, to collective and one sided ones.

3.2 Existing Networking Protocols for Sensor Networks

Existing wired network protocols, like X10, CAN, BacNet, ... are the standards. They were designed for home automation or industry processes without having in mind the Internet world developments. Numerous other industrial communication layers already exist for radio networks, but unfortunately almost one per hardware provider ! A strong effort was made by the ZigBee consortium [5] to discuss and formalize an industrial standard for networking automation. But if its resulting stack is very satisfactory for the interface description of devices needs at the network edges, its networking part is not in the same mature state nor fully adapted to the Internet down to the WSN nodes.

3.3 Existing OS

Among the open source projects, the TinyOS project from Berkeley [6] is the lighter memory footprint and most popular event driven OS dedicated for WSN. TinyDB, its companion module is also very important for the remote management of sensors. As well its programming interface and associated compile tool is also dedicated to WSN. On this TinyOS base, the ArchRock middleware includes a very practical web server interface. FreeRTOS [7] is a more heavy weight tool with a larger memory footprint and a full IP stack. Contiki [8] brings several nice functionalities and an IP stack with TCP for a remarkably small memory footprint. However, it is not dedicated to WSN but to the generic embedded systems world.

Some of these OSes were ported to several micro controller platforms, see TinyOS in the Table 1 for instance. Some others are very dependent on the hardware they were developed on or simply need too large resources that precludes some very light WSN hardwares. Table 1 presents the main parameters of a panel of WSN nodes along the, rather short, history of the field and involving several hardware and software technologies.

Table 1. WSN hardware platforms and available OSes

	Mica-x	TelosB	SensiNode	SunSPOT
μ controller	Atmega88-128	TI MSP430	TI 8051 (CC 2431)	ARM 920T
Bus	8-bit	16-bit	8-bit	32-bit
RAM	128kB	10kB	8kB	512kB
Flash	512k	48k	128k	4M
Radio	variable	CC2420	CC2420 (CC 2431)	CC2420
Onboard Sensors	none	humidity + temp. + 2xlight	temp. + light	temp. + light + accelero.
IO	yes	16 (8 ADC)	21 (8 ADC)	yes
Onboard Antenna	no	yes	yes	no
OS	TinyOS	TinyOS, Contiki	FreeRTOS, TinyOS	Java JMX
OS mem. footprint	0.4kB	0.4kB, 2kB	3kB , 0.4kB	80kB

4 The iWSN Project

The iWSN project aims to design an open source middleware for WSN micro-controllers including an IPv6 communication stack following the 6lowPAN IETF

RFC [9], and tools to provide an easy configuration and usage of the WSN platforms in the building energy saving domain.

Implementing generic IP, distributed functions and tools while preserving the advantages of a dedicated but very constrained WSN architecture is not that easy. In a very modular structure based on TinyOS, we investigate the implementation of core functions in a software stack, see Figure 1, that will provide enough flexibility for our building monitoring and control applications.

4.1 Mesh Routing Strategies

Several mesh-routing strategies are envisioned to answer the needs of the different applications on such a system. These strategies may be implemented in different network layers to optimize their behavior for the end-user applications.

Also, the objective function will be often linked to resource consumption, especially battery power. Thus the routing and scheduling of the communication functions may depend on the local states of the intermediate nodes in the mesh graph.

4.2 IPv6 over Low Power and Lossy Radio Networks

There are numerous reasons for the use of IPv6 [10] in such a system. On the networking side is the inheritance of the IP addressing schema, the Internet network architecture and tools for configuration, maintenance and security.

This open standard also gives provisions for a large adhesion of the rather diverse communities involved in the research, the technical usage and business of WSNs.

The web interface might be the most important end-user reason for providing IP to the WSN nodes. Such a simple remote way to control the data loggers and relays is mandatory to open the domain to a large community.

The standardization activities are ongoing and summarized in the IETF 6lowPAN working group. With the 6lowPAN charters, a few research teams and start-ups started networking stack developments for WSN. Some stacks are proprietary like ArchRock [11], which is based on an enhancement of TinyOS, some other are open source like the NanoStack from SensiNode [12]. We are currently developing such a 6lowPAN stack for TinyOS.

Furthermore, not taking into account the battery power spending, one could also provide MPI over IPv6 like in [13] since the 6lowPAN stack should provide a complete substitute of IPv6.

4.3 Active Message Implementation

The TinyOS communication stack is based on the Active Messages (AM) paradigm developed at Berkeley in the nineties [14]. This should allows for an implementation of MPI over AM, like in clusters, see for instance [15], but with very different optimization goals, namely power consumption, robustness or memory footprint.

Raw AM can be implemented very close to the hardware resources and thus can provide high performance MPI communication libraries. However, with only this communication layer, applications do not benefit from the easy web access interface. So direct AM usage should be limited to the internal WSN distributed functions calls.

4.4 Distributed Communication and Computation Functions

Automation transfer functions as well as associative aggregation must be computed on-line to increase batteries' lives by decreasing the communication volume.

For instance, the associative functions, such as MAX, MIN, SUM, AVERAGE, can be computed along spanning trees routed by a gateway. More complex transfer functions may needs dedicated communication graphs inside the WSN. A communication middleware like MPI could help in the design, mapping and scheduling of these complex algorithms inside the WSN architecture.

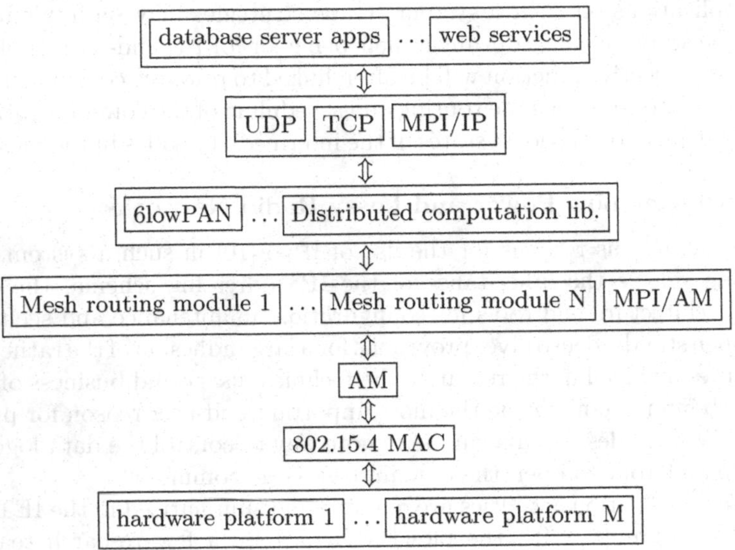

Fig. 1. iWSN software stack

5 Conclusions

While Ad-hoc networking is a mature domain, WSN is still in its infancy. There is a great deal of work to do to provide a communication system that will allow to optimize the autonomy of the WSN platform and to fulfill the needs of each WSN application domains.

A subset of MPI could be of great help to write distributed communication and computation functions modules within the WSN and inherit of library style programming in WSN component based applications. Moreover, even if this is a difficult task regarding the constrained resources of WSN, our first implementation approach is optimistic and depends on the options chosen in the iWSN software stack.

A complete IPv6 interface is mandatory for end-users and the WSN lightweight nodes may not be able to accommodate two communication system programming layers. Thus, the introduction of the ability to choose the layer target in IP encapsulated packet may be a win-win approach.

We may end up with two communication layers, one MPI-like to code distributed functions embedded in the kernel core of each WSN node and one IPv6-like for end user remote accesses during WSN operations.

References

1. Stojmenovic, I.: Handbook of Sensor Networks. Wiley, Chichester (2005)
2. Wireless Sensor Networks. Kluwer Academic Publishers, Dordrecht (2004)
3. Clarke, J.A.: Energy Simulation in Building Design. Butterworth Heinemann (2001)
4. The message passing interface (mpi) standard (1992-2008), `http://www-unix.mcs.anl.gov/mpi`
5. Zigbee alliance (2005-2008), `http://www.zigbee.org`
6. Hill, J.L.: System architecture for wireless sensor networks. PhD thesis, University of California, Berkeley, Adviser-David E. Culler (2003)
7. Freertos (2008), `http://www.freertos.org`
8. Dunkels, A., Grönvall, B., Voigt, T.: Contiki - a lightweight and flexible operating system for tiny networked sensors. In: EmNetS. IEEE Computer Society Press, Los Alamitos (2004)
9. Montenegro, G., Kushalnagar, N., Hui, J., Culler, D.: Transmission of ipv6 packets over ieee 802.15.4 networks. Technical Report 4944, IETF (2007)
10. Montenegro, G., Kushalnagar, N., Schumacher, C.: Ipv6 over low-power wireless personal area networks (6lowpans): Overview, assumptions, problem statement, and goals. Technical Report 4919, IETF (2007)
11. Archrock (2006-2008), `http://www.archrock.com`
12. Nanostack by sensinode (2007-2008) , `http://sensinode.com`
13. Knoth, A., Kauhaus, C., Fey, D., Schneidenbach, L., Schnor, B.: Challenges of mpi over ipv6. In: ICNS. IEEE Computer Society Press, Los Alamitos (2008)
14. Mainwaring, A.M., Culler, D.E.: Active messages: Organization and applications programming interface. Technical report, University of California at Berkeley (1995)
15. Ciaccio, G., Chiola, G.: Gamma and mpi/gamma on gigabit ethernet. In: Dongarra, J., Kacsuk, P., Podhorszki, N. (eds.) PVM/MPI 2000. LNCS, vol. 1908. Springer, Heidelberg (2000)

Author Index